GUN CONTROL

GUN CONTROL

Erin Brown

INFORMATION PLUS® REFERENCE SERIES
Formerly Published by Information Plus, Wylie, Texas

ISSN 1534-1909

Jefferson Madison
Regional Library
Charlottesville, Virginia

GALE
CENGAGE Learning

Farmington Hills, Mich • San Francisco • New York • Waterville, Maine
Meriden, Conn • Mason, Ohio • Chicago

GALE
CENGAGE Learning

Gun Control

Erin Brown

Kepos Media, Inc.: Steven Long and Janice Jorgensen, Series Editors

Project Editor: Laura Avery

Rights Acquisition and Management: Ashley Maynard, Carissa Poweleit

Composition: Evi Abou-El-Seoud, Mary Beth Trimper

Manufacturing: Rita Wimberley

© 2017 Gale, Cengage Learning

ALL RIGHTS RESERVED. No part of this work covered by the copyright herein may be reproduced or distributed in any form or by any means, except as permitted by U.S. copyright law, without the prior written permission of the copyright owner.

This publication is a creative work fully protected by all applicable copyright laws, as well as by misappropriation, trade secret, unfair competition, and other applicable laws. The authors and editors of this work have added value to the underlying factual material herein through one or more of the following: unique and original selection, coordination, expression, arrangement, and classification of the information.

For product information and technology assistance, contact us at
Gale Customer Support, 1-800-877-4253.
For permission to use material from this text or product, submit all requests online at www.cengage.com/permissions.
Further permissions questions can be e-mailed to permissionrequest@cengage.com

Cover photograph: © Stephanie Frey/Shutterstock.com.

While every effort has been made to ensure the reliability of the information presented in this publication, Gale, a part of Cengage Learning, does not guarantee the accuracy of the data contained herein. Gale accepts no payment for listing; and inclusion in the publication of any organization, agency, institution, publication, service, or individual does not imply endorsement of the editors or publisher. Errors brought to the attention of the publisher and verified to the satisfaction of the publisher will be corrected in future editions.

Gale
27500 Drake Rd.
Farmington Hills, MI 48331-3535

ISBN-13: 978-0-7876-5103-9 (set)
ISBN-13: 978-1-4103-2549-5

ISSN 1534-1909

This title is also available as an e-book.
ISBN-13: 978-1-4103-3270-7 (set)
Contact your Gale sales representative for ordering information.

Printed in the United States of America
1 2 3 4 5   21 20 19 18 17

TABLE OF CONTENTS

PREFACE vii

CHAPTER 1
The History of the Right to Bear Arms 1
  The right of the individual to keep and bear arms is rooted in English common law. Colonial Americans believed that individual ownership of guns was essential to the formation of a militia, and it was colonial militias that won America's freedom. In modern times the debate over gun control evokes strong emotions. It focuses on the meaning of the Second Amendment, gun violence, and whether placing greater restrictions on gun ownership will make society safer.

CHAPTER 2
How Many Guns Are There, and Who Owns Them? 9
  More than 375 million guns are either privately owned or available for sale in the United States. The Tiahrt Amendment, which restricts the ability of the Bureau of Alcohol, Tobacco, Firearms, and Explosives to report on sales of multiple handguns and firearms tracing statistics, is explained. Also included in this chapter is a discussion of so-called assault weapons, gun safety, and characteristics of gun owners.

CHAPTER 3
Firearm Laws, Regulations, and Ordinances 19
  The federal government became involved in gun regulation because of a wave of lawlessness that accompanied Prohibition, and Congress has since passed several laws regulating the sale and use of firearms. The major federal laws are the Gun Control Acts of 1968 and 1986, their amendments, and the Brady Handgun Violence Prevention Act. Most gun laws are passed at the state and local levels. This chapter provides an overview of major federal laws as well as the variations in state and local law regarding gun purchases, permitting, and carrying.

CHAPTER 4
Court Rulings on Firearms 41
  The rights of state and local governments to control gun ownership have consistently been upheld by the courts. This chapter looks at landmark and recent court rulings on gun regulation at the federal, state, and local levels.

CHAPTER 5
Firearms and Crime 57
  Guns and crime often go together. This chapter describes the types of crimes in which guns are involved, discusses the prevalence of gun use in various crimes, and considers trends in criminal gun use. The chapter also charts the history of mass shootings in the United States and details the rise of active shooter incidents in the 21st century.

CHAPTER 6
Gun-Related Injuries and Fatalities............... 83
  Guns cause thousands of injuries and fatalities each year. These incidents are a major public health issue. The public health establishment tracks weapons-related injuries and deaths to find ways to reduce them. The role of firearms in suicide versus homicide, the public costs of firearm fatalities, and the use of firearms in self-defense are addressed.

CHAPTER 7
Guns and Youth.......................... 99
  During the late 1980s and early 1990s, much of the gun violence was attributable to young people, as the rise of youth gangs and the drug trade coincided with the nationwide proliferation of handguns. By the mid-1990s and beyond, falling rates of gun violence among young people led to declining overall rates of gun violence. These historical developments, as well as ongoing concerns about school shootings and other forms of youth firearm violence, are delved into in this chapter.

CHAPTER 8
Public Attitudes toward Gun Control 123
  The focus of this chapter is on a sampling of poll results showing Americans' attitudes toward the strictness of gun laws and public responses to major firearm crimes, such as the Newtown, Connecticut, school shooting. The chapter also examines the demographics of gun ownership.

CHAPTER 9
There Should Be Stricter Gun Control Laws 131
  This chapter presents a sample of prominent expert opinion supporting increased regulation of firearms at the federal level.

CHAPTER 10
There Should Not Be Stricter Gun Control Laws..... 137
  This chapter presents a sample of prominent expert opinion opposed to increased regulation of firearms at the federal level.

APPENDIX: STATE CONSTITUTION ARTICLES CONCERNING WEAPONS................... 143

IMPORTANT NAMES AND ADDRESSES....... 147

RESOURCES............................. 149

INDEX 151

PREFACE

Gun Control is part of the *Information Plus Reference Series*. The purpose of each volume of the series is to present the latest facts on a topic of pressing concern in modern American life. These topics include the most controversial and studied social issues of the 21st century: abortion, capital punishment, care for the elderly, child abuse, energy, the environment, health care, immigration, national security, social welfare, women, youth, and many more. Although this series is written especially for high school and undergraduate students, it is an excellent resource for anyone in need of factual information on current affairs.

By presenting the facts, it is the intention of Gale, Cengage Learning, to provide its readers with everything they need to reach an informed opinion on current issues. To that end, there is a particular emphasis in this series on the presentation of scientific studies, surveys, and statistics. These data are generally presented in the form of tables, charts, and other graphics placed within the text of each book. Every graphic is directly referred to and carefully explained in the text. The source of each graphic is presented within the graphic itself. The data used in these graphics are drawn from the most reputable and reliable sources, such as from the various branches of the U.S. government and from private organizations and associations. Every effort has been made to secure the most recent information available. Readers should bear in mind that many major studies take years to conduct and that additional years often pass before the data from these studies are made available to the public. Therefore, in many cases the most recent information available in 2017 is dated from 2014 or 2015. Older statistics are sometimes presented as well, if they are landmark studies or of particular interest and no more-recent information exists.

Although statistics are a major focus of the *Information Plus Reference Series*, they are by no means its only content. Each book also presents the widely held positions and important ideas that shape how the book's subject is discussed in the United States. These positions are explained in detail and, where possible, in the words of their proponents. Some of the other material to be found in these books includes historical background, descriptions of major events related to the subject, relevant laws and court cases, and examples of how these issues play out in American life. Some books also feature primary documents or have pro and con debate sections that provide the words and opinions of prominent Americans on both sides of a controversial topic. All material is presented in an evenhanded and unbiased manner; readers will never be encouraged to accept one view of an issue over another.

HOW TO USE THIS BOOK

Gun control is a topic that elicits passionate emotions. On one side are those who believe the right to own guns should be subject to more stringent restrictions and regulations, given their propensity to be used to commit crimes and suicide. This side tends to see firearm violence as a public health issue and to advocate the kind of robust government response that any public health crisis, such as an epidemic of a disease, might elicit. On the other side are those who consider the Second Amendment to be one of the quintessential American freedoms. This side tends to see restrictions on the right to own guns as an infringement of basic liberty and as progress down a slippery slope that will end in the banning of all firearms.

Gun Control consists of 10 chapters and four appendixes. Each chapter is devoted to a particular aspect of gun control in the United States. For a summary of the information that is covered in each chapter, please see the synopses that are provided in the Table of Contents. Chapters generally begin with an overview of the basic facts and background information on the chapter's topic, then proceed to examine subtopics of particular interest. For example, Chapter 5: Firearms and Crime begins by explaining the current sources of the best government data on violent crime, then proceeds to look at the role firearms play in these crimes and the demographics of both offenders and victims. The chapter further discusses some of the deadliest mass shootings in American history, the increased frequency of active shooter incidents in the United States since 2000, and the extent to which police are fatally shot or injured in the line of duty by criminals with guns. In exploring the channels through which criminals gain access to firearms, the chapter also notes the lack of up-to-date, authoritative data on this subject, due to restrictions on government research that went into effect in 2003. Readers can find their way through a chapter by looking for the section and subsection headings, which are clearly set off from the text. They can also refer to the book's extensive Index if they already know what they are looking for.

Statistical Information

The tables and figures featured throughout *Gun Control* will be of particular use to readers in learning about this issue. These tables and figures represent an extensive collection of the most recent and important statistics on gun ownership, gun violence, and gun control legislation, as well as the trends—for example, graphics outline the percentage of Americans who own various types of firearms, the number and result of criminal background checks under the Brady gun control laws, and statistics on firearm violence. Gale, Cengage Learning, believes that making this information available to readers is the most important way to fulfill the goal of this book: to help readers understand the issues and controversies surrounding gun control in the United States and reach their own conclusions.

Each table or figure has a unique identifier appearing above it, for ease of identification and reference. Titles for the tables and figures explain their purpose. At the end of each table or figure, the original source of the data is provided.

To help readers understand these often complicated statistics, all tables and figures are explained in the text. References in the text direct readers to the relevant statistics. Furthermore, the contents of all tables and figures are fully indexed. Please see the opening section of the Index at the back of this volume for a description of how to find tables and figures within it.

Appendixes

Besides the main body text and images, *Gun Control* has four appendixes. The first appendix lists the 50 states and cites the specific articles of the 44 states that have constitutional provisions regarding weapons. The second appendix is the Important Names and Addresses directory. Here, readers will find contact information for a number of government and private organizations that can provide further information on aspects of gun control policy and gun rights. The third appendix is the Resources section, which can also assist readers in conducting their own research. In this section, the author and editors of *Gun Control* describe some of the sources that were most useful during the compilation of this book. The final appendix is the Index. It has been greatly expanded from previous editions and should make it even easier to find specific topics in this book.

COMMENTS AND SUGGESTIONS

The editors of the *Information Plus Reference Series* welcome your feedback on *Gun Control*. Please direct all correspondence to:

Editors
Information Plus Reference Series
27500 Drake Rd.
Farmington Hills, MI 48331-3535

CHAPTER 1
THE HISTORY OF THE RIGHT TO BEAR ARMS

The right to bear arms has deep roots in American legal and cultural traditions. From the time colonists settled on North American soil, Americans have held weapons to protect themselves. Armed citizen-soldiers won America's freedom from English rule more than two centuries ago. Partly because of this connection to the country's history, attempts to restrict a citizen's right to own a gun evoke strong emotions.

The modern debate over gun control erupted after a series of high-profile assassinations during the 1960s, including the assassinations of President John F. Kennedy (1917–1963), the civil rights leader Martin Luther King Jr. (1929–1968), and Senator Robert F. Kennedy (1925–1968; D-NY). In the decades that followed, the debate gained new urgency as gun-related violence among ordinary Americans increased. In the ongoing series *Homicide Trends in the United States* (2016, http://www.bjs.gov/index.cfm?ty=pbse&sid=31), the Bureau of Justice Statistics indicates that gun-related homicides rose sharply during the late 1970s and peaked in 1980 and that the number of homicides rose again beginning in the late 1980s and peaked in the mid-1990s. Although the annual numbers of gun-related homicides began dropping nationally after the mid-1990s' peak, mass shootings such as the 2012 killings of 20 first graders and six staff members at Sandy Hook Elementary School in Connecticut and the 2016 rampage that left 49 people dead and 53 wounded at the Pulse nightclub in Orlando, Florida (the deadliest mass shooting in U.S. history as of December 2016) have kept the issue of gun control in the forefront of the collective American consciousness.

WHAT IS GUN CONTROL?

In "Gun Control Explained" (NYTimes.com, October 7, 2015), Richard Pérez-Peña defines gun control as "a broad term that covers any sort of restriction on what kinds of firearms can be sold and bought, who can possess or sell them, where and how they can be stored or carried, what duties a seller has to vet a buyer, and what obligations both the buyer and the seller have to report transactions to the government." Additionally, Pérez-Peña points out that gun control may extend to the legality of certain types of magazines (cartridge holders that feed bullets automatically into a gun's chamber) and ammunition or may apply to technological restrictions, such as sensors that recognize the grip of the handler and prevent a gun from firing unless it is being held by its registered owner.

In the second decade of the 21st century the gun control debate is generally focused on issues pertaining to the vetting of prospective gun buyers through background checks, whether and where people should be allowed to carry firearms in public, and whether the right to bear arms should include semiautomatic assault rifles and high-capacity magazines. Often cited as one of the most polarizing issues in American politics, the gun control debate typically breaks down along party and ideological lines, with Republicans and those who identify as conservative staunchly in favor of unfettered gun rights, while Democrats and those who identify as liberal advocate for stricter gun legislation.

At the heart of the gun control debate is the interpretation of the Second Amendment to the U.S. Constitution. One side claims that gun ownership is an individual right guaranteed by the Second Amendment and that guns are vital for self-protection. The other side believes guns should be banned or restricted because many innocent lives are lost due to their misuse. Gun control advocates say there is no longer a compelling need for people to "keep and bear arms" as there was when the Constitution was ratified in 1788. They sometimes add the argument that the constitutionally guaranteed right was never meant to apply to individuals. This argument was supported by a consensus of legal experts and the U.S. Supreme Court until the 21st century, when an interpretation focusing on the individual right to bear arms won a majority of

TABLE 1.1

Poll respondents' views on the meaning of the Second Amendment, 2016

["Please read the following passage: 'A well-regulated militia, being necessary to the security of a free state, the right of the people to keep and bear arms, shall not be infringed.' Now, for each of the following, please indicate whether you feel the passage does or does not support it."]

Base: U.S. adults		Supports this	Does not support this
A state militia's right to own firearms	%	73	27
Any citizen's right to own firearms	%	70	30
A state's right to form a militia	%	66	34
A state's right to regulate firearm ownership among its citizens	%	54	46
A state's right to regulate the ownership of firearms among its militia	%	54	46
The federal government's right to regulate state militias	%	46	54
The federal government's right to regulate who can own firearms	%	44	56
A private citizen's right to form a militia	%	40	60

Note: Percentages may not add up to 100% due to rounding.

SOURCE: "Table 7. Perceived as Supported/Not Supported by Second Amendment," in *73% Favor Some Restrictions on Firearms Sales/Ownership; 16% Favor No Limitations*, The Harris Poll, January 13, 2016, http://www.theharrispoll.com/politics/Gun-Rights.pdf (accessed August 31, 2016)

adherents among Supreme Court justices. Regardless of the merits of either side's argument or of legal experts' interpretations, polls show that a majority (70%) of Americans understand that the Second Amendment upholds an individual's right to own firearms. (See Table 1.1.)

The right of the individual to keep and use weapons has a long tradition in Western civilization. The Greek philosopher Aristotle (384–322 BC) wrote in *Politics* that ownership of weapons was necessary for true citizenship and participation in the political system. By contrast, another Greek philosopher, Plato (428–347 BC), wrote in the *Republic* that he believed in a monarchy with few liberties and saw the disarming of the populace as essential to the maintenance of an orderly and autocratic system. In *De Officiis*, the Roman politician Marcus Tullius Cicero (106–43 BC) expressed support for the individual's right to own weapons for self-defense and for public defense against tyranny. Similarly, in *Discourse*, the Italian political philosopher Niccolò Machiavelli (1469–1527) advocated an armed populace of citizen-soldiers to keep headstrong rulers in line.

AN EARLY PRECEDENT: MILITIAS AND THE OWNERSHIP OF WEAPONS

One of the first documents to link the bearing of arms with a militia (an army composed of citizens called to action in time of emergency) was the English *Assize of Arms* of 1181, which directed every free man to have access to weaponry. Henry II of England (1133–1189) signed this law to enable the rapid creation of a militia when needed, but the law also permitted carrying arms in self-defense and forbade the use of arms only when the intention was to "terrify the King's subjects." In 1328, under the reign of King Edward III (1312–1377), Parliament enacted the Statute of Northampton, which prohibited the carrying of arms in public places but did not overrule the right to carry arms in self-defense.

EARLY GUN CONTROL LAWS

The 17th century was a period of great turmoil in England, as Parliament and the monarchy struggled for control of the government. A critical issue related to the civil wars that erupted in 1642 was whether the king or Parliament had the right to control the militia. When the wars ended in early 1660, England briefly fell under the control of a military government, which authorized its officers to search for and seize all arms owned by Catholics or any other people deemed dangerous. In late 1660, with the coronation of Charles II (1630–1685), the English monarchy was restored, but the battle between Parliament and the monarchy continued.

The Game Act of 1671, an early example of a gun control law, was enacted to keep the ownership of hunting lands and weaponry in the hands of the wealthy and to restrict hunting and gun ownership among the peasants. People without an annual income of at least 40 to 100 pounds could no longer keep weapons, even for self-defense. In 1689 Queen Mary II (1662–1694) and King William III (1650–1702) were installed as corulers of England. When the pair took their oaths of office, they were presented with a new Bill of Rights (2008, http://avalon.law.yale.edu/17th_century/england.asp), which outlined the relationship of Parliament and the monarchy to the people. This Bill of Rights included a specific right of "Protestants [to] arms for their defence suitable to their conditions and as allowed by law." It also condemned abuses committed by standing armies (armies maintained by the government on a long-term basis, even while at peace) and declared "that the raising or keeping a standing army within the kingdom in time of peace, unless it be with consent of Parliament, is against law." Furthermore, the Bill of Rights removed the word *guns* from the list of items the poor were forbidden to own by the Game Act of 1671. From this time on, the right to keep and bear arms belonged to all Englishmen, whether rich or poor.

THE AMERICAN MILITIA AND THE RIGHT TO BEAR ARMS

Much of U.S. law is rooted in the system of laws developed in England (called common law) because most American colonists came from England, bringing with them English values, traditions, and legal concepts. Many of the English were familiar with the famous judge William Blackstone (1723–1780), who listed in his *Commentaries*

(2000, http://press-pubs.uchicago.edu/founders/documents/v1ch16s5.html) the "right of having and using arms for self-preservation and defense." This right, brought to North America by the English, was exercised by the colonists against the English during the Revolutionary War (1775–1783) and was later incorporated into the U.S. Constitution.

During the mid-18th century an increasing British military presence in the colonies alerted colonists to the danger of a standing army. When British soldiers shot and killed five men on the streets of Boston in 1770, an event that became known as the Boston Massacre, colonists grew further concerned. The Boston Massacre became a milestone on the road to the American Revolutionary War. In 1775 the British army encountered the Massachusetts militia at Lexington—famously recalled in "Concord Hymn" by the essayist and poet Ralph Waldo Emerson (1803–1882) as "the shot heard round the world"—and the ensuing seizure of colonial arms and munitions convinced other colonies that a militia was necessary to achieve the "security of a free state."

Because the individual colonies did not have enough money to purchase weapons, each man was required to maintain a firearm so he could report immediately for duty and form a militia. It was taken for granted by the colonists that the right to individually possess and bear arms was inseparable from the right to form a militia—without these privileges, the right to organize a militia would have little meaning. Thomas Jefferson (1743–1826) stated, "No freeman shall be debarred the use of arms (within his own lands or tenements)," and Richard Henry Lee (1732–1794) of Virginia observed that "to preserve liberty, it is essential that the whole body of the people always possess arms."

In *The Federalist No. 29* ([January 9, 1788] November 22, 2016, http://www.constitution.org/fed/federa29.htm), one of a series of papers written after the Revolutionary War to convince the colonists to ratify the Constitution, Alexander Hamilton (1755?–1804) spoke of the right to bear arms in the sense of an unorganized militia, which consisted of the "people at large." He suggested that this militia could mobilize against a standing army if the army usurped the government's authority or if it supported a tyrannical government. Such a standing army, declared Hamilton, could "never be formidable to the liberties of the people while there is a large body of citizens, little, if at all, inferior to them in discipline and the use of arms, who stand ready to defend their own rights and those of their fellow-citizens."

James Madison (1751–1836) attributed the colonial victory to armed citizens. In *The Federalist No. 46* ([January 29, 1788] November 22, 2016, http://www.constitution.org/fed/federa46.htm), he wrote, "Besides the advantage of being armed, which the Americans possess over the people of almost every other nation, the existence of subordinate governments, to which the people are attached, and by which the militia officers are appointed, forms a barrier against the enterprises of ambition, more insurmountable than any which a simple government of any form can admit of. Notwithstanding the military establishments in the several kingdoms of Europe, which are carried as far as the public resources will bear, the governments are afraid to trust the people with arms."

THE U.S. CONSTITUTION, THE RIGHT TO BEAR ARMS, AND THE MILITIA

The Revolutionary War ended in 1783. In 1787, 39 men gathered in Philadelphia, Pennsylvania, to sign the newly written Constitution. Three refused to sign because the document did not include a Bill of Rights. One reluctant signer protested that, without a Bill of Rights, Congress "at their pleasure may arm or disarm all or any part of the freemen of the United States."

By 1791 James Madison had written the 10 amendments to the Constitution that became known as the Bill of Rights. He was influenced by state bills of rights and many amendment suggestions from the state conventions that ratified the Constitution. Overall, four basic beliefs were assimilated into the Second Amendment: the right of the individual to possess arms, the fear of a professional army, the dependence on militias regulated by the individual states, and the control of the military by civilians.

According to the Constitution Society, in *Documents on the First Congress Debate on Arms and Militia* (February 25, 2005, http://www.constitution.org/mil/militia_debate_1789.htm), the ratifying state conventions offered similar suggestions about the militia and the right to bear arms. Although New Hampshire did not mention the militia, it did state that "no standing Army shall be Kept up in time of Peace unless with the consent of three fourths of the Members of each branch of Congress, nor shall Soldiers in Time of Peace be Quartered upon private Houses without the consent of the Owners."

Another New Hampshire amendment read, "Congress shall never disarm any Citizen unless such as are or have been in Actual Rebellion." Maryland proposed five separate amendments, which Virginia consolidated by stating: "That the people have a right to keep and bear arms; that a well regulated Militia composed of the body of the people trained to arms is the proper, natural and safe defence of a free State. That standing armies in time of peace are dangerous to liberty, and therefore ought to be avoided, as far as the circumstances and protection of the Community will admit; and that in all cases the military should be under strict subordination to and governed by the Civil power."

The New York convention offered more than 50 amendments, including the following: "That the People have a right to keep and bear Arms; that a well regulated Militia, including the body of the People capable of bearing Arms, is the proper, natural and safe defence of a free State."

As early as 1776 Congress had advised the colonies to form new governments "such as shall best conduce to the happiness and safety of their constituents." Within a year after the Declaration of Independence was signed, nearly every state had drawn up a new constitution. The constitutions of several states guaranteed the rights of individuals to bear arms but forbade the maintenance of a standing army. In *State Constitutional Right to Keep and Bear Arms Provisions, by Date* (2006, http://www.law.ucla.edu/volokh/beararms/statedat.htm), Eugene Volokh of the University of California, Los Angeles, notes that North Carolina's constitution ensured "that the people have a right to bear arms, for the defence of the State; and, as standing armies, in time of peace, are dangerous to liberty, they ought not to be kept up; and that the military should be kept under strict subordination to, and governed by, the civil power."

Because of their fear of tyranny and repression by a standing army, the colonists preferred state militias to provide protection and order. Such militias could also act as counterbalances against any national standing army. Some people believe that the individual right to bear arms was guaranteed by state laws providing for a militia made up of people trained to use arms.

The Bill of Rights was adopted in 1791. The Second Amendment states: "A well regulated Militia, being necessary to the security of a free State, the right of the people to keep and bear Arms, shall not be infringed."

THE MODERN DEBATE

Efforts to control gun ownership are usually a response to gun-related violence, and most of these initiatives focus on weapons that are perceived to be designed specifically to kill other humans, rather than on guns designed for hunting. Table 1.2 lists the types of firearms that are used in crime and defines many of the types of weapons mentioned in this chapter.

An effort to decrease crime during the early 1930s led to an unsuccessful attempt by President Franklin D. Roosevelt (1882–1945) to pass legislation requiring the registration of handguns. Nonetheless, the National Firearms Act was passed in 1934, which imposed a tax on the manufacture and sale of "Title II weapons" and mandated the registration of those weapons. Title II weapons include machine guns, short-barreled rifles, short-barreled shotguns, destructive devices (such as grenades), and a catchall category that includes novelty devices such as pen guns and cane guns. It does not, however, include handguns.

The next major piece of gun control legislation—the Gun Control Act of 1968—was passed after the assassinations of King and Senator Kennedy. The act encompasses the Title I portion of the U.S. federal firearms laws, and the National Firearms Act described earlier contains the Title II portion. Title I has provisions that include requiring serial numbers on all guns, setting standards for gun dealers, prohibiting mail-order and interstate sales of firearms, prohibiting the importation of guns not used for sporting purposes, and setting penalties for carrying and using firearms in crimes of violence or drug trafficking. It also prohibits certain categories of people, such as convicted felons, drug addicts, illegal aliens, and minors, from buying or possessing firearms. However, there was no efficient national system for carrying out background checks until the Brady Handgun Violence Prevention Act was passed in 1993.

TABLE 1.2

Types of firearms used in crime

Types	
Handgun	A weapon designed to fire a small projectile from one or more barrels when held in one hand with a short stock designed to be gripped by one hand.
Revolver	A handgun that contains its ammunition in a revolving cylinder that typically holds five to nine cartridges, each within a separate chamber. Before a revolver fires, the cylinder rotates, and the next chamber is aligned with the barrel.
Pistol	Any handgun that does not contain its ammunition in a revolving cylinder. Pistols can be manually operated or semiautomatic. A semiautomatic pistol generally contains cartridges in a magazine located in the grip of the gun. When the semiautomatic pistol is fired, the spent cartridge that contained the bullet and propellant is ejected, the firing mechanism is cocked, and a new cartridge is chambered.
Derringer	A small single- or multiple-shot handgun other than a revolver or semiautomatic pistol.
Rifle	A weapon intended to be fired from the shoulder that uses the energy of the explosive in a fixed metallic cartridge to fire only a single projectile through a rifled bore for each single pull of the trigger.
Shotgun	A weapon intended to be fired from the shoulder that uses the energy of the explosive in a fixed shotgun shell to fire through a smooth bore either a number of ball shot or a single projectile for each single pull of the trigger.
Firing action	
Fully automatic	Capability to fire a succession of cartridges so long as the trigger is depressed or until the ammunition supply is exhausted. Automatic weapons are considered machine guns subject to the provisions of the National Firearms Act.
Semiautomatic	An autoloading action that will fire only a single shot for each single function of a trigger.
Machine gun	Any weapon that shoots, is designed to shoot, or can be readily restored to shoot automatically more than one shot without manual reloading by a single function of the trigger.
Submachine gun	A simple fully automatic weapon that fires a pistol cartridge that is also referred to as a machine pistol.
Ammunition	
Caliber	The size of the ammunition that a weapon is designed to shoot, as measured by the bullet's approximate diameter in inches in the United States and in millimeters in other countries. In some instances, ammunition is described with additional terms, such as the year of its introduction (.30/06) or the name of the designer (.30 Newton). In some countries, ammunition is also described in terms of the length of the cartridge case (7.62 × 63 mm).
Gauge	For shotguns, the number of spherical balls of pure lead, each exactly fitting the bore, that equals one pound.

SOURCE: Marianne W. Zawitz, "What Are the Different Types of Firearms?" in *Guns Used in Crime*, U.S. Department of Justice, Office of Justice Programs, Bureau of Justice Statistics, July 1995, http://www.bjs.gov/content/pub/pdf/GUIC.PDF (accessed August 31, 2016)

The Brady Handgun Violence Prevention Act (the Brady law) was named for James Brady (1940–2014), who was an official in the administration of President Ronald Reagan (1911–2004). Brady was shot during a 1981 assassination attempt on the president. The Brady law imposed a five-day waiting period on handgun purchases and a background check on buyers to determine whether they were illegal aliens or had a history of criminal behavior, mental illness, or drug use. It required state and local law enforcement agencies to carry out the background checks until a national system could be established. These early background checks were never effectively carried out because Congress did not provide the funds, and in 1997 the requirement was found unconstitutional by the U.S. Supreme Court under the 10th Amendment (states' rights). The court stated in *Printz v. United States* (521 U.S. 898) that the federal government had no right to order state and local law enforcement agencies to carry out federal programs.

The five-day waiting period and the background check requirement were eventually replaced by a national database for background checks, which became effective in November 1998. Known as the National Instant Criminal Background Check System (NICS), this computerized system is managed by the Federal Bureau of Investigation (FBI). It is used to perform background checks on people seeking to buy handguns or long guns from federal firearms licensees. (Commonly referred to as an FFL, a federal firearms license is usually required for anyone selling a firearm.) According to the FBI, in "Total NICS Firearm Background Checks: November 30, 1998–October 31, 2016" (October 2016, https://www.fbi.gov/file-repository/nics_firearm_checks_-_month_year.pdf), a total of 247.9 million background checks had been conducted between the introduction of the NICS and October 31, 2016.

"Collective Rights" versus "Individual Rights"

The modern debate over gun control is described by Robert J. Spitzer in *The Politics of Gun Control* (2016) as a split between a "collective" or "militia-based" interpretation and an "individualist" interpretation of the Second Amendment. Spitzer, a professor of political science at the State University of New York, Cortland, claims that it is essential to interrogate the meaning and consequences of the Second Amendment because it is a "touchstone of the gun debate."

Proponents of the collective interpretation favor stricter control of guns. They point to the opening words of the Second Amendment—"A well regulated militia, being necessary to the security of a free state"—as an indication that the amendment was intended to guarantee the right of states to maintain militias. They argue that since colonial times the concept of a citizens' militia has fallen from use, having been replaced by the National Guard. By this view, the general population does not need unfettered access to guns because it is no longer expected to form a militia during times of need.

In contrast, proponents of the individualist interpretation hold that individuals have a right to keep and bear arms. Opponents of extensive gun control add that the phrase "right of the people" is used in the Second Amendment, as well as in other amendments in the Bill of Rights, and in each case it refers to a right of individuals.

In November 2001 the U.S. Court of Appeals for the Fifth Circuit held in *United States v. Emerson* (No. 99-10331) that the Second Amendment protects the right of individuals to "privately possess and bear their own firearms." Conversely, in December 2002 the U.S. Court of Appeals for the Ninth Circuit ruled in *Silveira v. Lockyer* (No. 01-15098) that the Second Amendment does not grant Americans a personal right to carry firearms. The court's ruling said the purpose of the Second Amendment was to maintain effective state militias. In December 2003 the U.S. Supreme Court declined to hear a challenge of this ruling, leaving the question of gun ownership rights in limbo. Then in November 2007 the high court announced that it would hear an appeal involving the constitutionality of a District of Columbia law banning the use or possession of all handguns. *District of Columbia v. Heller* (554 U.S. 570) was argued in March 2008. In June 2008 the court ruled in a 5–4 decision that the Second Amendment guarantees individuals the right to bear arms. This ruling affected federal jurisdictions only, but in June 2010, with the ruling in *McDonald v. City of Chicago* (No. 08-1521 [2010]), the high court extended the same interpretation of the Second Amendment to include state and local jurisdictions. These rulings enshrined the individual's right to bear arms to a greater degree than any previous ruling since the beginning of the modern gun control era.

Is Gun Control Unconstitutional?

This strengthening of the individual's right to bear arms, however, did not preclude the ability of the federal and state governments to craft and enforce gun control legislation. The majority opinion in *Heller*, written by Justice Antonin Scalia (1936–2016), noted that "it is not a right to keep and carry any weapon whatsoever in any manner whatsoever and for whatever purpose." For example, the District of Columbia requires that firearms be registered and bans the carrying of guns in public, both openly and concealed. Keeping a loaded gun in one's home is legal. The provision for gun control legislation applies to state and local jurisdictions in the *McDonald* ruling as it does to federal jurisdictions in the *Heller* ruling.

Arguments for and against Gun Control

One of the key issues in the debate over gun control is whether placing greater restrictions on gun ownership will make society safer. Many opponents of extensive gun control think that access to guns makes it possible for law-abiding Americans to protect themselves and deter crime. By contrast, proponents of extensive gun control hold that Americans infrequently use guns for this purpose.

Handguns are a particular point of contention in the gun control debate because they are seen as the weapon of choice for criminals. However, advocates of gun rights point out that upstanding citizens use handguns for self-defense and argue that any attempt to control handgun use is unconstitutional. Those opposing this argument reply that no gun control legislation—including legislation affecting handguns—has ever been declared unconstitutional by the Supreme Court under the Second Amendment.

The statistics appear to indicate that millions of Americans use guns to defend themselves annually. Advocates of extensive gun control or gun prohibition argue that the defensive use of handguns does not offset the offensive use of handguns by criminals, which accounts for thousands of deaths and hundreds of thousands of injuries annually. Spitzer, a gun control advocate, notes that "on an individual level, a gun in the hand of a victim can thwart or stop a crime. On an aggregate level, however, more guns mean more gun problems, even though many citizens believe that guns make them safer."

Additionally, Spitzer and other gun control advocates maintain that the high rate of gun ownership in the United States contributes to the suicide rate, noting that guns are more effective and instantaneous than other modes of committing suicide, thus resulting in more deaths and fewer opportunities to intervene. Gun rights advocates tend to argue that suicidal individuals will find other methods for carrying out their plans. Similar disagreement exists on the topic of accidental injuries and fatalities from guns: gun control advocates maintain that these incidents represent preventable injuries and deaths, whereas gun rights advocates point out that many consumer goods (those not dealing with guns) also result in accidental injuries and deaths but that this does not justify burdening their owners with complicated regulatory systems.

Although many advocates for extensive gun control measures point out that the founding fathers could not have foreseen the existence of the highly effective handguns and assault weapons of the 21st century, the Second Amendment Foundation, a nonprofit group that promotes the right to bear arms, counters that the improved quality of modern weapons does not override the Second Amendment's insistence on the people's right to arm themselves. Likewise, the Second Amendment Foundation and other gun rights groups profess no faith in the ability of stricter gun control measures to keep weapons out of the hands of criminals. They argue that law-abiding citizens' access to guns through legal channels would become limited, whereas criminals would continue to acquire weapons through illegal means.

Spitzer argues that in the interest of national security a compromise must be reached between those who favor gun control and those who favor gun rights. He suggests that citizens should not have access to assault weapons, that access to handguns should be limited, and that ownership of hunting and sporting weapons should be protected. This position is reflected by most gun control advocates, few of whom call for the outright banning of all guns or even of all handguns. Gun rights advocates, meanwhile, often view compromise as a slippery slope that will eventually lead to the wholesale disarming of the U.S. population.

Expert Views

Gun control advocates typically maintain that guns are infrequently used in self-defense and that their widespread use in criminal acts and suicide necessitates stricter regulation. Gun rights advocates typically maintain that guns in the hands of law-abiding citizens make society safer by deterring crime and that further restrictions will only make it harder for these law-abiding citizens to obtain weapons. Experts, like laypeople who participate in the gun control debate, disagree about the degree to which guns are used as protection versus the degree to which they cause unnecessary death and suffering.

Estimates vary widely as to the number of times handguns are used in self-defense annually. Gun rights advocates often point to the research of Gary Kleck (1951–), a criminologist at Florida State University, who has maintained since the mid-1990s that guns are used in self-defense roughly 2 million to 2.5 million times per year. Kleck first published these findings in "Armed Resistance to Crime: The Prevalence and Nature of Self-Defense with a Gun" (*Journal of Law and Criminology*, vol. 86, no. 1, 1995), cowritten with Marc Gertz, and he has substantiated his estimates in subsequent works, most notably the book *Armed: New Perspectives on Gun Control* (2001), cowritten with Don B. Kates (1941–2016). If guns really are used in self-defense 2 million to 2.5 million times per year, then claims by gun rights advocates stating that guns prevent more loss of life than they cost would be more than validated. However, Kleck's findings have not been duplicated by other researchers, and they are considered to be extreme overestimates by most others in fields relating to gun use and safety.

Perhaps the most prominent of Kleck's critics is David Hemenway of the Harvard Injury Control Research Center (HICRC), one of the most reputable research centers in the United States associated with the societal effects of firearms. Hemenway has published a number of studies refuting Kleck's claims, beginning with "Survey Research and Self-Defense Gun Use: An Explanation of Extreme Overestimates" (*Journal of Criminal Law and Criminology*, vol. 87, no. 4, Summer 1997), a direct response to the first influential Kleck and Gertz paper. Hemenway maintains that Kleck and Gertz's original research was subject to a "huge overestimation bias" and a failure to validate survey results relative to other statistics.

For example, Hemenway points out that, when checked against FBI statistics regarding the number and circumstances of robberies annually, Kleck and Gertz's findings about defensive gun use in robberies would require one to believe that more than 100% of gun owners subjected to robbery attempts used guns in self-defense, although more than two-thirds of them were asleep at the time of the intruder's arrival and other studies at that time (based on police reports) showed that only around 1.5% of robbery victims used guns in self-defense. Similarly, Kleck and Gertz estimate that people who used guns in self-defense injured approximately 207,000 criminals each year during the early 1990s, although only around 100,000 people, very few of whom were perpetrators of firearm assaults, were treated for firearm injuries in emergency departments nationally.

Kleck and Kates vigorously reject such critiques, countering that studies conducted by Hemenway and others are the ones with major methodological flaws. Kleck and Kates argue that they are the only researchers to have conducted extensive surveys devoted exclusively to the topic of armed self-defense and that their large sample size and unique methodology makes them more reliable than other scholars. They also maintain that arguments such as Hemenway's, which compare specific subsets of crime data (such as robbery statistics) to the Kleck and Kates findings about the number of times guns are used in self-defense, are logically inconsistent. Because the total number of nonfatal gunshot wounds is unknown, Kleck and Kates argue, crime statistics do not capture the number of times crimes are attempted, let alone the number of times guns are used defensively. In their view, comparing their estimates of defensive gun use with incomplete crime statistics is not a sound means of determining whether their estimates are accurate.

As outlined by Hemenway and Matthew Miller in "Public Health Approach to the Prevention of Gun Violence" (*New England Journal of Medicine*, vol. 368, May 23, 2013), HICRC researchers typically treat firearm ownership and use as a public health issue, an approach that seeks to reduce firearm injuries and deaths by the same strategies that have been used successfully in recent decades to guard public health and safety against outbreaks of infectious disease, tobacco use, and motor vehicle collisions. Hemenway and Miller explain that key tenets of the public health approach include viewing the problem as one that is population-based, rather than as one involving specific individuals; focusing on strategies for prevention that address the environment in which the problem occurs, rather than on the individuals at the site of the incident (e.g., perpetrators or victims); creating change through a systems approach that reduces the likelihood of accidents and transgressions that lead to injuries and fatalities; exploring all possible interventions, from changing social norms to passing new legislation; and emphasizing shared responsibility, rather than blame, in an effort to engage as many people and institutions as possible to solve the problem.

Since the late 1980s HICRC researchers have conducted wide-ranging reviews of the scholarly literature on firearm ownership rates and violence as well as data analysis for all 50 states and for 26 developed countries and have published their findings in numerous scholarly journals. For example, in "Firearm Legislation and Firearm-Related Fatalities in the United States" (*Journal of the American Medical Association*, vol. 173, no. 9, May 13, 2013), Eric W. Fleegler et al. examine whether stricter state firearm laws correlate with lower firearm fatality rates. After more than two decades of investigation, the overarching finding of HICRC researchers is that higher rates of gun ownership consistently correlate with higher rates of gun deaths. This finding holds true in all countries, even when the United States (which has the highest rate of gun ownership among developed countries and is thus a statistical outlier) is excluded; and across U.S. states, it holds true for every age group of residents, even when controlling for poverty and urbanization.

CHAPTER 2
HOW MANY GUNS ARE THERE, AND WHO OWNS THEM?

OWNERSHIP BY PRIVATE CITIZENS
There Can Be Only Estimates

"How many guns are there?" is a question that cannot be answered with exact figures for the United States due to differences among the states. Each has its own system of counting and classifying guns. Some states do not require registration of guns, and unregistered guns cannot be included in an official count. In addition, some types of gun data are restricted from public access. The result is that there can be only estimates of the total number of guns that U.S. residents possess.

The ATF Estimates

The Bureau of Alcohol, Tobacco, Firearms, and Explosives (ATF) is a law enforcement organization in charge of reducing violent crime, among other tasks. One of its duties is to keep firearms out of the hands of criminals. The ATF is also responsible for estimating the total number of firearms in the United States. It does this by adding domestic firearms production and imports since 1899, then subtracting firearms exports during the same period. The ATF does not take into account guns that are destroyed or that no longer work. The ATF statistics also do not account for guns that are smuggled into or out of the United States or guns that are manufactured illegally.

According to the ATF, in *Firearms Commerce in the United States—2001/2002* (2002), between 1899 and 1999 an estimated 248 million guns became available for sale in the United States (not including those produced for the military). This number included more than 87 million rifles, 86 million handguns, and 72 million shotguns. The ATF estimates that there were 1.5 million guns produced in the United States in 1950, 3.7 million in 1970, 5.6 million in 1980, 3.8 million in 1990, and 4 million in 1999. The ATF data suggest that the number of imported rifles, shotguns, and handguns combined averaged 1 million per year during the 1990s, with handguns accounting for roughly half of that figure. Exports averaged fewer than 400,000 per year. Putting all estimates together, by the end of 1999 the total number of guns privately owned or available for sale in the United States came to approximately 260 million.

The publication of ATF data on gun sales was restricted by law between 2003 and 2008, as a result of an amendment authored by Representative Todd Tiahrt (1951–; R-KS) and attached to funding bills for the U.S. Departments of Commerce, Justice, and State. The Tiahrt Amendment also made it illegal for the agency to report on firearms tracing statistics (gun trace statistics). The Tiahrt Amendment restrictions were relaxed in 2008 and again in 2010, allowing for increased data sharing among law enforcement agencies and for the public release of some data that had been suppressed. However, provisions preventing the public disclosure of certain gun trace data, as well as data use in lawsuits against the firearms industry, remained in place as of December 2016.

With the resumption of regular ATF reporting on gun sales, it is possible to add to the cumulative gun totals between 1899 and 1999 to arrive at a rough estimate of guns either privately owned or available for sale in the United States as of 2014. Table 2.1, Table 2.2, and Table 2.3 provide statistics from the ATF on the number of firearms manufactured, imported, and exported since 1986. Using yearly totals from these tables, it is possible to determine that 77.8 million firearms were manufactured in the United States between 2000 and 2014 and that 3.6 million of these were exported to other countries, for a total of 74.2 million guns that were both manufactured and offered for sale in the United States. Additionally, 41.1 million firearms were imported from other countries to the United States, bringing the total number of firearms either privately owned or available for sale between 2000 and 2014 to 115.3 million. This figure can be added to the 260 million guns estimated to be in the United States as of

TABLE 2.1

Firearms manufactured, 1986–2014

Calendar year	Pistols	Revolvers	Rifles	Shotguns	Misc. firearms*	Total firearms
1986	662,973	761,414	970,507	641,482	4,558	3,040,934
1987	964,561	722,512	1,007,661	857,949	6,980	3,559,663
1988	1,101,011	754,744	1,144,707	928,070	35,345	3,963,877
1989	1,404,753	628,573	1,407,400	935,541	42,126	4,418,393
1990	1,371,427	470,495	1,211,664	848,948	57,434	3,959,968
1991	1,378,252	456,966	883,482	828,426	15,980	3,563,106
1992	1,669,537	469,413	1,001,833	1,018,204	16,849	4,175,836
1993	2,093,362	562,292	1,173,694	1,144,940	81,349	5,055,637
1994	2,004,298	586,450	1,316,607	1,254,926	10,936	5,173,217
1995	1,195,284	527,664	1,411,120	1,173,645	8,629	4,316,342
1996	987,528	498,944	1,424,315	925,732	17,920	3,854,439
1997	1,036,077	370,428	1,251,341	915,978	19,680	3,593,504
1998	960,365	324,390	1,535,690	868,639	24,506	3,713,590
1999	995,446	335,784	1,569,685	1,106,995	39,837	4,047,747
2000	962,901	318,960	1,583,042	898,442	30,196	3,793,541
2001	626,836	320,143	1,284,554	679,813	21,309	2,932,655
2002	741,514	347,070	1,515,286	741,325	21,700	3,366,895
2003	811,660	309,364	1,430,324	726,078	30,978	3,308,404
2004	728,511	294,099	1,325,138	731,769	19,508	3,099,025
2005	803,425	274,205	1,431,372	709,313	23,179	3,241,494
2006	1,021,260	385,069	1,496,505	714,618	35,872	3,653,324
2007	1,219,664	391,334	1,610,923	645,231	55,461	3,922,613
2008	1,609,381	431,753	1,734,536	630,710	92,564	4,498,944
2009	1,868,258	547,195	2,248,851	752,699	138,815	5,555,818
2010	2,258,450	558,927	1,830,556	743,378	67,929	5,459,240
2011	2,598,133	572,857	2,318,088	862,401	190,407	6,541,886
2012	3,487,883	667,357	3,168,206	949,010	306,154	8,578,610
2013	4,441,726	725,282	3,979,570	1,203,072	495,142	10,844,792
2014	3,633,454	744,047	3,379,549	935,411	358,165	9,050,626

*Miscellaneous firearms are any firearms not specifically categorized in any of the firearms categories defined on the ATF (Bureau of Alcohol, Tobacco, and Firearms) Form 5300.11 Annual Firearms Manufacturing Exportation Report (AFMER). (Examples of miscellaneous firearms would include pistol grip firearms, starter guns, and firearm frames and receivers.)

Notes: The AFMER report excludes production for the U.S. military but includes firearms purchased by domestic law enforcement agencies. The report also includes firearms manufactured for export. AFMER data is not published until one year after the close of the calendar year reporting period because the proprietary data furnished by filers is protected from immediate disclosure by the Trade Secrets Act. For example, calendar year 2012 data was due to ATF by April 1, 2013, but not published until January 2014.

SOURCE: "Exhibit 1. Firearms Manufactured (1986–2014)," in *Firearms Commerce in the United States: Annual Statistical Update, 2016*, U.S. Department of Justice, Bureau of Alcohol, Tobacco, Firearms, and Explosives, August 9, 2016, https://www.atf.gov/resource-center/docs/2016-firearms-commerce-united-states/download (accessed August 31, 2016)

1999, for a total of 375.3 million guns privately owned or available for sale in the United States by the end of 2014.

U.S. FIREARMS MANUFACTURING

Table 2.1 shows the numbers of pistols, revolvers, rifles, and shotguns that were manufactured in the United States between 1986 and 2014. The number of pistols manufactured annually rose from 662,973 in 1986 to 2,093,362 in 1993, an increase of 216%, and then dropped to a low of 626,836 in 2001. Thereafter, pistol production steadily increased, rising to an all-time high of 4,441,726 pistols in 2013. Rifle manufacture ranged between roughly 1 million and 1.5 million annually between 1986 and 2006, before increasing to a high of 2,248,851 in 2009. Rifle production fell to 1,830,556 weapons in 2010, a 19% decline from the previous year, before returning to record levels in 2011 (2,318,088) and then dramatically exceeding those levels in 2013 (3,979,570).

Excluded from the totals detailed in Table 2.1 are National Firearms Act (NFA) weapons. NFA weapons include machine guns, short-barreled rifles, short-barreled shotguns, silencers, and destructive devices. (See Chapter 1 for a description of the National Firearms Act of 1934. In addition, Table 1.2 in Chapter 1 lists and describes several types of firearms.) In 1986 Congress banned the manufacture of machine guns for private sale, but they can be manufactured for export, for the military, and for law enforcement personnel. Approximately 4.4 million NFA weapons were registered in the United States in 2016, including 575,602 machine guns. (See Table 2.4.) Texas (468,581), California (324,417), Virginia (281,356), and Florida (277,227) had the most NFA weapons registered in 2016.

The top firearms manufacturers in the United States are Sturm, Ruger & Co., Inc. (also known as Ruger), Remington Arms Company, LLC, and Smith & Wesson. Jade Molde and Russ Thurman indicate in "U.S. Firearms Industry Today 2016" (ShootingIndustry.com, 2016) that these three companies produced a combined total of 4,465,924 firearms in 2014. With headquarters in Connecticut and manufacturing facilities in New Hampshire, North Carolina, and Arizona, Ruger produced 722,029 pistols,

TABLE 2.2

Firearms imported, 1986–2015

Calendar year	Shotguns	Rifles	Handguns	Total
1986	201,000	269,000	231,000	701,000
1987	307,620	413,780	342,113	1,063,513
1988	372,008	282,640	621,620	1,276,268
1989	274,497	293,152	440,132	1,007,781
1990	191,787	203,505	448,517	843,809
1991	116,141	311,285	293,231	720,657
1992	441,933	1,423,189	981,588	2,846,710
1993	246,114	1,592,522	1,204,685	3,043,321
1994	117,866	847,868	915,168	1,880,902
1995	136,126	261,185	706,093	1,103,404
1996	128,456	262,568	490,554	881,578
1997	106,296	358,937	474,182	939,415
1998	219,387	248,742	531,681	999,810
1999	385,556	198,191	308,052	891,799
2000	331,985	298,894	465,903	1,096,782
2001	428,330	227,608	710,958	1,366,896
2002	379,755	507,637	741,845	1,629,237
2003	407,402	428,837	630,263	1,466,502
2004	507,050	564,953	838,856	1,910,859
2005	546,403	682,100	878,172	2,106,675
2006	606,820	659,393	1,166,309	2,432,522
2007	725,752	631,781	1,386,460	2,743,993
2008	535,960	602,364	1,468,062	2,606,386
2009	558,679	864,010	2,184,417	3,607,106
2010	509,913	547,449	1,782,585	2,839,947
2011	529,056	998,072	1,725,276	3,252,404
2012	973,465	1,243,924	2,627,201	4,844,590
2013	936,235	1,507,776	3,095,528	5,539,539
2014	648,339	791,892	2,185,037	3,625,268
2015	644,293	815,817	2,470,101	3,930,211

Note: Statistics prior to 1992 are for fiscal years; 1992 is a transition year with five quarters.

SOURCE: "Exhibit 3. Firearms Imports (1986–2015)," in *Firearms Commerce in the United States: Annual Statistical Update, 2016*, U.S. Department of Justice, Bureau of Alcohol, Tobacco, Firearms, and Explosives, August 9, 2016, https://www.atf.gov/resource-center/docs/2016-firearms-commerce-united-states/download (accessed August 31, 2016)

281,430 revolvers, 706,192 rifles, and 4,446 shotguns in 2014. Founded in 1816, Remington Arms is the oldest and second-largest firearms manufacturer in the United States. Headquartered in North Carolina, with plants in New York, Kentucky, and Arizona, the company concentrates heavily on long-gun manufacturing. Of the 1,409,129 firearms the company produced in 2014, 929,823 were rifles and 413,535 were shotguns. By contrast, Smith & Wesson, which was founded in 1852 and is headquartered in Massachusetts, with additional manufacturing facilities in New Hampshire and Maine, is predominantly a handgun manufacturer. Of the 1,342,698 firearms the company produced in 2014, 914,700 were pistols and 268,722 were revolvers.

FIREARMS IMPORTS

The United States imports a sizable number of guns each year. According to the Gun Control Act of 1968, imported firearms must be "generally recognized as particularly suitable for or readily adaptable to sporting purposes, excluding surplus military firearms." Gun import statistics from the U.S. International Trade Commission show that firearms imports for the U.S. civilian market more than quadrupled between 1986 and 1993, from 701,000 to 3,043,321. (See Table 2.2.) Firearms imports, however, declined by more than a million units the following year and remained below 1.5 million annually through 2001. The first decade of the 21st century saw an increase in the number of firearms imported into the United States, surpassing 2 million units in 2005 and continuing to increase in the years that followed. After reaching an all-time high of 5,539,539 in 2013, firearms imports declined to 3,930,211 in 2015, but this was still nearly double the number imported a decade earlier.

Table 2.5 shows the countries from which firearms were imported in 2015. The top-four source countries for imported firearms—Austria (926,534 total firearms), Brazil (602,449), Croatia (338,535), and Canada (337,877)—supplied 56% of all firearms imported into the United States. Handguns (2,470,101) were the most common type of imported firearm, accounting for 63% of total imports. Austria was the leading supplier of imported handguns (923,951), followed by Brazil (485,639), Croatia (338,535), and Germany (236,935). Canada (334,268) was by far the leading source country for imported rifles, while Turkey (220,310), Italy (199,266), and China (164,818) were the leading source countries for imported shotguns.

FIREARMS EXPORTS

Exports of most firearms and ammunition are regulated under the Arms Export Control Act of 1976, under which prospective exporters are generally required to obtain a license from the Directorate of Defense Trade Controls, an office within the U.S. Department of State. Table 2.3 shows the number of firearms that were exported by U.S. manufacturers between 1986 and 2014. Between 1986 and 1993 the export of firearms nearly doubled from 217,448 to 431,204. From that peak, however, firearms exports began a period of decline, reaching a low of 139,920 in 2004. After 2004 firearms exports increased significantly in 2005 (194,682) and then nearly doubled in 2006 (367,521). After several years of reduced numbers, U.S. firearms exports surpassed 2006 levels in 2013 and reached an all-time high of 420,932 in 2014.

The Small Arms Survey, an independent research center based at the Graduate Institute of International and Development Studies in Geneva, Switzerland, monitors the global circulation and use of small arms (predominantly guns and ammunition, but also including rocket and grenade launchers), providing research and analysis to support the reduction of armed violence and illegal arms trafficking around the world. The organization's annual report provides data for major small arms exporters (those with annual exports of at least $10 million). In *Small Arms Survey 2015* (April 2015, http://www.smallarmssurvey.org/publications/by-type/year

TABLE 2.3

Firearms exported, 1986–2014

Calendar year	Pistols	Revolvers	Rifles	Shotguns	Misc. firearms*	Total firearms
1986	16,511	104,571	37,224	58,943	199	217,448
1987	24,941	134,611	42,161	76,337	9,995	288,045
1988	32,570	99,289	53,896	68,699	2,728	257,182
1989	41,970	76,494	73,247	67,559	2,012	261,282
1990	73,398	106,820	71,834	104,250	5,323	361,625
1991	79,275	110,058	91,067	117,801	2,964	401,165
1992	76,824	113,178	90,015	119,127	4,647	403,791
1993	59,234	91,460	94,272	171,475	14,763	431,204
1994	93,959	78,935	81,835	146,524	3,220	404,473
1995	97,969	131,634	90,834	101,301	2,483	424,221
1996	64,126	90,068	74,557	97,191	6,055	331,997
1997	44,182	63,656	76,626	86,263	4,354	275,081
1998	29,537	15,788	65,807	89,699	2,513	203,344
1999	34,663	48,616	65,669	67,342	4,028	220,318
2000	28,636	48,130	49,642	35,087	11,132	172,627
2001	32,151	32,662	50,685	46,174	10,939	172,611
2002	22,555	34,187	60,644	31,897	1,473	150,756
2003	16,340	26,524	62,522	29,537	6,989	141,912
2004	14,959	24,122	62,403	31,025	7,411	139,920
2005	19,196	29,271	92,098	46,129	7,988	194,682
2006	144,779	28,120	102,829	57,771	34,022	367,521
2007	45,053	34,662	80,594	26,949	17,524	204,782
2008	54,030	28,205	104,544	41,186	523	228,488
2009	56,402	32,377	61,072	36,455	8,438	194,744
2010	80,041	25,286	76,518	43,361	16,771	241,977
2011	121,035	23,221	79,256	54,878	18,498	296,888
2012	128,313	19,643	81,355	42,858	15,385	287,554
2013	167,653	21,236	131,718	49,766	22,748	393,121
2014	126,316	25,521	207,934	60,377	784	420,932

*Miscellaneous firearms are any firearms not specifically categorized in any of the firearms categories define on the Bureau of Alcohol, Tobacco, and Firearms (ATF) Form 5300.11 Annual Firearms Manufacturing and Exportation Report (AFMER). (Examples of miscellaneous firearms would include pistol grip firearms, starter guns, and firearm frames and receivers.)
Notes: The AFMER report excludes production for the U.S. military but includes firearms purchased by domestic law enforcement agencies. This exhibit does not include statistics related to the National Firearms Act (NFA).

SOURCE: "Exhibit 2. Firearms Manufacturers' Exports (1986–2014)," in *Firearms Commerce in the United States: Annual Statistical Update, 2016*, U.S. Department of Justice, Bureau of Alcohol, Tobacco, Firearms, and Explosives, August 9, 2016, https://www.atf.gov/resource-center/docs/2016-firearms-commerce-united-states/download (accessed August 31, 2016)

book/small-arms-survey-2015.html), the organization notes that in 2012 the world's top-earning small arms exporters included Italy, whose exports were valued at $544 million, Germany ($472 million), Brazil ($374 million), and Austria ($293 million). Even so, the United States easily eclipsed all these countries, with small arms exports valued at $935 million. The Small Arms Survey indicates that the top importers were the United States, Canada, Germany, Australia, France, the United Kingdom, Thailand, and Indonesia.

AUTOMATIC AND SEMIAUTOMATIC FIREARMS

The Firearms Owners' Protection Act, which amended the Gun Control Act of 1968, was signed into law in May 1986 and banned private ownership of any machine gun not already lawfully owned. Machine guns are fully automatic weapons, which means they can fire a steady stream of bullets. (See Table 1.2 in Chapter 1.)

Semiautomatic guns manufactured to look like machine guns, commonly referred to as "assault weapons," are frequently confused with machine guns by the media and the public. The term *assault weapons* generally refers to military-style semiautomatic firearms. These weapons look like guns used in the military, but because they are semiautomatic (able to fire only a single shot for each pull of the trigger), they are not actual military-grade weapons. Nevertheless, the term *assault weapons* is often used by the media and the public to refer to machine guns, which are military weapons.

Although assault weapons do not rise to the lethality level of military-grade machine guns, they do typically enable greater lethality than firearms used for hunting, especially when they are outfitted with accessories such as high-capacity magazines (cartridge holders that feed bullets automatically into a gun's chamber). Many mass shooters have used assault weapons with magazines, enabling them to shoot many times without having to pause for reloading. Gun control advocates often argue that semiautomatic assault weapons are inappropriately effective killing tools suitable only for criminal purposes and that these weapons should be banned accordingly. Defenders of the right to own assault weapons typically consider assault weapons to be no more deadly than hunting rifles, and they point to the applicability of assault weapons for hunting, informal target shooting, and competitive target shooting.

TABLE 2.4

National Firearms Act registered weapons, by state, February 2016

State	Any other weapon[a]	Destructive device[b]	Machinegun[c]	Silencer[d]	Short barreled rifle[e]	Short barreled shotgun[f]	Total
Alabama	1,166	77,283	19,390	30,849	3,765	2,234	134,687
Alaska	322	4,651	1,652	4,213	1,333	1,268	13,439
Arkansas	604	45,813	5,189	14,370	2,464	1,060	69,500
Arizona	1,185	86,309	16,318	28,942	10,351	2,143	145,248
California	3,884	256,420	29,516	11,702	9,472	13,423	324,417
Colorado	961	44,598	7,111	18,689	4,859	1,660	77,878
Connecticut	708	12,220	37,939	8,352	3,374	983	63,576
District of Columbia	69	39,134	4,496	308	787	1,101	45,895
Delaware	32	2,748	583	324	184	537	4,408
Florida	3,448	151,672	34,373	61,015	18,409	8,310	277,227
Georgia	1,929	66,433	28,497	49,357	8,132	11,175	165,523
Hawaii	34	6,863	410	143	59	62	7,571
Iowa	876	15,398	3,495	983	521	965	22,238
Idaho	641	17,172	4,039	18,176	2,437	459	42,924
Illinois	979	101,572	30,563	1,717	2,278	1,716	138,825
Indiana	1,615	41,343	18,880	28,395	4,709	8,816	103,758
Kansas	704	22,580	3,667	7,715	2,276	966	37,908
Kentucky	1,088	30,580	13,568	22,238	2,914	1,811	72,199
Louisiana	543	51,268	6,686	25,203	3,603	1,712	89,015
Massachusetts	831	15,028	6,705	6,772	2,312	981	32,629
Maryland	998	49,865	26,074	12,340	4,102	4,068	97,447
Maine	578	3,320	4,858	2,467	1,725	454	13,402
Michigan	1,155	24,561	12,913	10,146	2,203	1,244	52,222
Minnesota	2,654	42,890	9,264	5,655	2,404	1,132	63,999
Missouri	1,378	30,433	9,472	15,052	4,080	2,499	62,914
Mississippi	426	9,628	4,220	10,589	1,747	798	27,408
Montana	444	3,675	2,373	7,928	1,042	438	15,900
North Carolina	922	84,705	11,947	24,346	6,697	2,922	131,539
North Dakota	204	1,926	1,581	4,836	642	271	9,460
Nebraska	763	6,291	2,217	6,576	1,354	810	18,011
New Hampshire	447	4,253	9,325	12,120	3,194	482	29,821
New Jersey	426	42,674	7,730	1,130	1,144	2,568	55,672
New Mexico	302	79,180	3,847	6,159	2,080	683	92,251
Nevada	812	35,251	10,336	14,928	5,504	954	67,785
New York	1,632	41,055	12,210	3,466	4,594	7,449	70,406
Ohio	1,872	78,639	19,952	30,714	6,137	5,920	143,234
Oklahoma	1,169	16,087	9,136	32,192	3,804	1,675	64,063
Oregon	1,549	20,825	6,415	17,537	4,158	1,439	51,923
Pennsylvania	2,115	152,262	17,989	29,742	7,779	12,835	222,722
Rhode Island	41	3,187	639	29	119	113	4,128
South Carolina	678	33,365	8,302	23,451	3,495	3,922	73,213
South Dakota	352	3,761	1,718	7,320	547	194	13,892
Tennessee	1,599	40,466	13,855	19,736	5,438	6,010	87,104
Texas	6,740	224,498	34,848	165,499	29,509	7,487	468,581
Utah	455	16,117	6,699	26,039	3,725	1,332	54,367
Virginia	2,854	193,728	33,199	31,205	12,665	7,705	281,356
Vermont	223	2,548	1,122	405	295	130	4,723
Washington	1,855	43,287	4,229	21,797	3,967	822	75,957
Wisconsin	760	30,096	7,292	11,156	3,169	1,224	53,697
West Virginia	446	16,928	6,774	5,020	1,349	1,023	31,540
Wyoming	303	120,899	1,774	3,744	675	392	127,787
Other US territories	6	359	215	18	12	97	707
Total	**57,777**	**2,545,844**	**575,602**	**902,805**	**213,594**	**140,474**	**4,436,096**

In 1989 the ATF issued an order banning the importation of 43 models of semiautomatic assault-type guns following a schoolyard shooting in Stockton, California, that killed five people and wounded 29 others. The Violent Crime Control and Law Enforcement Act of 1994 banned the sale and possession of 19 assault-type firearms and copycat models, including the Uzi, the TEC-9, and the Street Sweeper. The act also limited the capacity of newly manufactured magazines to 10 bullets. (Chapter 3 describes this law in more detail.)

The assault weapons ban expired in September 2004. Bills were introduced in the U.S. House of Representatives in 2007 and 2008 to reinstate and extend the assault weapons ban, but neither bill became law. During the presidential campaigns of 2008 and 2012 the possibility of reinstating the ban on assault weapons was raised. These attempts took on new urgency following the Newtown, Connecticut, mass shooting in December 2012, in which 20 first graders and six school staff members were fatally shot with a semiautomatic assault weapon. However, the Newtown shooting did not substantially change the political feasibility of reinstating the ban. Senator Dianne Feinstein's (1933–; D-CA) Assault Weapons Ban of 2013 won only 40 votes in the U.S. Senate in April of that year. In December 2015—in the wake of a terrorist mass shooting in San Bernardino,

TABLE 2.4

National Firearms Act registered weapons, by state, February 2016 [CONTINUED]

[a]The term "any other weapon" means any weapon or device capable of being concealed on the person from which a shot can be discharged through the energy of an explosive, a pistol or revolver having a barrel with a smooth bore designed or redesigned to fire a fixed shotgun shell, weapons with combination shotgun and rifle barrels 12 inches or more, less than 18 inches in length, from which only a single discharge can be made from either barrel without manual reloading, and shall include any such weapon which may be readily restored to fire. Such term shall not include a pistol or a revolver having a rifled bore, or rifled bores, or weapons designed, made, or intended to be fired from the shoulder and not capable of firing fixed ammunition.
[b]Destructive device generally is defined as (a) any explosive, incendiary, or poison gas (1) bomb, (2) grenade, (3) rocket having a propellant charge of more than 4 ounces, (4) missile having an explosive or incendiary charge of more than one-quarter ounce, (5) mine, or (6) device similar to any of the devices described in the preceding paragraphs of this definition; (b) any type of weapon (other than a shotgun or a shotgun shell which the director finds is generally recognized as particularly suitable for sporting purposes) by whatever name known which will, or which may be readily converted to, expel a projectile by the action of an explosive or other propellant, and which has any barrel with a bore of more than one-half inch in diameter; and (c) any combination of parts either designed or intended for use in converting any device into any destructive device described in paragraph (a) or (b) of this section and from which a destructive device may be readily assembled. The term shall not include any device which is neither designed nor redesigned for use as a weapon; any device, although originally designed for use as a weapon, which is redesigned for use as a signaling, pyrotechnic, line throwing, safety, or similar device; surplus ordnance sold, loaned, or given by the Secretary of the Army pursuant to the provisions of section 4684(2), 4685, or 4686 of title 10, United States Code; or any other device which the director finds is not likely to be used as a weapon, is an antique, or is a rifle which the owner intends to use solely for sporting, recreational, or cultural purposes.
[c]Machinegun is defined as any weapon which shoots, is designed to shoot, or can be readily restored to shoot, automatically more than one shot, without manual reloading, by a single function of the trigger. The term shall also include the frame or receiver of any such weapon, any part designed and intended solely and exclusively, or combination of parts designed and intended, for use in converting a weapon into a machinegun, and any combination of parts from which a machinegun can be assembled if such parts are in the possession or under the control of a person.
[d]Silencer is defined as any device for silencing, muffling, or diminishing the report of a portable firearm, including any combination of parts, designed or redesigned, and intended for the use in assembling or fabricating a firearm silencer or firearm muffler, and any part intended only for use in such assembly or fabrication.
[e]Short-barreled rifle is defined as a rifle having one or more barrels less than 16 inches in length, and any weapon made from a rifle, whether by alteration, modification, or otherwise, if such weapon, as modified, has an overall length of less than 26 inches.
[f]Short-barreled shotgun is defined as a shotgun having one or more barrels less than 18 inches in length, and any weapon made from a shotgun, whether by alteration, modification, or otherwise, if such weapon as modified has an overall length of less than 26 inches.

SOURCE: "Exhibit 8. National Firearms Act Registered Weapons by State (Feb 2016)," in *Firearms Commerce in the United States: Annual Statistical Update, 2016*, U.S. Department of Justice, Bureau of Alcohol, Tobacco, Firearms, and Explosives, August 9, 2016, https://www.atf.gov/resource-center/docs/2016-firearms-commerce-united-states/download (accessed August 31, 2016)

California, in which 14 people were killed and 22 were seriously injured by attackers whose weapons included a semiautomatic assault weapon—Representative David Cicilline (1961–; D-RI) introduced the Assault Weapons Ban of 2015. The bill received broad backing from House Democrats, but as of December 2016 it remained in committee and was widely believed to have no chance of passing under the Republican-controlled Congress.

GUNS IN THE HOME
Percentages of Adults Having Guns in the Home

The Gallup Organization, an independent, nonpartisan polling group, has followed trends in U.S. gun ownership over the decades. In *Guns* (2016, http://www.gallup.com/poll/1645/guns.aspx), Gallup indicates that the percentage of adult Americans answering "yes" to the question "Do you have a gun in your home?" has varied since 1960, although the general trend has been downward. During the 1960s about half of all respondents reported having a gun in their home, and by the early 1980s only 40% answered "yes" to the question. The percentage answering "yes" rose through the late 1980s and reached 51% in the early 1990s, before settling in a range between 34% and 43% through the late 1990s and into the second decade of the 21st century. In 2014, 41% of adult Americans reported having a gun in their home.

Table 2.6, which shows the results of the General Social Surveys conducted between 1973 and 2014 by the National Opinion Research Center at the University of Chicago, yields a similar picture of decreasing firearms prevalence in American homes. The center's surveys find that the percentage of adults who had a gun in the home decreased from 47% in 1973 to 31% in 2014, the lowest rate in four decades. Tom W. Smith and Jaesok Son of the National Opinion Research Center explain in *General Social Survey Final Report: Trends in Gun Ownership in the United States, 1972–2014* (2015, http://www.norc.org/PDFs/GSS%20Reports/GSS_Trends%20in%20Gun%20Ownership_US_1972-2014.pdf) that the significant decline in household gun ownership is largely attributable to a nationwide decline in the popularity of hunting. Whereas 31.6% of American adults reported being a hunter, or married to a hunter, in 1973, this number had fallen to just 15.4% in 2014.

Research on Guns in the Home and Safety

Does the presence of guns in the home increase or decrease the overall safety of a household? This question has long been a point of contention between gun rights and gun control advocates. Guns in the home can be used to fend off intruders, and gun rights advocates typically maintain that the increased safety brought about by their use in such circumstances outweighs any negative safety effects they may have. However, guns can also be used in conflicts between household members, be accidentally discharged, and be used to commit suicide. Numerous studies have attempted to quantify the effect on safety of having guns in the home, and most have found that the injury or death of a household member or other innocent person is a much more likely outcome than the successful protection of the home against an intruder.

TABLE 2.5

Firearms imported, by country, 2015

	Handguns	Rifles	Shotguns	Total firearms
Austria	923,951	1,867	716	926,534
Brazil	485,639	78,585	38,225	602,449
Croatia	338,535	0	0	338,535
Canada	3,417	334,268	192	337,877
Italy	107,940	27,222	199,266	334,428
Turkey	80,939	339	220,310	301,588
Germany	236,935	17,651	1,536	256,122
China[a]	0	0	164,818	164,818
Czech Republic	76,029	28,181	109	104,319
Philippines	79,457	5,603	6,400	91,460
Japan	0	87,012	904	87,916
Serbia	18,066	62,308	0	80,374
Belgium	18,679	54,497	713	73,889
Finland	0	50,481	0	50,481
Argentina	42,304	0	0	42,304
Romania	9,460	17,982	0	27,442
Spain	239	25,371	839	26,449
Israel	15,618	4,302	0	19,920
United Kingdom	7,419	4,687	578	12,684
Bulgaria	6,267	5,100	0	11,367
Poland	10,783	527	0	11,310
Russia	0	4,406	5,150	9,556
Switzerland	3,971	2,463	0	6,434
Portugal	0	2,117	4,175	6,292
Hungary	1,521	27	0	1,548
Slovak Republic	1,075	0	0	1,075
Slovenia	1,058	0	0	1,058
Other[b]	799	821	362	1,982
Total	**2,470,101**	**815,817**	**644,293**	**3,930,211**

[a]On May 26, 1994, the United States instituted a firearms imports embargo against China. Sporting shotguns, however, are exempt from the embargo.
[b]Imports of fewer than 1,000 per country.
Notes: Imports from Afghanistan, Belarus, Burma, China, Cuba, Democratic Republic of Congo, Haiti, Iran, Iraq, Libya, Mongolia, North Korea, Rwanda, Somalia Sudan, Syria, Unita (Angola), Vietnam, may include surplus military curio and relic firearms that were manufactured in these countries prior to becoming proscribed or embargoed and had been outside those proscribed countries for the preceding five years prior to import. Imports may also include those that obtained a waiver from the U.S. State Department. Imports from Georgia, Kazakstan, Kyrgyzstan, Moldova, Russian Federation, Turkmenistan, Ukraine, Uzbekistan are limited to firearms enumerated on the Voluntary Restraint Agreement (VRA).

SOURCE: "Exhibit 5. Firearms Imported into the United States by Country of Manufacture 2015," in *Firearms Commerce in the United States: Annual Statistical Update, 2016*, U.S. Department of Justice, Bureau of Alcohol, Tobacco, Firearms, and Explosives, August 9, 2016, https://www.atf.gov/resource-center/docs/2016-firearms-commerce-united-states/download (accessed August 31, 2016)

TABLE 2.6

Percentage of U.S. households with guns, 1973–2014

	% of households with guns			
	Gun	No gun	Refused	Don't know	Missing
1973	47.0	51.4	1.0	0.0	0.6
1974	46.1	52.9	0.7	0.1	0.3
1976	46.5	52.0	1.1	0.0	0.4
1977	50.4	48.9	0.1	0.1	0.5
1980	47.3	51.8	0.1	0.1	0.7
1982	45.3	52.9	1.3	0.2	0.3
1984	44.9	54.0	0.6	0.1	0.4
1985	44.2	54.9	0.7	0.0	0.3
1987	46.0	53.3	0.5	0.0	0.1
1988	39.8	58.4	1.0	0.0	0.7
1989	46.0	53.7	0.0	0.0	0.3
1990	42.2	56.7	0.0	0.1	1.0
1991	39.6	58.7	1.0	0.0	0.7
1993	42.0	57.1	0.7	0.0	0.2
1994	40.6	58.1	1.0	0.2	0.2
1996	40.1	59.4	0.4	0.1	0.0
1998	34.8	64.6	0.3	0.1	0.3
2000	32.4	66.1	1.2	0.1	0.2
2002	33.5	65.5	1.0	0.0	0.0
2004	34.7	62.9	1.6	0.2	0.6
2006	33.1	65.2	1.6	0.1	0.0
2008	34.0	64.6	1.1	0.3	0.0
2010	31.1	65.0	3.5	0.5	0.0
2012	33.1	64.7	2.0	0.2	0.0
2014	31.0	65.7	3.2	0.2	0.0

SOURCE: Adapted from Tom W. Smith and Jaesok Son, "Table 1. Trends in Household Gun Ownership," in *General Social Survey Final Report: Trends in Gun Ownership in the United States, 1972–2014*, National Opinion Research Center, University of Chicago, 2015, http://www.norc.org/PDFs/GSS%20Reports/GSS_Trends%20in%20Gun%20Ownership_US_1972-2014.pdf (accessed September 7, 2016)

In "Injuries and Deaths Due to Firearms in the Home" (*Journal of Trauma*, vol. 45, no. 2, August 1998), a landmark study on the relationship between guns in the home and overall safety, Arthur L. Kellerman et al. report on their analysis of police, medical examiner, emergency medical, and hospital records pertaining to shootings in Memphis, Tennessee; Seattle, Washington; and Galveston, Texas. Of the 626 shootings that occurred in these three cities during the study period, 54 were unintentional shootings, 118 were attempted or completed suicides, and 438 were assaults or homicides. By contrast, only 13 shootings were legally justifiable or motivated by self-defense, and three of those involved law enforcement officers. The researchers accordingly conclude that "guns kept in homes are more likely to be involved in a fatal or nonfatal accidental shooting, criminal assault, or suicide attempt than to be used to injure or kill in self-defense."

Subsequent studies have reached varying quantitative conclusions, but most have been in broad agreement regarding their overall conclusions. For example, Garen J. Wintemute of the University of California, Davis, School of Medicine frames in "Guns, Fear, the Constitution, and the Public's Health" (*New England Journal of Medicine*, vol. 358, April 3, 2008) the presence of guns in the home as a public health issue and summarizes findings from a variety of researchers to quantify the risks that accompany firearm ownership. He states, "Living in a home where there are guns increases the risk of homicide by 40 to 170% and the risk of suicide by 90 to 460%. Young people who commit suicide with a gun usually use a weapon kept at home, and among women in shelters for victims of domestic violence, two thirds of those who come from homes with guns have had those guns used against them."

In *U.S. Health in International Perspective: Shorter Lives, Poorer Health* (Steven H. Woolf and Laudan Aron, eds., 2013), a joint report by the National Research Council and the Institute of Medicine, the "American health-wealth paradox" is examined, wherein the U.S. population demonstrates relatively poor health status and lower life expectancy than the populations of comparatively poorer countries. The report considers this paradox from many different perspectives, including social, cultural, and environmental factors, government policies, the health care system, and individual behaviors. With regard to the latter, the report posits, "One behavior that probably explains the excess lethality of violence and unintentional injuries in the United States is the widespread possession of firearms and the common practice of storing them (often unlocked) at home." The report notes that the United States holds between 35% and 50% of all civilian-owned firearms in the world. Furthermore, among 23 countries in the Organisation for Economic Co-operation and Development (a group consisting primarily of countries with advanced industrialized economies), 80% of all firearms deaths occurred in the United States in 2003.

According to most researchers, children in homes where guns are present are at particular risk of injury or death. In the policy statement "Firearm-Related Injuries Affecting the Pediatric Population" (*Pediatrics*, vol. 130, no. 5, November 1, 2012), the American Academy of Pediatrics (AAP) notes that "although rates have declined since the American Academy of Pediatrics (AAP) issued the original policy statement in 1992, firearm-related deaths continue as 1 of the top 3 causes of death in American youth." The AAP goes on to maintain, as part of its official recommendations, that "the most effective measure to prevent suicide, homicide, and unintentional firearm-related injuries to children and adolescents is the absence of guns from homes and communities."

Nevertheless, the National Rifle Association and gun rights advocates maintain the dissenting view that guns in the hands of law-abiding citizens are effective and necessary for deterring violence. This argument continues to rest heavily on research that was conducted during the late 1990s by Gary Kleck (1951–), a criminologist at Florida State University. As is discussed in Chapter 1, Kleck and coauthor Don B. Kates (1941–2016; Kates later became a research fellow with the Independent Institute in Oakland, California) argue in *Armed: New Perspectives on Gun Control* (2001) that guns are used for defense 2 million to 2.5 million times per year in the United States. They posit that this estimate has been confirmed many times "by all surveys of comparable technical merit." They also contend that the study of guns and violence in general, and the study of guns and self-defense in particular, are plagued by "advocacy scholarship" and "junk science."

CHARACTERISTICS OF FIREARMS OWNERS

In *Why Own a Gun? Protection Is Now Top Reason* (March 12, 2013, http://www.people-press.org/files/legacy-pdf/03-12-13%20Gun%20Ownership%20Release.pdf), the Pew Research Center provides an overview of the characteristics of gun owners in the United States based on surveys conducted in February 2013. Pew finds that 37% of survey respondents reported the presence of a gun in their home, 24% of whom reported personally owning the gun and 13% of whom reported that someone else in the household owned the gun. (See Table 2.7.) Nearly half (45%) of the men surveyed reported a gun in the household, and 37% of them claimed to own the gun personally, compared with 30% and 12% of women, respectively. Respondents aged 50 to 64 years (40%) and 65 years and older (40%) owned a gun at a higher rate than those between the ages of 18 and 29 years (35%) and 30 to 49 years (35%). There appeared to be no conclusive correlation between education levels and household gun ownership. Likewise, there were no dramatic differences in the gun-owning rates of parents versus nonparents.

By contrast, the demographic indicators that had the strongest correlations with gun ownership were race and Hispanic origin, U.S. region, and urban/suburban/rural location. (See Table 2.7.) Racial differences in gun ownership rates were pronounced: 46% of white survey respondents reported the presence of a gun in the home, compared with 21% of African American and 17% of Hispanic respondents. Likewise, respondents in the Midwest (45%) and the South (42%) were much more likely

TABLE 2.7

Gun owners by selected demographic characteristics, 2013

	Household owns gun %	Personally own gun %	Someone else
Total	37	24	13
Men	45	37	8
Women	30	12	17
18–29	35	16	19
30–49	35	24	11
50–64	40	30	10
65+	40	29	11
White	46	31	15
Black	21	15	7
Hispanic	17	11	6
College grad+	37	24	12
Post-graduate	33	20	13
College grad	40	27	13
Some college	43	28	15
High school or less	33	22	11
Northeast	25	17	8
Midwest	45	27	18
South	42	29	13
West	30	21	9
Urban	28	18	10
Suburban	36	24	12
Rural	59	39	20
Married	45	31	14
Not married	30	19	11
Parents	34	23	11
Men	45	40	5
Women	24	8	16
Non-parents	39	26	13
Men	45	36	9
Women	33	15	18

SOURCE: "Demographics of Gun Ownership," in *Why Own a Gun? Protection Is Now Top Reason*, Pew Research Center, March 12, 2013, http://www.people-press.org/files/legacy-pdf/03-12-13%20Gun%20Ownership%20Release.pdf (accessed September 22, 2016)

TABLE 2.8

Reasons for owning a gun, 1999 and 2013

What is main reason you own a gun?	1999 Protection %	1999 Hunting %	1999 Other %	2013 Protection %	2013 Hunting %	2013 Other %	2013 N
All gun owners	26	49	26	48	32	18	421
Men	21	55	24	42	36	20	303
Women	43	26	31	65	21	13	118
18–49	27	47	25	52	29	18	157
50+	22	52	26	45	36	18	260
College grad+	25	49	26	45	28	25	154
Some college	34	45	21	53	31	14	128
High school grad or less	22	51	27	45	37	17	138
Rep/lean Rep	23	50	27	45	39	14	216
Dem/lean Dem	28	49	23	53	28	19	155

SOURCE: "Why Do You Own a Gun?" in *Why Own a Gun? Protection Is Now Top Reason*, Pew Research Center, March 12, 2013, http://www.people-press.org/files/legacy-pdf/03-12-13%20Gun%20Ownership%20Release.pdf (accessed September 7, 2016)

TABLE 2.9

Attitudes of non-gun households toward having a gun in the home, by comfort or discomfort, 2013

Among those without a gun in household	Comfortable %	Uncomfortable %	Don't know %
Total	40	58	3 = 100
Men	49	47	4 = 100
Women	33	65	2 = 100
18–29	55	41	4 = 100
30–49	39	58	2 = 100
50–64	28	69	3 = 100
65+	36	62	2 = 100
White	47	50	2 = 100
Black	36	61	3 = 100
Hispanic	24	73	3 = 100
College grad+	36	61	3 = 100
Some college	47	50	2 = 100
High school or less	37	60	3 = 100
Republican	58	38	4 = 100
Democrat	30	69	2 = 100
Independent	44	53	3 = 100
Parents	36	61	3 = 100
Non-parents	41	56	3 = 100

Note: Based on those without a gun in the household. Whites and blacks include only those who are not Hispanic; Hispanics are of any race.

SOURCE: "How Would You Feel about Having a Gun in Your Home?" in *Why Own a Gun? Protection Is Now Top Reason*, Pew Research Center, March 12, 2013, http://www.people-press.org/files/legacy-pdf/03-12-13%20Gun%20Ownership%20Release.pdf (accessed September 7, 2016)

to report the presence of a gun in the household than those living in the West (30%) or Northeast (25%). A solid majority of rural respondents (59%) reported having a gun in their home, compared with 36% of suburban respondents and 28% of urban respondents.

Comparing its 2013 survey results with results from a similar survey that was conducted by ABC News and the *Washington Post* in 1999, Pew finds that gun owners' reasons for owning a gun have changed significantly. (See Table 2.8.) In 1999 roughly half (49%) of all gun owners reported owning a gun primarily for the purpose of hunting, and approximately a quarter (26%) gave protection as the primary motivation for gun ownership. By 2013 the proportions of respondents giving these reasons for gun ownership had reversed: 48% said protection was the primary reason they owned a gun and 32% said hunting was the primary reason.

Meanwhile, among those Americans who did not keep a gun in their home, 58% reported that they would be uncomfortable doing so. (See Table 2.9.) Whereas men without a gun in their home were about evenly split between those who would (49%) and would not (47%) be comfortable doing so, the disparity among women was much more pronounced. Among women who did not keep a gun in their home, 33% indicated they would be comfortable doing so, while 65% indicated they would not.

CHAPTER 3
FIREARM LAWS, REGULATIONS, AND ORDINANCES

FEDERAL GOVERNMENT REGULATION OF GUNS

Americans have long debated the issue of federal regulation of firearms. Those in favor of regulation argue that only federal firearm laws can limit access by criminals, juveniles, and other high-risk people, thereby reducing violent crime. Supporters of regulation also contend that without federal laws, states with few firearms restrictions will supply guns illegally to states with stiffer restrictions. Opponents of federal oversight advance Second Amendment arguments against any kind of gun control. They also argue that federal gun laws only create extra burdens for law-abiding citizens who seek to buy and sell firearms.

Until the 1920s the states made their own decisions about whether and how to regulate firearms. The federal government stepped in following ratification of the 18th Amendment (1919), which made it illegal to manufacture, transport, and sell "intoxicating liquors" in the United States. Known as Prohibition, the ban sparked a national crime wave as gangsters built empires on the illegal manufacture and sale of alcohol. For the first time, crime was seen as a national problem, and people debated solutions at the federal level.

The first federal law regulating firearms was passed in 1927 (18 USC 1715). The law outlawed mailing any firearm other than a shotgun or rifle through the U.S. Postal Service, except for firearms shipped for official law enforcement purposes. This law, which was still in effect as of 2016, was intended to curb the mail-order business in handguns, and it prompted many states to pass their own regulations regarding the sale and use of handguns. Supporters of the federal law tried to get a law passed forbidding the shipment of handguns across state lines by all commercial carriers, but they failed. Instead, commercial carriers other than the U.S. Postal Service were allowed to carry handguns across state lines.

The next federal regulation was the National Firearms Act of 1934, which was designed to make it more difficult to acquire especially dangerous "gangster-type" weapons such as machine guns, sawed-off shotguns, and silencers. The legislation placed heavy taxes on all aspects of the manufacture and distribution of these firearms and required registration of each firearm through the entire production, distribution, and sales process. This law was still in force, with amendments, as of 2016.

Ordinary firearms were the subject of the Federal Firearms Act of 1938. The act required any manufacturer or dealer who sent or received firearms across state lines to have a federal firearms license and to keep a record of the names and addresses of people purchasing firearms. Firearms could not be sent to anyone who was a fugitive from justice or had been convicted of a felony. It became illegal to transport stolen guns from which the manufacturer's mark had been rubbed out or changed. This act was replaced with the passage of the Gun Control Act of 1968.

THE GUN CONTROL ACT OF 1968

The Gun Control Act of 1968 was passed in the wake of the assassinations of President John F. Kennedy (1917–1963), the civil rights leader Martin Luther King Jr. (1929–1968), and Senator Robert F. Kennedy (1925–1968; D-NY). It repealed the Federal Firearms Act of 1938 and amended the National Firearms Act of 1934 by adding bombs and other destructive devices to machine guns and sawed-off shotguns as items strictly controlled by the government. Its purpose was to assist federal, state, and local law enforcement agencies in reducing crime and violence.

The Gun Control Act of 1968 had two major sections. Title I required anyone dealing in firearms or ammunition—whether locally or across state lines—to be federally licensed under tough new standards and to keep records of all commercial gun sales. Title I also prohibited the interstate mail-order sale of all firearms and ammunition, the interstate sale of handguns generally, and the interstate sale of long guns (rifles and shotguns), except under certain conditions. It forbade sales to minors and those with criminal records, outlawed the importation of nonsporting firearms, and established special penalties for the use or carrying of a firearm while committing a crime of violence or drug trafficking. However, Title I did not forbid the importation of unassembled weapons parts, and some individuals and companies were suspected of importing separate firearms parts and then reassembling them into a complete weapon as a means of getting around the law.

Title II, the National Firearms Act, reenacted the 1934 National Firearms Act and extended the Gun Control Act to cover private ownership of destructive devices such as submachine guns, bombs, and grenades. Enforcement of the federal laws became the responsibility of the U.S. Department of the Treasury, which created the Bureau of Alcohol, Tobacco, and Firearms (ATF) in 1972. The ATF has since come under the control of the U.S. Department of Justice and has been renamed the Bureau of Alcohol, Tobacco, Firearms, and Explosives.

Efforts to Amend the Gun Control Act of 1968

After the passage of the Gun Control Act of 1968, Congress found itself under siege from both pro-gun and antigun groups. Firearms owners and dealers complained about ATF enforcement efforts, which seemed to target law-abiding citizens while neglecting criminals with firearms and people selling firearms illegally. Others voiced concern that the act penalized sportsmen. In contrast, those urging tighter control on guns believed the act did not go far enough in keeping firearms out of the hands of criminals. Every Congress from 1968 to 1986 introduced dozens of pieces of legislation to strengthen, repeal, or diminish the requirements of the 1968 act. In 1986 the gun control advocates were successful.

THE FIREARMS OWNERS' PROTECTION ACT OF 1986

In 1986, nearly 20 years after the passage of the Gun Control Act of 1968, Congress passed major legislation that amended the 1968 law. The Firearms Owners' Protection Act of 1986, commonly referred to as the Gun Control Act of 1986, was still in effect as of 2016. The battle over this piece of legislation was fierce. David T. Hardy describes in "The Firearm Owners' Protection Act of 1986: A Historical and Legal Perspective" (*Cumberland Law Review*, vol. 17, 1986) the reactions of those in favor and those opposed to its passage. Those in favor of the act called its passage "necessary to restore fundamental fairness and clarity to our nation's firearms laws." Those opposed called it an "almost monstrous idea" and a "national disgrace." The following sections compare the 1968 act with the current standards under the 1986 act.

Changes Implemented with the Gun Control Act of 1986 and Further Amendments

PROHIBITED PEOPLE. The 1968 legislation identified the categories of people to whom firearms could not be sold by a federally licensed firearms dealer, also called a federal firearms licensee (FFL). These included convicted felons, drug abusers, and the mentally ill. The 1986 legislation clarified some inconsistencies in the 1968 act and made it unlawful for anyone, whether licensed or not, to sell a gun to a high-risk individual. The definition of the term *high risk* has been refined over the years and includes groups such as felons, fugitives, illegal aliens, and those subject to a restraining order.

The 1996 enactment of the Domestic Violence Offender Gun Ban (generally known as the Lautenberg Amendment after its sponsor Senator Frank Lautenberg [1924–2013; D-NJ]) to the Gun Control Act of 1968 expanded the group of people prohibited from legally obtaining a gun. The Lautenberg Amendment made it a felony for anyone convicted of a misdemeanor crime of domestic violence (e.g., assault or attempted assault on a family member) to ship, transport, possess, or receive firearms or ammunition. There was no exception for military personnel or law enforcement officers engaged in official duties unless their record has been expunged (erased). The amendment also made it a felony for anyone to sell or issue a firearm or ammunition to a person with such a conviction. The Lautenberg Amendment was ruled unconstitutional in 1999, but that ruling was reversed two years later. It was challenged and upheld again in 2009 in *United States v. Hayes* (555 U.S. 415), and, as of 2016, the Lautenberg Amendment was still in force.

The Omnibus Consolidated and Emergency Supplemental Appropriations Act of 1999 also amended the Gun Control Act of 1968 by prohibiting aliens admitted under a nonimmigrant visa from obtaining firearms. This included people traveling temporarily in the United States, those studying in the United States who maintained a residence abroad, and some foreign workers. Exceptions included people entering the United States for lawful hunting or sporting purposes, official representatives of a foreign government, and foreign law enforcement officers from friendly foreign governments who entered the United States on official business. In addition, the U.S. attorney general had the authority to waive the prohibition on submission of a petition by an alien.

PURCHASING FIREARMS AND AMMUNITION. The Gun Control Act of 1968 stated that firearms and ammunition could be purchased only on the premises of an FFL and that FFLs were required to record all firearms and ammunition transactions. The 1986 law made the purchase of ammunition and gun components by mail legal. Under the new law firearms dealers were required to keep records only of armor-piercing ammunition sales. Establishments selling only ammunition (no firearms) would not have to be licensed, thus allowing stores that previously may not have carried ammunition because of licensing requirements to do so.

The Treasury and General Government Appropriations Act of 2000 amended the Gun Control Act to require that firearms owners who pawn their weapons must undergo a background check when they seek to redeem the weapons.

CARRYING A GUN BETWEEN JURISDICTIONS. The 1968 act did not address the effects of state and local regulations concerning the intrastate (within a state) transportation of weapons. The 1986 act made it legal to transport any legally owned gun through a jurisdiction where it would otherwise be illegal, provided the possession and transporting of the weapon were legal at the point of origin and the point of destination. In addition, the gun had to be unloaded and placed in a locked container or in the trunk of a vehicle during transport.

FEDERAL CRIMES. The 1968 legislation prohibited the carrying or use of a firearm during or in relation to a federal crime of violence. In addition, anyone convicted of such a crime would be subject to a minimum penalty over and above any sentence received for the primary offense. The 1986 act added serious drug offenses to the category of prohibited crimes, and it doubled the existing penalty for use of a machine gun or a gun equipped with a silencer.

FORFEITURE OF FIREARMS AND AMMUNITION. Before 1986 any firearm or ammunition involved in, used in, or intended to be used in violation of the Gun Control Act or other federal criminal law could be taken away from the gun owner. However, under the Gun Control Act of 1986, forfeiture is no longer automatic. For some offenses, a willful element must be demonstrated; for others, knowledge of an offense is enough. In the case of a firearm being "intended for use" in a violation, "clear and convincing evidence" of the intent must be shown. In addition, only specified crimes now justify forfeiture, including crimes of violence, drug-related offenses, and certain violations of the Gun Control Act. In all cases forfeiture proceedings must begin within 120 days of seizure, and the court will award attorney fees to the owner if the owner wins the case.

CRIMINAL PENALTIES. The 1968 law stated that a demonstration of willfulness was not needed as an element of proof of violation of any provision of the act, whereas the Gun Control Act of 1986 required proof of willful violations and/or knowing violations to prosecute. The 1986 act also reduced licensee record-keeping violations from felonies to misdemeanors.

LEGAL DISABILITIES. Under the 1968 law, any person who had been convicted of a crime and sentenced to prison for more than one year was restricted from shipping, transporting, or receiving a firearm and could not be granted an FFL. State pardons could not erase the conviction for federal purposes. However, a convicted felon could make a special request to the U.S. secretary of the treasury to be allowed to possess a firearm. The secretary of the treasury had to certify that the possession of a firearm by the convicted felon was not contrary to public interest and safety and that the applicant did not commit a crime involving a firearm or violate federal gun control laws.

Under the 1986 act, however, state pardons can erase convictions for federal purposes, unless the person is specifically denied the right to possess or receive firearms. The act allows those who violated the law to appeal—even those whose crime involved the use of a firearm or the violation of federal gun control laws.

MACHINE GUN FREEZE. The National Firearms Act of 1934 imposed production and transfer taxes and registration requirements on firearms typically associated with criminal activity. These restrictions applied specifically to machine guns; destructive devices such as bombs, missiles, and grenades; and firearm silencers. The definition of a machine gun included "any combination of parts designed and intended for use in converting a weapon to a machine gun."

The Firearms Owners' Protection Act of 1986 made it "unlawful for any person to transfer or possess a machine gun" unless it was manufactured and legally owned before May 19, 1986. In addition, the definition was revised to include any combination of parts "designed and intended solely and exclusively for use in conversion." By refusing to review a lower court's decision in *Farmer v. Higgins* (907 F.2d 1041 [11th Cir. 1990]), the U.S. Supreme Court upheld the ban on machine-gun ownership.

THE LAW ENFORCEMENT OFFICERS' PROTECTION ACT OF 1986

The Law Enforcement Officers' Protection Act of 1986 also amended the 1968 Gun Control Act. This act banned the manufacture or importation of certain varieties of armor-piercing ammunition. Called "cop-killers," these bullets are capable of penetrating police officers' bulletproof vests. The law defined the banned ammunition as handgun bullets made of specific hard metals: tungsten alloys, steel, brass, bronze, iron, beryllium copper, or depleted uranium. (Standard ammunition is made from lead.) It also decreed that the licenses of dealers who knowingly sold such ammunition should be revoked.

In 1994 the Violent Crime Control and Law Enforcement Act broadened the ban to include other metal-alloy ammunition. Both laws limit the sale of such bullets to the U.S. military or to the police.

The legislation banning armor-piercing ammunition was politically significant because it polarized two traditionally allied groups: the National Rifle Association of America (NRA) and the police. The NRA had long assisted in training police officers in marksmanship. The police saw the ammunition ban as a personal issue, because the cop-killer bullets were intended to harm them. A police lobby, the Law Enforcement Steering Committee (LESC), was formed to get this and other legislation passed. The NRA opposed the legislation as originally written because, it said, the legislation would have also banned most of the types of ammunition used for hunting and target shooting. The final version of the legislation exempted bullets made for rifles and sporting purposes.

THE UNDETECTABLE FIREARMS ACT OF 1988

Highly publicized aircraft hijackings during the 1980s prompted Congress to begin hearings in 1986 to determine if plastic firearms represented a danger to airline passengers if used by terrorists. The passage of the Undetectable Firearms Act of 1988 followed. It banned the manufacture, import, sale, transfer, or possession of a plastic firearm. The act also stated: "If the major parts of the firearms do not permit an accurate X-ray picture of the gun's shape, the firearm is [also defined as] a plastic firearm, even if the firearm contains more than 3.7 ounces of electromagnetically detectable metal."

The NRA called the proposed legislation unnecessary and the first step toward a total ban on handguns. The lobbying efforts of police officers on the LESC convinced the U.S. attorney general Edwin Meese (1931–) that plastic weapons posed an unacceptable hazard to public safety and that this bill was a necessary piece of legislation. The ban was renewed and made permanent in December 2003.

THE "TOY GUN LAW" OF 1988

A law requiring that a "toy, look-alike, or imitation firearm shall have as an integral part, permanently affixed, a blaze orange plug inserted in the barrel of such toy, look-alike, or imitation firearm" was passed as section four of the Federal Energy Management Improvement Act of 1988. The law provided for alternative markings if the orange plug could not be used. However, the plug could be removed and the markings painted over.

Ending Sales of Look-Alike Toy Guns

U.S. sales of realistic-looking toy guns generated almost $250 million in 1993. Even so, in November 1994 three major toy retailers—Kay-Bee Toy Stores, Toys "R" Us, and Bradlees—announced that they would stop selling the look-alike toy guns because the toys had led to tragic consequences. On September 27, 1994, in two separate incidents in Brooklyn, New York, 13-year-old Nicholas Heyward Jr. and 16-year-old Jamiel Johnson were shot by police officers who mistook the boys' toy guns for real ones. Heyward died from his gunshot wounds, while Johnson was seriously wounded. Besides these tragic police errors, a report by the New York Police Department revealed that realistic-looking toy guns had been used in 534 felonies up until October 1994. In 1999 the New York City council passed a law requiring toy guns to be made with certain pastel or fluorescent colors so they would be easily distinguishable to law enforcement officers.

However, despite the law and retailers' pledges sales of realistic-looking toy guns did not stop. In 2004, after a New York City council investigation found that 20% of all toy and discount stores in the city were still selling the realistic-looking toy guns, Mayor Michael Bloomberg (1942–) launched a renewed effort to eliminate the look-alike guns from local stores' shelves. The city began levying large fines against retailers that violated the city law. Between December 2009 and January 2010 the city conducted an advertising campaign designed to illustrate how difficult it was to distinguish between the look-alike toy guns and real, deadly weapons.

Although New York City has made national headlines for its aggressive stance on the issue of look-alike toy guns (also commonly referred to as replica guns), across the country the imitation firearms have proved extremely difficult to regulate, due in large part to strenuous opposition from the NRA, makers and vendors of the toy guns, and other gun rights advocates. In "Fatal Texas Shooting Highlights Struggle to Regulate Replica Guns" (January 13, 2012, https://www.publicintegrity.org/2012/01/13/7870/fatal-texas-shooting-highlights-struggle-regulate-replica-guns), Susan Ferriss of the Center for Public Integrity explains that in the absence of strong federal restrictions, a "confusing patchwork of laws to regulate the sale, use, and color of replicas has developed among states and communities," but the replica guns remain widely available to juveniles, and shooting incidents in which the toys are mistaken for real weapons continue to occur with alarming frequency. Among the many such incidents that have occurred in the United States in recent years, perhaps none galvanized more public outcry than the 2014 shooting death of 12-year-old Tamir Rice in Cleveland, Ohio. Responding to a 911 call in which an onlooker had reported seeing Rice in a public park carrying what appeared to be a handgun, a police officer fatally shot the boy. As it turned out, Rice had been carrying a replica gun that had its orange safety indicator removed.

In 2015 the *Guardian* newspaper launched "The Counted," a project to track the number and demographics of people killed in the United States by law enforcement and the circumstances under which they were killed. Ciara McCarthy and Jon Swaine report in "At Least 28 People Holding BB or Pellet Guns Were Killed by Police in US in 2015" (Guardian.com, December 29, 2015) that in 2015 U.S. law enforcement killed at least 28 people who were carrying replica guns much like the one in the Rice case. The most recent report on this issue by the Department of

Justice, *Toy Guns: Involvement in Crime & Encounters with Police* (http://www.bjs.gov/content/pub/pdf/tg-icep.pdf), was published in June 1990.

THE GUN-FREE SCHOOL ZONES ACT OF 1990

Congressional passage of the Gun-Free School Zones Act, which was part of the Crime Control Act of 1990, stipulated that it was unlawful for anyone to knowingly possess firearms in school zones. Considered quite strict, the law made it illegal to carry even unloaded firearms in an unlocked case or bag while on public sidewalks in designated school zones. This applied to the sidewalk in front of the firearm owner's residence as well, if that portion of the walkway was within 1,000 feet (305 m) of the grounds of any public or private school—whether or not school was in session. Gun control advocates thought the legislation was needed to send a message to teachers and law enforcement personnel that the federal government was behind them in the effort to get guns out of schools.

Gun rights advocates condemned this attempt to restrict their rights. The act was challenged as unconstitutional and was struck down by the Supreme Court in *United States v. Lopez* (514 U.S. 549 [1995]). The court upheld state and local authority in the regulation of schools and ruled that Congress had exceeded its power in passing the original act. In response to the court's ruling, Congress approved a slightly revised version of the Gun-Free School Zones Act in 1996. The focus of the act was changed from possessing a firearm in a school zone to possessing a firearm "that has moved in or that otherwise affects interstate or foreign commerce" in a school zone. The amended law has since been upheld numerous times in U.S. circuit courts.

THE BRADY HANDGUN VIOLENCE PREVENTION ACT OF 1993

In 1981 a mentally disturbed young man named John W. Hinckley Jr. (1955–)—armed with a handgun—shot and seriously wounded the presidential press secretary James Brady (1940–2014) during an assassination attempt on President Ronald Reagan (1911–2004). After the attempt, Hinckley was quickly apprehended, placed in custody, tried, and found not guilty by reason of insanity. His violent act was the impetus behind the Brady Handgun Violence Prevention Act, which was passed by Congress in November 1993 after years of debate.

"Interim Brady": The Five-Day Waiting Period

The Brady Handgun Violence Prevention Act (commonly referred to as the Brady law), which went into effect on February 28, 1994, established a national five-day waiting period for a person to purchase a handgun. This waiting period allowed time for local law enforcement officers to conduct background checks on handgun purchasers for mental instability or criminal records. The waiting time also provided a cooling-off period to reduce gun-related crimes of passion. Provisions for a waiting period and for background checks by local law enforcement officials were interim measures of the Brady law that could be dropped after a computerized national instant criminal identification system was developed. Such a system became operational in the United States in 1998.

Before 1998, within the first day of the waiting period, licensed firearms dealers were required to send a copy of the purchaser's sworn statement to the chief law enforcement officer where the purchaser resided. The purchaser was asked many questions, including whether he or she was a fugitive from justice, addicted to drugs, illegally in the United States, or convicted of a crime. The dealer could complete the sale after five business days unless the police notified the dealer that the sale would violate the law. It also required the police to respond within 20 business days to any request for a written explanation of a request denial. If the sale was not denied, local law enforcement officials were required to destroy the sworn statement and any other record of the transaction within 20 business days.

In 1997 the Brady law's interim provision requiring law enforcement officers to conduct background checks was ruled unconstitutional by the Supreme Court. Consequently, gun purchasers were no longer required to fill out the Brady Handgun Purchase Form. However, the five-day waiting period remained in place until November 30, 1998, when it expired. It was replaced by a mandatory computerized criminal background check through the National Instant Criminal Background Check System (NICS) before any firearm purchase from a federally licensed firearms dealer.

People Who May Not Purchase Firearms under the Brady Law

The Brady law prohibits firearms sales to any person who:

- Is charged with a crime punishable by imprisonment for more than one year or has been convicted of such a crime

- Is a fugitive from justice

- Is an unlawful user of a controlled substance

- Has been judged mentally ill or has been committed to a mental institution

- Has renounced U.S. citizenship

- Is subject to a court order restraining him or her from harassing, stalking, or threatening an intimate partner or a child

- Has been convicted of domestic violence

- Is an illegal alien

- Has been dishonorably discharged from the military

"PERMANENT BRADY": THE NICS

On November 30, 1998, the five-day waiting period for handgun purchasers was replaced by the NICS, which was designed to quickly screen purchasers of both handguns and long guns. As a result, all FFLs must verify the identity of a customer and receive authorization for the sale from the NICS, which usually takes less than a minute. This is a permanent provision of the Brady law.

How the NICS Works

Under the NICS, FFLs call a toll-free telephone number and provide information on prospective firearms buyers. The Federal Bureau of Investigation (FBI) checks the NICS, which has access to four databases: an interstate criminal records database containing nearly 72 million criminal history records as of December 2016; a database containing information on people who are the subject of protection orders, criminal warrants, and other violations; the NICS's own database, which is not shared with other government agencies and which contained 15.6 million active records as of October 31, 2016; and a database maintained by the U.S. Department of Homeland Security's immigration enforcement section, which allows for background searches of noncitizens who are attempting to receive firearms. From 1999 to 2005 the number of background checks conducted annually under the NICS fluctuated between 8.5 million and 9.1 million. (See Table 3.1.) Since 2006, when 10 million checks were conducted, the annual number of checks has more than doubled, reaching 23.1 million in 2015. As of August 31, 2016, the NICS had processed 243.6 million background checks since its introduction in 1998.

In many states some or all inquiries are directed to the state's designated point of contact (POC), which will then conduct a background check through the NICS. In states without a designated POC, gun dealers contact the FBI directly, and the FBI runs the check through the NICS. Figure 3.1 shows the components of the national firearm check system and how the NICS fits into the system. (The firearm transferee is the person buying the gun, and the federal firearm licensee is the person licensed to sell the gun.) The FFL must initiate a background check on the person wanting to buy a gun. Depending on the state, the FFL does this by contacting either the NICS directly or a state-designated POC. In the POC states, the states act as intermediaries between the FFLs and the NICS.

As of 2014, 30 states, five territories, and the District of Columbia were non-POC entities, which means the FBI conducted all the NICS checks. (See Figure 3.2.) Thirteen states were full POC states, which means they maintained their own Brady NICS Program and conducted their own background checks by electronically accessing the NICS. The rest of the states were partial POC states, in that they conducted NICS checks on handgun transfers and had the FBI conduct checks on long gun transfers.

When the NICS check is initiated, one or more of the following things can happen:

- No disqualifying information is found (the search takes 30 seconds or less) and the transaction can proceed immediately.

TABLE 3.1

Total National Instant Criminal Background Check System checks, November 30, 1998–August 31, 2016

Year	Jan	Feb	Mar	Apr	May	Jun	Jul	Aug	Sep	Oct	Nov	Dec	Totals
1998											21,196	871,644	892,840
1999	591,355	696,323	753,083	646,712	576,272	569,493	589,476	703,394	808,627	945,701	1,004,333	1,253,354	9,138,123
2000	639,972	707,070	736,543	617,689	538,648	550,561	542,520	682,501	782,087	845,886	898,598	1,000,962	8,543,037
2001	640,528	675,156	729,532	594,723	543,501	540,491	539,498	707,288	864,038	1,029,691	983,186	1,062,559	8,910,191
2002	665,803	694,668	714,665	627,745	569,247	518,351	535,594	693,139	724,123	849,281	887,647	974,059	8,454,322
2003	653,751	708,281	736,864	622,832	567,436	529,334	533,289	683,517	738,371	856,863	842,932	1,008,118	8,481,588
2004	695,000	723,654	738,298	642,589	542,456	546,847	561,773	666,598	740,260	865,741	890,754	1,073,701	8,687,671
2005	685,811	743,070	768,290	658,954	557,058	555,560	561,358	687,012	791,353	852,478	927,419	1,164,582	8,952,945
2006	775,518	820,679	845,219	700,373	626,270	616,097	631,156	833,070	919,487	970,030	1,045,194	1,253,840	10,036,933
2007	894,608	914,954	975,806	840,271	803,051	792,943	757,884	917,358	944,889	1,025,123	1,079,923	1,230,525	11,177,335
2008	942,556	1,021,130	1,040,863	940,961	886,183	819,891	891,224	956,872	973,003	1,183,279	1,529,635	1,523,426	12,709,023
2009	1,213,885	1,259,078	1,345,096	1,225,980	1,023,102	968,145	966,162	1,074,757	1,093,230	1,233,982	1,223,252	1,407,155	14,033,824
2010	1,119,229	1,243,211	1,300,100	1,233,761	1,016,876	1,005,876	1,069,792	1,089,374	1,145,798	1,368,184	1,296,223	1,521,192	14,409,616
2011	1,323,336	1,473,513	1,449,724	1,351,255	1,230,953	1,168,322	1,157,041	1,310,041	1,253,752	1,340,273	1,534,414	1,862,327	16,454,951
2012	1,377,301	1,749,903	1,727,881	1,427,343	1,316,226	1,302,660	1,300,704	1,526,206	1,459,363	1,614,032	2,006,919	2,783,765	19,592,303
2013	2,495,440	2,309,393	2,209,407	1,714,433	1,435,917	1,281,351	1,283,912	1,419,088	1,401,562	1,687,599	1,813,643	2,041,528	21,093,273
2014	1,660,355	2,086,863	2,488,842	1,742,946	1,485,259	1,382,975	1,402,228	1,546,497	1,456,032	1,603,469	1,803,397	2,309,684	20,968,547
2015	1,772,794	1,859,584	2,012,488	1,711,340	1,580,980	1,529,057	1,600,832	1,745,410	1,795,102	1,976,759	2,243,030	3,314,594	23,141,970
2016	2,545,802	2,613,074	2,523,265	2,145,865	1,870,000	2,131,485	2,197,169	1,853,815					17,880,475
Total													243,558,967

Note: These statistics represent the number of firearm background checks initiated through the National Instant Criminal Background Check System (NICS). They do not represent the number of firearms sold. Based on varying state laws and purchase scenarios, a one-to-one correlation cannot be made between a firearm background check and a firearm sale.

SOURCE: "Total NICS Background Checks: November 30, 1998–August 31, 2016," in *National Instant Criminal Background Check System*, Federal Bureau of Investigation, August 31, 2016, https://www.fbi.gov/file-repository/nics_firearm_checks_-_month_year.pdf/view (accessed September 7, 2016)

FIGURE 3.1

Components of the National Instant Criminal Background Check System

Note: POC is point of contact.

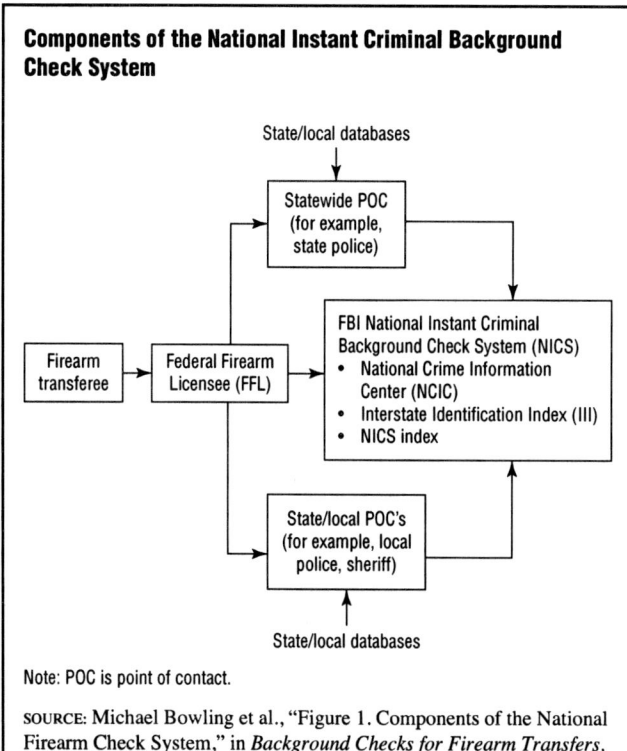

SOURCE: Michael Bowling et al., "Figure 1. Components of the National Firearm Check System," in *Background Checks for Firearm Transfers, 2002*, U.S. Department of Justice, Office of Justice Programs, Bureau of Justice Statistics, September 2003, http://www.bjs.gov/content/pub/pdf/bcft02.pdf (accessed September 14, 2016)

- The transaction is briefly delayed because it is necessary for the FBI to look outside the NICS for data.

- The FBI contacts state or local law enforcement for information because it is not possible to conduct an NICS check electronically. The FBI has three business days to complete its background check. If the check cannot be completed in three days, the transaction may still occur, even though potentially disqualifying information might exist in the NICS. This is called a "default proceed" transaction. The dealer does not, however, have to complete the sale, and the FBI will continue to review the case for two more weeks. If disqualifying information is discovered after three business days, the FBI then contacts the dealer to find out whether the gun was transferred (sold) under the default proceed rule. Default proceed transactions are a matter of particular concern because potentially disqualifying information on a gun buyer may well exist. This sometimes occurs when information about a person's recent arrest or conviction has not been entered into the national database.

- The dealer transfers a gun to a prohibited person in a default proceed transaction. If this occurs, the FBI may later notify local law enforcement agencies and the ATF in an attempt to retrieve the gun and take appropriate action, if any, against the buyer.

- The NICS check returns disqualifying information on the buyer and the transfer is denied.

Figure 3.3 illustrates the NICS procedures as they unfolded per 100 applicants in 2014. For every 100 applicants, 66 proceeded immediately and were cleared for purchase, while 34 were transferred to the FBI's NICS section for analysis. Of those 34, 25 were immediately cleared for approval and nine were held for additional review. Ultimately, only one out of every 100 firearm transactions led to a denial by the NICS.

Table 3.2 shows the reasons for NICS denials of gun transfers between November 30, 1998, and August 31, 2016. The most common reason, which accounted for 731,723 (54.3%) denials over this period, was the applicant's previous conviction of a felony punishable by more than one year in prison or a misdemeanor punishable by more than two years. Other leading reasons for rejection were an applicant's status as a fugitive from justice (161,821, or 12%), an applicant's prior conviction of a misdemeanor crime of domestic violence (126,169, or 9.3%), and an applicant's unlawful use of or addiction to a controlled substance (117,339, or 8.7%).

As of August 31, 2016, there were 15.2 million active records in the NICS system, or 15.2 million individuals known to the FBI to be legally prohibited from owning a firearm. (See Table 3.3.) More than 6.9 million (45.5%) were ineligible for firearm ownership due to their unlawful immigration status, and 4.5 million (29.8%) were ineligible due to their mental health status. Although people convicted of crimes were the most likely individuals to be denied permits, they represented only 2.6 million (17.4%) active NICS records, suggesting that they were far more likely to attempt to purchase guns illegally than other prohibited people.

Impact of the Brady Law on Gun Violence

In "The Limited Impact of the Brady Act: Evaluation and Implications" (Daniel W. Webster and Jon S. Vernick, eds., *Reducing Gun Violence in America: Informing Policy with Evidence and Analysis*, 2013), Philip J. Cook and Jens Ludwig analyze their findings that the Brady law's restrictions on gun purchases do not result in discernable reductions in gun-related deaths, either by homicide or suicide. One of the most compelling reasons for this lack of impact is the so-called private sale loophole. They explain, "Most crime guns are obtained from people who are not licensed FFLs through private transactions that are largely unregulated under existing federal law—that is, these crime guns are obtained in the off-the-books secondary gun market." An additional shortcoming of the current NICS system is that the criminal background check is not strict enough. That is, by screening potential gun buyers only for felony and misdemeanor domestic violence convictions, but not for violent misdemeanors, the NICS allows gun sales to too many adults whose histories indicate an inclination toward violent crime. Lastly, the databases used in the NICS are often incomplete or insufficiently networked between local, state, and federal authorities, thereby allowing many gun

FIGURE 3.2

National Instant Criminal Background Check System participation map, 2014

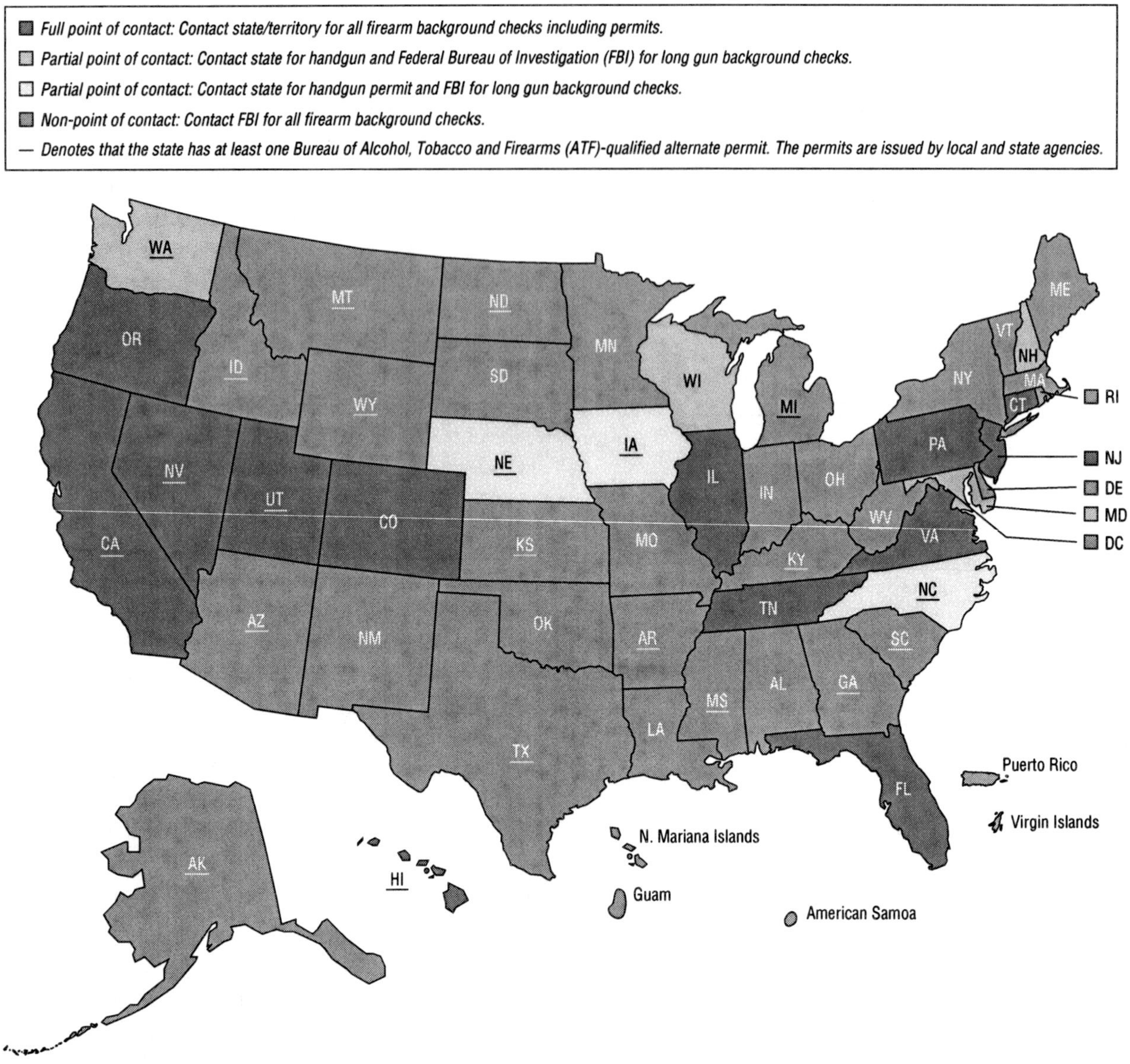

Full point of contact: Contact state/territory for all firearm background checks including permits.
Partial point of contact: Contact state for handgun and Federal Bureau of Investigation (FBI) for long gun background checks.
Partial point of contact: Contact state for handgun permit and FBI for long gun background checks.
Non-point of contact: Contact FBI for all firearm background checks.
— Denotes that the state has at least one Bureau of Alcohol, Tobacco and Firearms (ATF)-qualified alternate permit. The permits are issued by local and state agencies.
WA
OR
CA
NV
ID
MT
WY
UT
CO
AZ
NM
ND
SD
NE
KS
OK
TX
MN
IA
MO
AR
LA
WI
IL
MI
IN
OH
KY
TN
MS
AL
GA
FL
SC
NC
VA
WV
PA
NY
VT
NH
ME
MA
CT
RI
NJ
DE
MD
DC
AK
HI
N. Mariana Islands
Guam
American Samoa
Puerto Rico
Virgin Islands

SOURCE: "NICS Participation Map," in *National Instant Criminal Background Check System (NICS) Operations, 2014*, Federal Bureau of Investigation, 2015, https://www.fbi.gov/file-repository/2014-nics-ops-report-050115.pdf (accessed September 7, 2016)

purchasers whose records indicate that they should be disqualified to slip through the system and obtain firearms. Still, in spite of their statistical findings, Cook and Ludwig clarify that they do not rule out the possibility that the Brady law has saved hundreds of lives over the years and therefore that it has been worthwhile.

Controversies over Brady Provisions

Major ongoing Brady law controversies center on four issues:

- The elimination of the five-day waiting period, beginning on November 30, 1998
- The length of time the FBI can retain records of gun transactions
- The provision of the law that exempts certain people from background checks
- The regulation of sales at gun shows and on the Internet

ELIMINATION OF THE WAITING PERIOD. The elimination of the waiting period is probably the thorniest issue connected to the Brady law. Gun control advocates support legislation to establish a minimum three-day waiting period. Gun rights advocates claim that 24 hours

FIGURE 3.3

National Instant Criminal Background Check System denial process per 100 applicants, 2014

- Potential gun purchasers: 100
- Proceeds: 66
- Immediate determinations: 25
- Transferred to the NICS section: 34
- Delayed for additional review: 9
- Transactions denied by the NICS section in 2014: 1.10

NICS = National Instant Criminal Background Check System.

SOURCE: "Untitled," *National Instant Criminal Background Check System (NICS) Operations, 2014*, Federal Bureau of Investigation, 2015, https://www.fbi.gov/file-repository/2014-nics-ops-report-050115.pdf (accessed September 7, 2016)

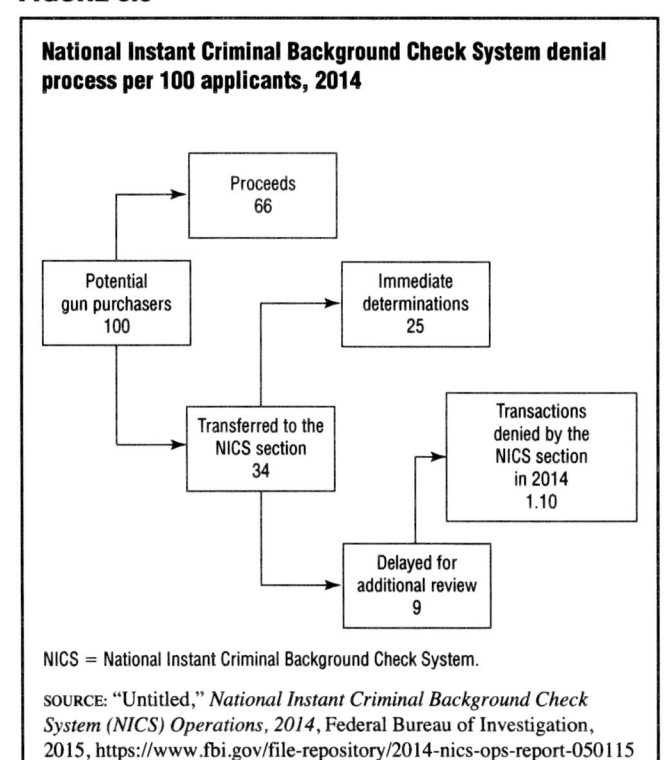

TABLE 3.2

National Instant Criminal Background Check System firearms denials, by reason, November 30, 1998–August 31, 2016

Rank	Prohibited category description	Total
1	Convicted of a crime punishable by more than one year or a misdemeanor punishable by more than two years	731,723
2	Fugitive from Justice	161,821
3	Misdemeanor crime of domestic violence conviction	126,169
4	Unlawful User/Addicted to a Controlled Substance	117,339
5	State prohibitor	69,341
6	Protection/Restraining Order for Domestic Violence	52,895
7	Under indictment/information	38,765
8	Adjudicated Mental Health	24,728
9	Illegal/unlawful alien	18,023
10	Federally Denied Persons File	5,939
11	Dishonorable discharge	1,036
12	Renounced U.S. Citizenship	73
	Total federal denials	**1,347,852**

SOURCE: "Federal Denials" in *National Instant Criminal Background Check System*, Federal Bureau of Investigation, August 31, 2016, https://www.fbi.gov/file-repository/federal_denials.pdf (accessed September 7, 2016)

TABLE 3.3

Active records in the National Instant Criminal Background Check System Index, August 31, 2016

Rank	Prohibited category description	Total
1	Illegal/unlawful alien	6,900,683
2	Adjudicated mental health	4,527,497
3	Convicted of a crime punishable by more than one year or a misdemeanor punishable by more than two years	2,639,888
4	Fugitive from justice	508,907
5	State prohibitor	276,866
6	Misdemeanor crime of domestic violence conviction	136,802
7	Protection/restraining order for domestic violence	66,934
8	Under indictment/information	40,604
9	Renounced U.S. citizenship	36,630
10	Unlawful user/addicted to a controlled substance	25,441
11	Dishonorable discharge	11,005
	Total active records in the NICS index	**15,171,257**

NICS = National Instant Criminal Background Check System.

SOURCE: "Active Records in the NICS Index," in *National Instant Criminal Background Check System*, Federal Bureau of Investigation, August 31, 2016, https://www.fbi.gov/file-repository/active_records_in_the_nics-index.pdf (accessed September 7, 2016)

is adequate. Gun control advocates maintain that a longer waiting period allows local police to search records that might not be in the NICS and that a waiting requirement of at least three days allows a "cooling-off" period, thereby preventing many crimes of passion. They worry that the speed of the background checks may help some of the targets evade the Brady law. This could happen if state and local records have not made it into the national database. For example, the NICS may miss individuals who have only recently been convicted of crimes. However, some states have mandatory waiting periods that must be respected before a person can purchase a firearm.

RETENTION OF RECORDS. Soon after the NICS started operations in 1998, the NRA filed a lawsuit arguing, among other things, that records relating to a background check should be immediately destroyed to prevent the formation of a central registry of gun buyers—a registry the NRA considers an invasion of privacy. The Department of Justice noted that the records would be kept, generally, for 90 days but for no more than six months to conduct audits to make sure gun dealers were following the law and to check for identity fraud and other abuses of the system. In July 2000 the federal court of appeals in the District of Columbia upheld in *NRA v. Reno* (No. 99-5270) the legality of the regulation, allowing the FBI to conduct periodic security audits of NICS records. In June 2001 the Supreme Court declined to hear the NRA's appeal of that decision, thereby allowing the lower court's decision to stand.

Only days later the U.S. attorney general John D. Ashcroft (1942–), a gun rights advocate and the head of the Department of Justice, proposed destroying the critical NICS records within one business day. Congress asked the U.S. General Accounting Office (GAO; now the U.S. Government Accountability Office) to investigate the implications of such a quick destruction of records. In July 2002 the GAO published *Gun Control: Potential Effects of Next-Day Destruction of NICS Background Check Records* (http://www.gao.gov/new.items/d02653.pdf). The GAO stated that 97% of the guns retrieved from illegal purchasers who had been incorrectly approved to buy guns between July 2001 and January 2002 would not have been detected under the attorney general's one-day destruction plan.

Two years later, under an amendment sponsored by Representative Todd Tiahrt (1951–; R-KS), the FBI was mandated to destroy the NICS records of a gun sale within 24 hours of allowing the sale to proceed. The amendment was approved as part of legislation that provided funding for the Departments of Commerce, Justice, and State and made it illegal for the agency to report on sales of multiple handguns and gun trace statistics. The amendment, which included numerous other measures meant to protect the privacy of gun owners and prohibit firearms purchasing records from being used in lawsuits against gun manufacturers, was loosened somewhat in 2008, but the requirement to destroy records within 24 hours remained in effect as of 2016.

EXEMPTION FROM BACKGROUND CHECKS. The Brady law exempts gun buyers from NICS background checks if they have a state permit that meets certain criteria established by the ATF, including holding a right-to-carry permit. However, states that issue permits are expected to incorporate an NICS check into their permit process. Buyers who do not have a permit must have an NICS check, and people renewing their permit must undergo an NICS check at that time.

GUN SHOW AND ONLINE REGULATION. The background check requirement on potential gun buyers applies to federally licensed gun dealers, manufacturers, importers, and pawnshop brokers. People who are not gun dealers and cannot be described as "engaged in the business" of selling firearms do not have to be licensed by the federal government to sell guns, nor do they need to perform background checks on their buyers. This means that someone who is not a gun dealer but who offers a gun or guns for sale at gun shows or flea markets, through classified ads, or online is not subject to the Brady law. According to the Brady Campaign to Prevent Gun Violence, in "Background Checks" (2016, http://www.bradycampaign.org/our-impact/campaigns/background-checks), these unregulated gun sales represent approximately 40% of total gun purchases.

As with sales of many other consumer products, online sales of guns have proliferated in the years since the Brady law was passed. Like gun shows, online marketplaces for guns allow unlicensed individuals to sell guns without conducting background checks; unlike gun shows, online marketplaces never close. Both Craigslist and eBay prohibit sales of firearms and related items, but online gun transactions are common on sites such as Armslist (http://www.armslist.com). Michael Luo, Mike McIntire, and Griff Palmer explain in "Seeking Gun or Selling One, Web Is a Land of Few Rules" (NYTimes.com, April 17, 2013) that "Armslist and similar sites function as unregulated bazaars, where the essential anonymity of the Internet allows unlicensed sellers to advertise scores of weapons and people legally barred from gun ownership to buy them." The reporters analyzed more than 170,000 Armslist ads over a period of three months and determined that 94% were posted by private, unlicensed sellers. They also contacted individuals attempting to buy and sell guns on the site. Luo, McIntire, and Palmer found numerous instances in which those intending to buy guns were convicted felons, fugitives from justice, domestic violence offenders, or others prohibited from owning guns. They also found cases in which individuals offered dozens of guns for sale over the three-month period, calling into question their status as individuals not engaged in the business of selling firearms.

In "Study Finds Vast Online Marketplace for Guns without Background Checks" (WashingtonPost.com, August 5, 2013), Philip Rucker notes that a study by the center-left advocacy group Third Way came to similar conclusions. The group analyzed Armslist listings in 10 states whose senators had recently voted against expanding background checks to change the gun show and Internet loophole. Rucker explains, "At any given time, more than 15,000 guns were for sale in those states, according to the study, and more than 5,000 of them were semi-automatic weapons. Nearly 2,000 ads were from prospective buyers asking to purchase specifically from private sellers, where no background checks are required." Rucker quotes Lanae Erickson Hatalsky, the director of social policy at Third Way, who said, "Nobody has any ability to stop these people who are looking for private sellers—and the only reason to do that is to evade the background check system."

The NRA and other gun rights advocates typically contest such descriptions of gun shows and online sales. The Institute for Legislative Action (ILA), the lobbying segment of the NRA, argues in "Background Checks for Guns" (August 8, 2016, https://www.nraila.org/issues/background-checks-nics/) that universal background checks would not prevent criminals from obtaining firearms because most perpetrators of firearm crimes obtain their guns through so-called straw purchasers (legally qualified buyers who purchase for someone who is not legally qualified) or "through theft, on the black market, from a drug dealer or 'on the street,' or from family members and friends." The ILA also contests the Brady Campaign's finding that 40% of gun purchases are unregulated, maintaining that the actual figure is much lower.

Still, in spite of the NRA's opposition, numerous national polls conducted between 2014 and 2016 indicate that a sizable majority of the U.S. public—consistently in the range of 80% to 90%—supports expanding background checks to cover all gun sales. Indeed, the Pew Research Center reports in *Opinions on Gun Policy and the 2016 Campaign* (August 26, 2016, http://www.people-press.org/

files/2016/08/08-26-16-Gun-policy-release.pdf) that even in the sharply polarized atmosphere of the 2016 presidential campaign between the Democratic nominee Hillary Rodham Clinton (1947–) and the Republican nominee Donald Trump (1946–), 83% of voters favored the extension of background checks to all private and gun show sales. Moreover, Pew found that support for background checks was widespread across all demographic groups, even in gun-owning households.

Legislation seeking to close the gun show and/or Internet loopholes, or to make background checks universal, has been introduced in every session of Congress since 2001, but none of these bills has been passed. In the absence of federal legislation, many states have closed the loopholes through state-level legislation by requiring either universal background checks or expanding background checks to include purchases at gun shows. According to the Law Center to Prevent Gun Violence, in "Universal Background Checks" (http://smartgunlaws.org/gun-laws/policy-areas/background-checks/universal-background-checks/), as of 2016 background checks for all handgun sales were required in California, Colorado, Connecticut, Delaware, Hawaii, Illinois, Iowa, Maryland, Massachusetts, Michigan, Nebraska, New Jersey, New York, North Carolina, Pennsylvania, Rhode Island, Washington, and the District of Columbia.

THE VIOLENT CRIME CONTROL AND LAW ENFORCEMENT ACT OF 1994

Banning Assault Weapons

Federal laws have banned the possession of automatic-fire guns (machine guns) since 1934 and their importation and manufacture for private use since 1986. A machine gun shoots a stream of bullets when the trigger is pulled, instead of a single shot.

In 1989 a semiautomatic firearm was used by Patrick E. Purdy (1964–1989) to kill five children at an elementary school in Stockton, California. This type of firearm shoots only one bullet each time the trigger is pulled, but the shooter does not have to do anything to "ready" the next shot; therefore, the shooter can fire as fast as he or she can pull the trigger. Some semiautomatic guns can be converted to automatic firearms.

In response to Purdy's crime, several states passed laws banning the sale and possession of semiautomatic assault-style weapons. It also led to the passage of the Violent Crime Control and Law Enforcement Act of 1994, which included a subsection commonly known as the Semiautomatic Assault Weapons Ban. This act banned the manufacture, transfer, and possession of semiautomatic assault-type firearms but did not outlaw such firearms lawfully possessed before its enactment.

According to the act, banned weapons were defined as guns with a detachable magazine (a cartridge holder that feeds bullets automatically into a gun's chamber) and two or more of the following: a bayonet lug (a metal mount for a thrusting knife), a flash suppressor, a protruding pistol grip, a folding stock (which allows the gun to be stored in a smaller space), or a threaded muzzle (used to attach other devices). The act also banned the making or sale of large-capacity ammunition magazines capable of holding more than 10 rounds (bullets). The federal law listed 19 types of banned semiautomatic firearms, including the Uzi, the TEC-9, the Street Sweeper, and their copycats, all of which are often referred to as "assault weapons." The definition of this term is not clear-cut, but an assault weapon is most frequently defined as a semiautomatic rifle, shotgun, or pistol with a combination of any or all the characteristics and accessories banned by the Violent Crime Control and Law Enforcement Act.

The act exempted at least 650 different sporting rifles. Furthermore, it was legal under this law to buy the accessories to convert these semiautomatic guns to automatic firearms, but the conversion itself was against federal law.

Although the act had several different provisions pertaining to guns, the focus of the debate over its passage was a proposed ban on semiautomatic military-style weapons. Pro-gun supporters repeatedly stressed that the use of the term *assault weapons* to describe the weapons banned under the law was misleading. They complained that the legislation was intended to ban military-style weapons but actually encompassed some ordinary rifles used in hunting and target shooting—weapons that are seldom used to commit crimes. The Semiautomatic Assault Weapons Ban Subsection of the Violent Crime Control and Law Enforcement Act expired on September 13, 2004. Despite broad support for its renewal at that time, it was not renewed; and despite ongoing calls for its reinstatement in the wake of numerous mass shootings carried out with weapons that were formerly banned under the law, it had not been renewed as of 2016. Chapter 9 offers congressional testimony about renewing the Semiautomatic Assault Weapons Ban.

Banning Juveniles from Possessing Handguns or Ammunition

The 1994 Violent Crime Control and Law Enforcement Act also prohibited the possession of a handgun or ammunition by a juvenile under 18 years of age. In addition, it prohibited the sale or private transfer of a handgun or ammunition to juveniles. Exemptions included cases in which the juvenile temporarily used the handgun for employment, with the permission of the owner, such as in ranching or farming, which sometimes require workers to use guns to kill predatory animals that threaten livestock.

Other exemptions included using a handgun for target practice, hunting, or a course of instruction on the safe and lawful use of a handgun.

Implementing New Regulations for Obtaining FFLs

Table 3.4 shows the number of active firearms licensees between 1975 and 2015. Even before the changes that came when the 1994 Violent Crime Control and Law Enforcement Act was implemented, the ATF realized that it needed to strengthen the procedures for obtaining an FFL. Between late 1992 and early 1993 news stories, such as Josh Sugarmann's "Gun Market Is Wide Open in America" (CSMonitor.com, April 23, 1993), revealed just how easy it was to get an FFL. The charge for a three-year license was a mere $10, and the only check was a short computer criminal history query.

In the years that followed, the ATF took steps to curtail the number of active FFLs. The Brady law raised the fee to $200 for a three-year license and $90 for a three-year renewal. In March 1994 the ATF sent out new application forms requiring each applicant to submit fingerprint cards and a photograph. These new regulations, as well as an extra step to ensure that the person applying for an FFL complied with state and local laws, were included as part of the Violent Crime Control and Law Enforcement Act. The act also required each licensee to report the theft or loss of a firearm within 48 hours.

There are 11 types of FFLs, including the Type 3 license—collector of curio and relic firearms. Many of the types of FFLs are specific to importers and manufacturers of firearms, their components, and their ammunition. The Type 1 FFL is for a dealer or gunsmith, and the Type 2 FFL is for a pawnbroker who deals in guns.

TABLE 3.4

Federal firearms licensees, 1975–2015

Fiscal year	Dealer	Pawn-broker	Collector	Manufacturer of Ammunition	Manufacturer of Firearms	Importer	Destructive device Dealer	Destructive device Manufacturer	Destructive device Importer	Total
1975	146,429	2,813	5,211	6,668	364	403	9	23	7	161,927
1976	150,767	2,882	4,036	7,181	397	403	4	19	8	165,697
1977	157,463	2,943	4,446	7,761	408	419	6	28	10	173,484
1978	152,681	3,113	4,629	7,735	422	417	6	35	14	169,052
1979	153,861	3,388	4,975	8,055	459	426	7	33	12	171,216
1980	155,690	3,608	5,481	8,856	496	430	7	40	11	174,619
1981	168,301	4,308	6,490	10,067	540	519	7	44	20	190,296
1982	184,840	5,002	8,602	12,033	675	676	12	54	24	211,918
1983	200,342	5,388	9,859	13,318	788	795	16	71	36	230,613
1984	195,847	5,140	8,643	11,270	710	704	15	74	40	222,443
1985	219,366	6,207	9,599	11,818	778	881	15	85	45	248,794
1986	235,393	6,998	10,639	12,095	843	1,035	16	95	52	267,166
1987	230,888	7,316	11,094	10,613	852	1,084	16	101	58	262,022
1988	239,637	8,261	12,638	10,169	926	1,123	18	112	69	272,953
1989	231,442	8,626	13,536	8,345	922	989	21	110	72	264,063
1990	235,684	9,029	14,287	7,945	978	946	20	117	73	269,079
1991	241,706	9,625	15,143	7,470	1,059	901	17	120	75	276,116
1992	248,155	10,452	15,820	7,412	1,165	894	15	127	77	284,117
1993	246,984	10,958	16,635	6,947	1,256	924	15	128	78	283,925
1994	213,734	10,872	17,690	6,068	1,302	963	12	122	70	250,833
1995	158,240	10,155	16,354	4,459	1,242	842	14	118	71	191,495
1996	105,398	9,974	14,966	3,144	1,327	786	12	117	70	135,794
1997	79,285	9,956	13,512	2,451	1,414	733	13	118	72	107,554
1998	75,619	10,176	14,875	2,374	1,546	741	12	125	68	105,536
1999	71,290	10,035	17,763	2,247	1,639	755	11	127	75	103,942
2000	67,479	9,737	21,100	2,112	1,773	748	12	125	71	103,157
2001	63,845	9,199	25,145	1,950	1,841	730	14	117	72	102,913
2002	59,829	8,770	30,157	1,763	1,941	735	16	126	74	103,411
2003	57,492	8,521	33,406	1,693	2,046	719	16	130	82	104,105
2004	56,103	8,180	37,206	1,625	2,144	720	16	136	84	106,214
2005	53,833	7,809	40,073	1,502	2,272	696	15	145	87	106,432
2006	51,462	7,386	43,650	1,431	2,411	690	17	170	99	107,316
2007	49,221	6,966	47,690	1,399	2,668	686	23	174	106	108,933
2008	48,261	6,687	52,597	1,420	2,959	688	29	189	113	112,943
2009	47,509	6,675	55,046	1,511	3,543	735	34	215	127	115,395
2010	47,664	6,895	56,680	1,759	4,293	768	40	243	145	118,487
2011	48,676	7,075	59,227	1,895	5,441	811	42	259	161	123,587
2012	50,848	7,426	61,885	2,044	7,423	848	52	261	169	130,956
2013	54,026	7,810	64,449	2,353	9,094	998	57	273	184	139,244
2014	55,431	8,132	63,301	2,596	9,970	1,133	66	287	200	141,116
2015	56,181	8,152	60,652	2,603	10,498	1,152	66	315	221	139,840

Note: Data is based on active firearms licenses and related statistics as of the end of each fiscal year.

SOURCE: "Exhibit 10. Federal Firearms Licensees Total (1975–2015)," in *Firearms Commerce in the United States: Annual Statistical Update, 2015*, U.S. Department of Justice, Bureau of Alcohol, Tobacco, Firearms, and Explosives, August 9, 2016, https://www.atf.gov/resource-center/docs/2016-firearms-commerce-united-states/download (accessed September 7, 2016)

In *Federal Firearms Licensees: Various Factors Have Contributed to the Decline in the Number of Dealers* (March 1996, http://www.gao.gov/archive/1996/gg96078.pdf), the GAO states that beginning in 1994, due to the revisions in the law, the number of FFLs and the number of FFL applications started to drop sharply. According to the ATF, in *Firearms Commerce in the United States—2001/ 2002* (2002), this decline is reflected in the number of active FFLs in 1993 (283,193), compared with 1996 (124,286). However, beginning in 1997 the decline appeared to level off, with the overall number of licenses stabilizing between 1997 (106,710) and 2001 (104,840); thereafter, the number of licenses began to increase.

Table 3.5 shows the number of active FFLs by state in 2015. Unsurprisingly, the states with the most FFLs were among the largest U.S. states. Texas (the second-largest state by population) had the most FFLs, at 10,910, followed by California (the largest state), at 8,261, and Florida (the fourth-largest state), at 7,507. New York (the third-largest state) and Illinois (the fifth-largest state) had comparatively few FFLs given their population size, at 4,081 and 5,295, respectively. There were 139,840 FFLs nationally, including those in the District of Columbia and the U.S. territories, in 2015.

Other Provisions of the Violent Crime Control and Law Enforcement Act

The Violent Crime Control and Law Enforcement Act imposed stiffer penalties for using a gun during a violent crime or drug felony. Amendments to the act and policies enacted by the ATF prohibited the possession of firearms by people guilty of domestic abuse. It also tightened rules and regulations for firearms dealers.

GUN LAWS PASSED IN RESPONSE TO SEPTEMBER 11, 2001

Some new gun laws were passed in the wake of the September 11, 2001, terrorist attacks on the United States. In November 2002 Congress passed the Arming Pilots against Terrorism Act, which allows airline pilots to carry handguns in the cockpit. The Law Enforcement Officers Safety Act, signed into law in 2004, allows off-duty or retired police officers with firearms training to travel the country with a concealed weapon.

THE NICS IMPROVEMENT AMENDMENTS ACT OF 2007

The NICS Improvement Amendments Act of 2007 was signed into law by President George W. Bush (1946–) in January 2008. This legislation mandates that information that would prohibit individuals from possessing or purchasing firearms be transmitted from state and local governments and federal agencies to the NICS. States are provided financial assistance to do this work and receive penalties if they fail to comply.

TABLE 3.5

Federal firearms licenses, by state, 2015

State	FFL* population
Alabama	2,385
Alaska	952
Arizona	3,202
Arkansas	2,018
California	8,261
Colorado	2,967
Connecticut	1,826
Delaware	340
District of Columbia	28
Florida	7,507
Georgia	3,807
Hawaii	288
Idaho	1,429
Illinois	5,295
Indiana	2,994
Iowa	2,172
Kansas	1,952
Kentucky	2,507
Louisiana	2,176
Maine	967
Maryland	3,417
Massachusetts	4,039
Michigan	4,262
Minnesota	2,676
Mississippi	1,560
Missouri	5,091
Montana	1,543
Nebraska	1,165
Nevada	1,368
New Hampshire	1,187
New Jersey	550
New Mexico	1,154
New York	4,081
North Carolina	4,819
North Dakota	666
Ohio	5,027
Oklahoma	2,483
Oregon	2,481
Pennsylvania	6,347
Rhode Island	610
South Carolina	2,240
South Dakota	803
Tennessee	3,482
Texas	10,910
Utah	1,361
Vermont	551
Virginia	4,374
Washington	2,857
West Virginia	1,497
Wisconsin	3,141
Wyoming	901
Other territories	124
Total	**139,840**

*Federal firearms licenses.
Note: Data is based on active firearms licenses and related statistics as of the end of the fiscal year.

SOURCE: "Exhibit 11. Federal Firearms Licenses by State 2015," in *Firearms Commerce in the United States: Annual Statistical Update, 2016*, U.S. Department of Justice, Bureau of Alcohol, Tobacco, Firearms, and Explosives, August 9, 2016, https://www.atf.gov/resource-center/docs/2016-firearms-commerce-united-states/download (accessed September 7, 2016)

The law was enacted in response to Seung-Hui Cho's (1984–2007) shooting rampage on April 16, 2007, at Virginia Polytechnic Institute and State University (Virginia Tech) in Blacksburg, Virginia. Cho had been ordered by a judge to receive mental health treatment, but that information was never entered into the NICS database. It would have disqualified Cho from purchasing the 9mm semiautomatic handgun and .22-caliber pistol he used to kill 32 students and teachers, and then himself. Although this law had been proposed in Congress since 2002, it languished until the Virginia Tech tragedy brought the issue to wide public attention.

STATE FIREARMS CONTROL LAWS

Laws and constitutional provisions relating to the purchase, ownership, and use of firearms differ dramatically from state to state as well as from county to county and town to town. The Law Center to Prevent Gun Violence (http://smartgunlaws.org/) maintains a comprehensive database of federal and state firearms laws. In most states the right to gun ownership is protected in the state constitution. (See Table 3.6.) As of 2016, only six states—California, Iowa, Maryland, Minnesota, New Jersey, and New York—did not have such a provision. Only 12 required a permit for the purchase of a firearm: California, Connecticut, Hawaii, Iowa,

TABLE 3.6

Firearm purchase laws, by state, 2016

State	Constitutional provision	Permit required to purchase - Long guns	Permit required to purchase - Handguns	Waiting period - Long guns	Waiting period - Handguns	Registration - Long guns	Registration - Handguns
Alabama	Article 1, Section 26	No	No	None	None	No	No[a]
Alaska	Article 1, Section 19	No	No	None	None	No	No
Arizona	Article 2, Section 26	No	No	None	None	No	No
Arkansas	Article 2, Section 5	No	No	None	None	No	No
California	None	Yes	Yes	10 days	10 days	No	Yes
Colorado	Article 2, Section 13	No	No	None	None	No	No[a]
Connecticut	Article 1, Section 15	Yes	Yes	None	None	Yes[b]	No[a]
Delaware	Article 1, Section 20	No	No	None	None	No	No[a]
Florida	Article 1, Section 8	No	No	None	3 days	No	No
Georgia	Article 1, Section 1	No	No	None	None	No	No
Hawaii	Article 1, Section 15	Yes	Yes	14 days	14 days	Yes	Yes
Idaho	Article 1, Section 11	No	No	None	None	No	No
Illinois	Article 1, Section 22	No	No	24 hours	72 hours	No	No
Indiana	Article 1, Section 32	No	No	None	None	No	No
Iowa	None	No	Yes	None	3 days	No	No
Kansas	Bill of Rights, Section 4	No	No	None	None	No	No
Kentucky	Article 1, Section 1	No	No	None	None	No	No
Louisiana	Article 1, Section 11	No	No	None	None	No[e]	No
Maine	Article 1, Section 16	No	No	None	None	No	No
Maryland	None	No	Yes	None	7 days	No	Yes
Massachusetts	Part 1, Article 17	Yes	Yes	None	None	No[a]	No[a]
Michigan	Article 1, Section 6	No	Yes	None	None	No	Yes
Minnesota	None	No	No	7 days[b]	7 days	No	No
Mississippi	Article 3, Section 12	No	No	None	None	No	No
Missouri	Article 1, Section 23	No	No	None	None	No	No
Montana	Article 2, Section 12	No	No	None	None	No	No
Nebraska	Article 1, Section 1	No	Yes	None	2 days	No	No
Nevada	Article 1, Section 11	No	No	None	None	No	No
New Hampshire	Part 1, Article 2-a	No	No	None	None	No	No
New Jersey	None	Yes	Yes	None	None	No	No[a]
New Mexico	Article 2, Section 6	No	No	None	None	No	No
New York	None	No[d]	Yes	None	None	No[d]	Yes
North Carolina	Article 1, Section 30	No	Yes	None	None	No	No[a]
North Dakota	Article 1, Section 1	No	No	None	None	No	No
Ohio	Article 1, Section 4	No	No	None	None	No	No
Oklahoma	Article 2, Section 26	No	No	None	None	No	No
Oregon	Article 1, Section 27	No	No	None	None	No	No[a]
Pennsylvania	Article 1, Section 21	No	No	None	None	No	No[a]
Rhode Island	Article 1, Section 22	No	Yes	7 days	7 days	No	No[e]
South Carolina	Article 1, Section 20	No	No	None	None	No	No
South Dakota	Article 6, Section 24	No	No	None	None	No	No
Tennessee	Article 1, Section 26	No	No	None	None	No	No
Texas	Article 1, Section 23	No	No	None	None	No	No
Utah	Article 1, Section 6	No	No	None	None	No	No
Vermont	Chapter 1, Article 16	No	No	None	None	No	No
Virginia	Article 1, Section 13	No	No	None	None	No	No
Washington	Article 1, Section 24	No	No	None	5 days[c]	No	No[a]
West Virginia	Article 3, Section 22	No	No	None	None	No	No
Wisconsin	Article 1, Section 25	No	No	None	None	No	No
Wyoming	Article 1, Section 24	No	No	None	None	No	No
District of Columbia	N/A	No	No	10 days	10 days	Yes[e]	Yes[e]

[a]Though not a complete registry of ownership, varying levels of documentation of transfers or permits are maintained in these states.
[b]For assault weapons.
[c]While not a mandatory waiting period, Washington allows five days to complete a background check prior to delivery of the handgun.
[d]Stricter gun control laws have been enacted in New York City, including requiring a permit to purchase a rifle or shotgun and restricting permission to carry firearms, even during transport to or from a target range.
[e]Safety course required.

SOURCE: Compiled by Laurie DiMauro and Erin Brown for Gale, © 2016.

Maryland, Massachusetts, Michigan, Nebraska, New Jersey, New York, North Carolina, and Rhode Island. In four of these states (California, Connecticut, Hawaii, and New Jersey) permits were required for all guns, including handguns and long guns (shotguns, rifles, and other long guns that fall under the category of "assault weapons"); and in seven states (Iowa, Maryland, Michigan, Nebraska, New York, North Carolina, and Rhode Island) permits were required for handguns but not for long guns. Although New York state law required permits only for handgun purchases, laws specific to New York City required permits for all gun purchases, among other restrictions.

All states had instant background checks under the terms of the Brady law in 2016, but only 10 and the District of Columbia required a waiting period between the purchase and receipt of a firearm: California, Florida, Hawaii, Illinois, Iowa, Maryland, Minnesota, Nebraska, Rhode Island, and Washington. (See Table 3.6.) Some of these waiting periods varied depending on whether a long gun or handgun was being purchased, and their length varied from 24 hours (for long guns in Illinois) to 14 days (for all guns in Hawaii). Registration (a record of the transfer or ownership of a specific firearm) was required (in various forms) in only six states (California, Connecticut, Hawaii, Maryland, Michigan, and New York) and the District of Columbia.

The Brady Campaign annually publishes a state "scorecard" that ranks states according to the strength or weakness of their gun laws. The state scorecard rankings for 2015 are shown in Figure 3.4 and Figure 3.5. Each state can earn up to 100 points, as in standard academic grading, and the measures evaluated include background checks, reporting of lost or stolen guns, and procedures for keeping guns out of the hands of dangerous people, such as those with records of domestic violence. States with the strongest gun control laws in these and other areas earn the most points, and states with laws considered the most likely to promote firearm violence earn the least points.

Figure 3.4 shows the Brady Campaign's ranking of the 15 states with the loosest gun laws in 2015. With a score of −39, Arizona had the most permissive gun laws in the nation, making it a criminal's number-one "choice" state for acquiring firearms. Following Arizona, states with the loosest gun laws were Alaska (−30), Wyoming (−28), Louisiana (−27), Montana (−25), Arkansas, (−24), Virginia (−22.5), Kentucky (−22), Florida (−20.5), Nevada (−20.5), Maine (−20), Mississippi (−19.5), Idaho (−19), New Mexico (−19), and Alabama (−18). Conversely, Figure 3.5 shows the Brady Campaign's ranking of the 10 states with the strictest gun laws in 2015. With a score of 76, California was considered to be the most difficult place in the country for criminals to buy, carry, or traffic guns. The other strictest states for firearms laws were Connecticut (73), Massachusetts (70), New Jersey (69), New York (65.5), Hawaii (62), Maryland (56), Rhode Island (55), Delaware (41), and Illinois (40.5).

In *The Brady Campaign State Scorecard* (March 2015, http://www.crimadvisor.com/data/Brady-State-Scorecard-2014.pdf), the Brady Campaign explains that although efforts to strengthen federal gun laws remain thwarted by political gridlock, significant legislative action at the state level suggests a growing national appetite for what it describes as "commonsense" gun policy reform. Indeed, the Brady Campaign notes an unprecedented level of legal change between 2013 and 2015, when 37 states passed a total of 99 laws that strengthened gun regulations.

These contrasting political dynamics remained much the same in mid-2016, according to the Law Center to Prevent Gun Violence, in "Gun Law Trendwatch" (August 1, 2016, http://smartgunlaws.org/wp-content/uploads/2016/08/Trendwatch-Mid-Year-2016-FINAL.pdf). In the wake of the horrific mass shooting at the Pulse nightclub in Orlando, Florida, on June 12, 2016, congressional Democrats resorted to a 15-hour filibuster (in the U.S. Senate) and a 25-hour sit-in protest (in the U.S. House of Representatives) to try to pass requirements for universal background checks and a ban on gun sales to anyone being monitored by the government for possible links to terrorism. However, as of December 2016, no such federal legislation had been passed. Meanwhile, several states passed new gun safety legislation in 2016. For example, Connecticut, Hawaii, Tennessee, and Virginia passed a variety of measures that made it more difficult for anyone with a history of domestic violence to obtain and carry firearms.

California enacted particularly extensive gun law reform in 2016, with laws that restricted magazine capacity to no more than 10 bullets, required background checks for ammunition purchases, and curbed straw purchasing (in which a person buys a gun legally to sell or give to someone else). Bill Chappell reports in "6 New Gun Control Laws Enacted in California, as Gov. Brown Signs Bills" (NPR.org, July 1, 2016) that the new law to restrict straw purchasing makes it a misdemeanor to file a false report about a gun being lost or stolen and imposed a 10-year wait before the purchaser can buy another gun. Although these bills were signed into law less than a month after the Orlando massacre, they were introduced months earlier, in response to the mass shooting that took place in San Bernardino in December 2015, in which the shooters were believed to have acquired some of their guns from a relative.

Although mass shootings often lead to legislative action for stricter gun laws, these tragic events can also spur gun rights activists to seek reforms that loosen gun laws, with the idea that an armed citizenry is better empowered to "take down" a mass shooter, thereby reducing the number of casualties the shooter can inflict. For example, in "NRA: 'Only Thing That Stops a Bad Guy with a Gun Is a Good Guy with a Gun'" (NPR.org, December 21, 2012),

FIGURE 3.4

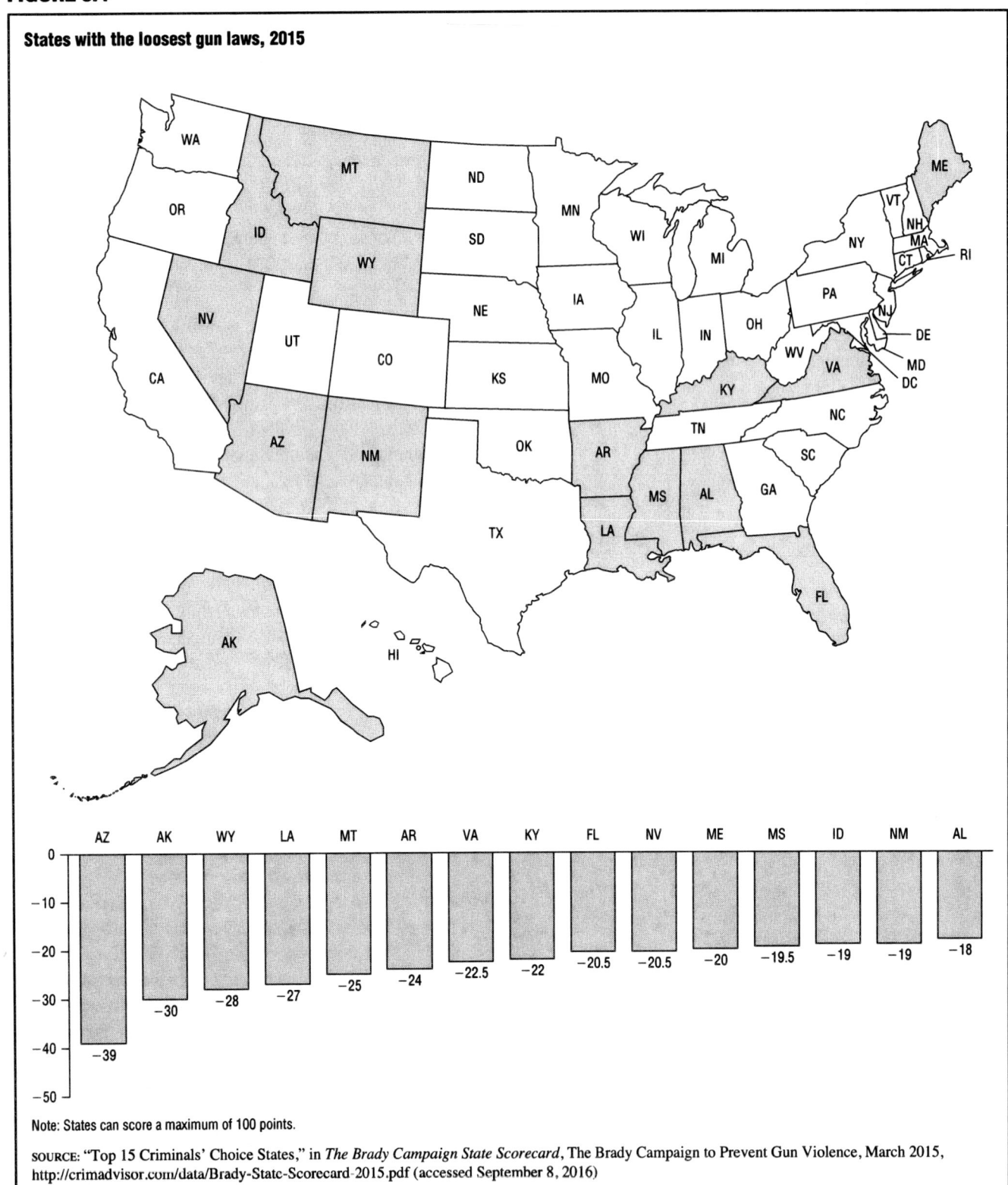

States with the loosest gun laws, 2015

Note: States can score a maximum of 100 points.

SOURCE: "Top 15 Criminals' Choice States," in *The Brady Campaign State Scorecard*, The Brady Campaign to Prevent Gun Violence, March 2015, http://crimadvisor.com/data/Brady-State-Scorecard-2015.pdf (accessed September 8, 2016)

Peter Overby interviewed Wayne LaPierre, the chief executive officer of the NRA, shortly after the Newtown, Connecticut, massacre. LaPierre encapsulated this logic, asserting, "The only thing that stops a bad guy with a gun, is a good guy with a gun." As chronicled by the Law Center to Prevent Gun Violence, in "Gun Law Trendwatch," 2016 also saw the passage of state laws to expand gun rights, especially by removing restrictions against carrying firearms in public places, such as churches, schools, and government buildings. A variety of such laws were passed in Georgia, Kansas, Mississippi, Oklahoma, and Tennessee, among other states.

FIGURE 3.5

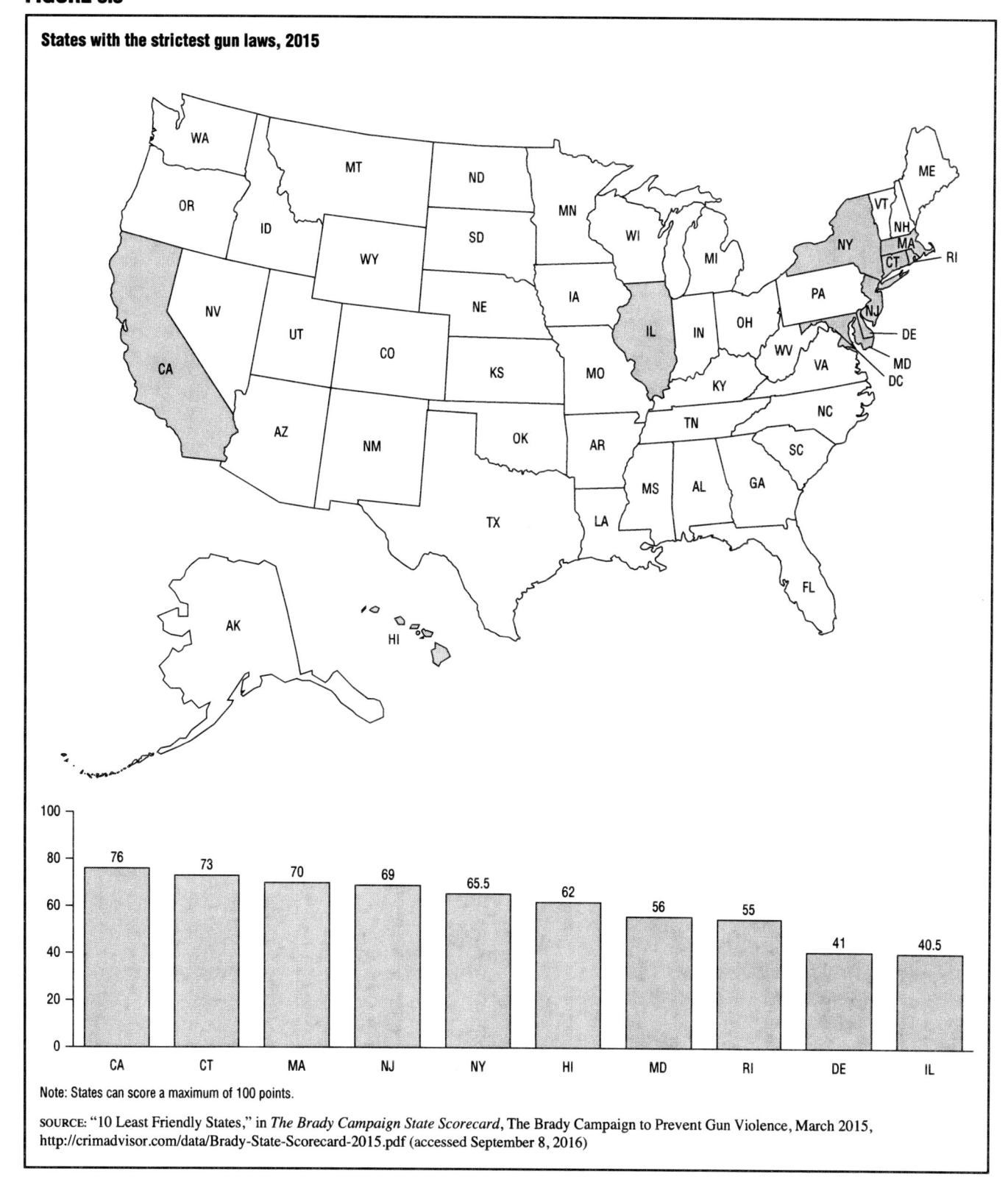

States with the strictest gun laws, 2015

Note: States can score a maximum of 100 points.

SOURCE: "10 Least Friendly States," in *The Brady Campaign State Scorecard*, The Brady Campaign to Prevent Gun Violence, March 2015, http://crimadvisor.com/data/Brady-State-Scorecard-2015.pdf (accessed September 8, 2016)

Concealed Weapons

According to the ILA, in "Gun Laws" (http://www.nraila.org/gun-laws.aspx), as of 2016 there were 41 right-to-carry (RTC) states, which means those states allowed individuals to carry concealed firearms for protection, either with or without a permit. (See Table 3.7.)

Of the 41 RTC states, 34 had shall-issue laws. These laws require local officials to issue a concealed handgun carry permit to anyone who applies, unless the applicant is prohibited by law from carrying a weapon. An additional six states (Alaska, Arizona, Maine, Mississippi, Vermont, and Wyoming) did not require a permit for carrying a

Gun Control   Firearm Laws, Regulations, and Ordinances   35

TABLE 3.7

Right-to-carry weapons laws, 2016

Shall issue	No permit required	Discretionary issue	No right to carry
Alabama	Alaska	Connecticut	California
Arkansas	Arizona		Delaware
Florida	Maine		Hawaii
Georgia	Mississippi		Maryland
Idaho	Vermont		Massachusetts
Illinois	Wyoming		New Jersey
Indiana			New York
Iowa			Rhode Island
Kansas			
Kentucky			
Louisiana			
Michigan			
Minnesota			
Missouri			
Montana			
Nebraska			
Nevada			
New Hampshire			
New Mexico			
North Carolina			
North Dakota			
Ohio			
Oklahoma			
Oregon			
Pennsylvania			
South Carolina			
South Dakota			
Tennessee			
Texas			
Utah			
Virginia			
Washington			
West Virginia			
Wisconsin			

SOURCE: Created by Erin Brown for Gale, © 2016.

concealed firearm. Connecticut had a discretionary-issue rather than a shall-issue law, which means its government has some discretion over the issuance of permits. In practical terms, however, Connecticut generally grants permits and is therefore considered to be an RTC state.

The remaining eight states were non-RTC states: California, Delaware, Hawaii, Maryland, Massachusetts, New Jersey, New York, and Rhode Island. (See Table 3.7.) These states generally did not allow handguns to be carried in public areas, but they had very limited and restrictive issue laws that allowed concealed handgun carry permits to be granted in certain circumstances. The District of Columbia prohibited individuals from carrying guns, as did New York City.

In some states the possession of a concealed weapons permit exempts gun purchasers from background check requirements that otherwise apply under the Brady law. Table 3.8 shows state policies regarding concealed weapons permits relative to the Brady background check as of 2016. In Alaska, for example, concealed weapons permits marked "NICS-exempt" qualify as alternatives to the background check requirements of the Brady law. In other states either a concealed weapons permit or a permit to purchase a handgun after a certain date entitle a purchaser to forgo a new background check.

Supporters of gun rights typically claim that research establishes a positive correlation between the authorization of concealed-carry and decreases in crime. This contention is generally based on the findings of John R. Lott Jr. and David B. Mustard in "Crime, Deterrence, and Right-to-Carry Concealed Handguns" (*Journal of Legal Studies*, vol. 26, no. 1, 1997), and expanded on in Lott's 1998 book *More Guns, Less Crime: Understanding Crime and Gun Control Laws*. Lott analyzed U.S. crime and socioeconomic data and determined that the falling crime rate of the 1990s and beyond was crucially linked to the adoption of shall-issue laws, which have been enacted in 34 states since 1991. (See Table 3.7.)

This assertion, however, has been contested by other researchers, including a panel assembled by the National Research Council of the National Academies, whose 2004 government-funded report, *Firearms and Violence: A Critical Review*, examined the raw data analyzed by Lott and found that no correlation between gun violence and RTC laws was evident. Numerous other researchers have also attacked Lott's findings. Lott responds to the many criticisms of his work in "What a Balancing Test Will Show for Right-to-Carry Laws" (*Maryland Law Review*, vol. 71, no. 4, 2012), noting that of 29 studies analyzing the issue as of 2012, 18 supported his findings, 10 found that shall-issue laws had no significant effect on crime, and one found that shall-issue laws brought about temporary increases in aggravated assaults.

In 2014 Abhay Aneja, John J. Donohue III, and Alexandria Zhang brought new findings to this discussion with the publication of *The Impact of Right to Carry Laws and the NRC Report: The Latest Lessons for the Empirical Evaluation of Law and Policy* (December 1, 2014, http://papers.ssrn.com/sol3/papers.cfm?abstract_id=2443681). In this report, the researchers reevaluate and update the methodology of the National Research Council's 2004 report, while also extending their data collection and analysis to cover 1979 to 2010 (whereas the original study looked at data from 1979 to 2000). According to Clifton B. Parker, in "Right-to-Carry Gun Laws Linked to Increase in Violent Crime, Stanford Research Shows" (Stanford.edu, November 14, 2014), the report by Aneja, Donohue, and Zhang provides "the most convincing evidence to date that right-to-carry laws are associated with an increase in violent crime," including rape, robbery, auto theft, burglary, larceny, and particularly aggravated assault. Indeed, between 1979 and 2010 aggravated assault increased by an estimated 8% in RTC states. Moreover, Aneja, Donohue, and Zhang attest, their data reflect "not even a hint of any crime declines."

Open Carry

The term *open carry* refers to the carrying of a firearm in plain view while in public. Advocates of open

TABLE 3.8

Permanent Brady permit chart, 2016

State/territory	Qualifying permits
Alabama	None
Alaska	Concealed weapons permits marked NICS-exempt
American Samoa	None
Arizona	Concealed weapons permits qualify.
Arkansas	Concealed weapons permits issued on or after April 1, 1999 qualify.*
California	Entertainment firearms permit only
Colorado	None
Connecticut	None
Delaware	None*
District of Columbia	None*
Florida	None*
Georgia	Georgia firearms licenses qualify.
Guam	None*
Hawaii	Permits to acquire and licenses to carry qualify.
Idaho	Concealed weapons permits qualify.
Illinois	None
Indiana	None
Iowa	Permits to acquire and permits to carry concealed weapons qualify.
Kansas	Concealed handgun licenses issued on or after July 1, 2010 qualify as alternatives to the background check.
Kentucky	Concealed weapons permits issued on or after July 12, 2006 qualify.
Louisiana	None*
Maine	None*
Maryland	None*
Massachusetts	None*
Michigan	Licenses to purchase a pistol qualify. Concealed pistol licenses (CPLs) issued on or after November 22, 2005, qualify as an alternative to a National Instant Criminal Background Check System (NICS) check. CPLs issued prior to November 22, 2005 and temporary concealed pistol licenses do not qualify as NICS alternative.
Minnesota	None*
Mississippi	License to carry concealed pistol or revolver issued to individuals under Miss. Stat. Ann. § 45-9-101 qualify. (Note: security guard permits issued under Miss. Stat. Ann. § 97-37-7 do not qualify).
Missouri	None*
Montana	Concealed weapons permits qualify.
Nebraska	Concealed handgun permit qualifies as an alternative. Handgun purchase certificates qualify.
Nevada	Concealed carry permit issued on or after July 1, 2011, qualify.
New Hampshire	None
New Jersey	None
New Mexico	None
New York	None
North Carolina	Permits to purchase a handgun and concealed handgun permits qualify.
North Dakota	Concealed weapons permits issued on or after December 1, 1999 qualify.*
Northern Mariana Islands	None
Ohio	None
Oklahoma	None*
Oregon	None*
Pennsylvania	None
Puerto Rico	None

TABLE 3.8

Permanent Brady permit chart, 2016 [CONTINUED]

State/territory	Qualifying permits
Rhode Island	None
South Carolina	Concealed weapons permits qualify.
South Dakota	None*
Tennessee	None
Texas	Concealed weapons permits qualify.
U.S. Virgin Islands	None
Utah	Concealed weapons permits qualify.
Vermont	None
Virginia	None
Washington	Concealed pistol license issued on or after July 22, 2011 qualify.
West Virginia	Concealed handgun license issued on or after June 4, 2014 qualify.
Wisconsin	None
Wyoming	Concealed weapons permits qualify.

*While certain permits issued in these states prior to November 30, 1998, were "grandfathered" as Brady alternatives, none of these grandfathered permits would still be valid under state law as of November 30, 2003.
NICS = National Instant Criminal Background Check System.
CPL = Concealed pistol licenses.
Note: Notwithstanding the dates set forth above, permits qualify as alternatives to the background check requirements of the Brady law for no more than five years from the date of issuance. The permit must be valid under state law in order to qualify as a Brady alternative.

SOURCE: "Permanent Brady Permit Chart," in *Firearms*, U.S. Department of Justice, Bureau of Alcohol, Tobacco, Firearms, and Explosives, June 10, 2014, https://www.atf.gov/rules-and-regulations/permanent-brady-permit-chart (accessed September 8, 2016)

---

carry, who are among the staunchest defenders of the right to bear arms in the United States, typically promote the practice with the goal of normalizing the presence of guns in public. The practice surged in popularity beginning in 2008, and open-carry demonstrations at political events, restaurants, and other venues have drawn controversy and been extensively covered by the media.

Businesses in open-carry states often find themselves caught in the middle between their customers who favor open-carry and those who do not. For example, in "Open Carry Group Holds Rally at Home Depot in North Richland Hills" (DallasNews.com, June 1, 2014), Doug J. Swanson reports that the home-improvement chain Home Depot decided in May 2014 to allow open carry within its stores. That same month 150 open-carry activists held a rally outside of a Home Depot store in North Richland Hills, Texas. In contrast, restaurant chains such as Sonic and Chili's, after taking time to review their corporate policies on the issue, eventually settled on a policy of prohibiting open carry in their locations. Other corporations that eventually joined the ranks of businesses asking customers not to openly carry firearms in their stores included the restaurant chain Chipotle and the retail chain Target. Open-carry activists in various states commonly maintain online listings of "OTC-friendly" stores as well as those that prohibit open carry, and they often call for boycotts of such stores.

Dan Solomon reports in "How Open Carry Forced Businesses in Texas to Take a Side in the Culture War" (TexasMonthly.com, January 5, 2016) that when open-carry handguns became legal in Texas in January 2016, businesses were required to display prominent signs indicating whether or not they allowed guns on their premises. "In other words," Solomon writes, "the mere existence of the expanded open carry law in Texas means that businesses whose main objective is selling hamburgers or craft supplies [have] to pick a side" and risk alienating their pro-gun customers or those who want to be able to shop, dine, and conduct other public business without having to confront the sight of armed citizens.

Federal law does not address open carry, with the exception of laws prohibiting firearms on certain federal properties. According to the Law Center to Prevent Gun Violence, in "Open Carrying" (http://smartgunlaws.org/gun-laws/policy-areas/firearms-in-public-places/open-carrying/), as of 2016 three states (California, Florida, and Illinois) and the District of Columbia prohibit the open carrying of any weapon, two states (New York and South Carolina) prohibit the open carrying of a handgun but not a long gun, and three states (Massachusetts, Minnesota, and New Jersey) prohibit the open carrying of a long gun but not a handgun. Fifteen states require a permit for open carrying of handguns: Connecticut, Georgia, Hawaii, Indiana, Iowa, Maryland, Massachusetts, Minnesota, Missouri, New Jersey, Oklahoma, Rhode Island, Tennessee, Texas, and Utah. Thirty-one states allow open carry of handguns without a permit, although some place certain restrictions on open carry. The open carrying of long guns is allowed without a permit in 44 states, but some states place restrictions on this right (such as stipulating that the gun must not be loaded). States that allow open carry generally prohibit it in certain locations, including schools, some state-owned properties, bars and other places serving alcohol, and on public transportation.

Minors

During the 1990s Americans became concerned with what the U.S. surgeon general called an epidemic of youth violence. In *Youth Violence: A Report of the Surgeon General* (January 2001, http://www.surgeongeneral.gov/library/youthviolence/default.htm), the surgeon general describes the sharply rising arrest rates of young people for violent crimes between 1983 and 1993–94. The surgeon general also notes that juvenile homicides increased 65% between 1987 and 1993. The number of older juveniles killed with firearms accounted for nearly all the growth.

Although arrest rates later declined, the surgeon general explains that some states passed tougher laws prohibiting juveniles from possessing firearms and/or punishing those who supplied them with guns. These state laws are in addition to the federal law that prohibits FFLs from selling or delivering handguns to people under the age of 21 years (18 U.S.C. 922[b][l]) and the federal law that prohibits people under the age of 18 years from possessing handguns (18 U.S.C. Sect. 922[x]). Federal law allows the sale of handguns to people between the ages of 18 and 21 years at gun shows, unless specifically prohibited by state law.

LOCAL ORDINANCES

"Preemption" is a legal doctrine under which regulatory authority at a higher level of government (federal or state) preempts (takes precedence over) regulatory authority at a lower level (county, city, or town). For example, although states have considerable authority to enact their own tax laws, federal law is considered preemptive on most issues relating to immigration and labor. As such, states cannot enact immigration or labor laws that conflict with federal law.

In 2016, 43 states maintained broad preemption statutes preventing town, city, and county governments from passing local ordinances that restrict firearms possession, transportation, sale, transfer, ownership, manufacturing, or repair. The Law Center to Prevent Gun Violence explains in "Local Authority to Regulate Firearms" (2016, http://smartgunlaws.org/gun-laws/policy-areas/other-laws/local-authority/) that these states' statutes vary in their exact terms, "but each one expressly preempts all, or substantially all, aspects of local firearms and/or ammunition regulation."

Only five states—Connecticut, Hawaii, Massachusetts, New Jersey, and New York—had no preemption statute in place regarding firearms regulation in 2016. It is this lack of state preemption that has enabled New York City to pass gun control measures that are considerably stricter than those at the state level. For example, the city does not recognize firearm permits from any other jurisdiction, including permits issued by the state of New York. Other restrictions include prohibiting possession of a gun by anyone under 21 years of age, requiring a license to purchase a firearm, and requiring handguns being transported in vehicles in New York City to be unloaded and in a locked container, with ammunition in a separate locked container.

Two additional states—California and Nebraska—grant local governments broad authority to regulate firearms and ammunition, with a few exceptions where state law is preemptive. In Nebraska, for example, state laws concerning the registration and concealed carrying of handguns are preemptive, such that city governments are prevented from passing more restrictive local measures. California state law is preemptive with regard to gun registration and licensing, as well as with the manufacture, possession, and sale of imitation guns and restrictions on where concealed weapons can be carried.

Gun rights advocates generally favor state preemption statutes. In the absence of state preemption, they argue, the existence of myriad different local firearms regulations places undue burden on responsible gun owners to understand and comply with laws that are not consistent from one county or town to the next. By contrast, gun control advocates generally oppose broad state preemption. According to the Law Center to Prevent Gun Violence, preemption statutes effectively "[mandate] a one-size-fits-all approach to firearms regulation," preventing local governments from addressing the particular gun violence challenges that exist within their communities. Pointing to California, where local governments have passed more than 300 new gun laws since the mid-1990s, the organization argues that local innovation is the key to discovering and developing effective gun violence reduction strategies that can be applied at the state level.

CHAPTER 4
COURT RULINGS ON FIREARMS

The U.S. Constitution and most state constitutions guarantee the right to bear arms, but the courts have ruled that this right may be strictly controlled. Many laws and regulations have been enacted at the local, state, and federal levels to regulate firearms. When these laws have been challenged, state and federal courts have consistently upheld the right of governments to require the registration of firearms, to determine how these weapons may be carried, and even to forbid the sale or use of some weapons under certain circumstances. Courts have also been asked to decide if manufacturers, dealers, or sometimes even relatives of the gun carrier should be held responsible when guns are used to commit crimes.

Table 4.1 presents a list of reasons an individual may be denied the right to bear arms. It has long been generally accepted that gun ownership will be denied to convicted felons, individuals who are not of sound mind, and individuals regarded as incapable because of a mental condition. Even these restrictions, however, have been challenged in court, and as new prohibited classes of people have been added, there have been more court cases clarifying the prohibitions' constitutionality. The following selection of court cases includes landmark decisions and more recent rulings on gun rights and regulations at the federal, state, and local levels.

SECOND AMENDMENT INTERPRETATIONS
The Bill of Rights: Federal versus State Protections

The "right of the people to keep and bear arms" is the essence of the Second Amendment, which is part of the Bill of Rights (ratified 1791) to the U.S. Constitution. For many years, when deciding cases based on any guarantee granted by the Bill of Rights, including the Second Amendment, the courts relied on the 1833 U.S. Supreme Court decision in *Barron v. City of Baltimore* (32 U.S. 243). In that case the court ruled that the Bill of Rights does not apply to or restrict the states. This meant that the Bill of Rights was a limitation to the federal government only; its protections applied to federal laws.

Under this interpretation, a state did not have to allow people "to keep and bear arms" unless the state's constitution guaranteed that right. If the state constitution did not ensure that right, then state or local authorities could arrest a person who possessed a firearm; nevertheless, federal law enforcement officials could not because of the Bill of Rights. Thus, state constitutions—rather than the Second Amendment of the U.S. Constitution—decided most gun cases. This situation, however, changed during the years of Chief Justice Earl Warren (1891–1974). Warren was the chief justice from 1953 to 1969, when most of the guarantees of the Bill of Rights were held to apply to the states through the 14th Amendment, which states, "No state shall make or enforce any law which shall abridge the privileges or immunities of citizens."

The Individual Rights Interpretation versus the Collective Rights Interpretation

Gun rights advocates interpret the Second Amendment as a guarantee to individuals of the right to keep and bear arms without governmental interference. Favoring stricter control of guns, proponents of the collective rights argument disagree and point to the opening words of the Second Amendment—"A well regulated militia, being necessary to the security of a free state"—as an intention to guarantee the right of states to maintain militias. The modern militia of the United States is the National Guard. Thus, collective rights advocates argue that the general population does not need unfettered access to guns because it is no longer expected to form a militia during times of need.

From 1939 to 2008 legal interpretations of the Second Amendment were based on the Supreme Court's unanimous ruling in *United States v. Miller* (307 U.S. 174 [1939]), which was widely regarded as an unclear opinion. Although many gun control advocates claim the ruling

TABLE 4.1

Firearm prohibitive criteria under the National Instant Criminal Background Check System

18 U.S.C. §922 (g) (1)
Has been convicted in any court of a crime punishable by imprisonment for a term exceeding one year.

18 U.S.C. §922 (g) (2)
Is a fugitive from justice.

18 U.S.C. §922 (g) (3)
Is an unlawful user of or addicted to any controlled substance.

18 U.S.C. §922 (g) (4)
Has been adjudicated as a mental defective or committed to a mental institution.

18 U.S.C. §922 (g) (5)
Is illegally or unlawfully in the United States.

18 U.S.C. §922 (g) (6)
Has been discharged from the Armed Forces under dishonorable conditions.

18 U.S.C. §922 (g) (7)
Having been a citizen of the United States, has renounced U.S. citizenship.

18 U.S.C. §922 (g) (8)
Is subject to a court order that restrains the person from harassing, stalking, or threatening an intimate partner or child of such intimate partner.

18 U.S.C. §922 (g) (9)
Has been convicted in any court of a misdemeanor crime of domestic violence.

18 U.S.C. §922 (n)
Is under indictment/information for a crime punishable by imprisonment for a term exceeding one year.

SOURCE: "Federal Prohibitors," in *National Instant Criminal Background Check System (NICS) Operations, 2014*, Federal Bureau of Investigation, 2015, https://www.fbi.gov/file-repository/2014-nics-ops-report-050115.pdf (accessed September 8, 2016)

declared that the Second Amendment's "obvious purpose" is to "assure the continuation and render possible the effectiveness of" state militias, thus reinforcing a collective rights interpretation, gun rights advocates suggest the ruling established an individual right to bear arms. In any case, during the seven decades that *Miller* served as precedent, lower courts typically upheld the constitutionality of restrictions on the individual's right to bear arms. Thus, although *Miller* has been used to support both sides in the gun control debate, its effect on subsequent legal rulings has been to privilege the collective rights interpretation of the Second Amendment.

The landmark Supreme Court ruling in *District of Columbia v. Heller* (554 U.S. 570 [2008]) established the current status quo in gun control legislation by taking a clear position on the individual right to bear arms. In the *Heller* decision, which concerned a challenge to the District of Columbia's decades-old ban on handguns and its restrictions regarding the storage of firearms in the home, the court unambiguously interpreted the Second Amendment as upholding an individual's right to possess firearms and keep them in the home for the purposes of self-defense. As a result of this new interpretation, the District of Columbia's ban on handguns and its storage requirements for guns were declared unconstitutional. The court did, however, establish the government's right to regulate firearm possession and sales within certain bounds.

FEDERAL COURT CASES

United States v. Cruikshank: Right to Bear Arms?

The first major federal case dealing with the Second Amendment was *United States v. Cruikshank* (92 U.S. 542 [1875]). The defendants were members of the Ku Klux Klan, a white supremacist group, and were convicted of conspiracy to deprive two African American men of their right of assembly and free speech and their right to keep and bear arms as guaranteed by the U.S. Constitution. The Supreme Court ruled, "'Bearing arms for a lawful purpose.' This is not a right granted by the Constitution. Neither is it in any manner dependent upon that instrument for its existence. The second amendment declares that it shall not be infringed; but this ... means no more than that it shall not be infringed by Congress. This is one of the amendments that has no other effect than to restrict the powers of the national government."

In this ruling, the Supreme Court agreed with a lower court ruling that the right to keep and bear arms is a birthright. It is not a right created or conferred by the Constitution. The Constitution, however, guarantees that this right shall not be impaired by the state or federal government. In addition, it is the duty of the state to protect and enforce this right.

United States v. Miller: Possession of Gangster-Type Weapons

The Supreme Court first addressed the meaning of the Second Amendment during the late 1930s, in a case involving a violation of the National Firearms Act of 1934, a federal law designed to make it more difficult to acquire especially dangerous "gangster-type" weapons. Jack Miller and Frank Layton were arrested by federal agents in 1938 and charged with traveling with an unregistered, gangster-type, sawed-off shotgun. A federal district court judge dismissed the case on the grounds that the National Firearms Act violated the Second Amendment. The U.S. government appealed to the Supreme Court in *United States v. Miller*.

The federal government argued that if the Second Amendment protected an individual's right to keep and bear arms, the only arms protected were those suitable to military purposes, not weapons such as sawed-off shotguns that "constitute the arsenal of the 'public enemy' and the 'gangster'"—weapons that the National Firearms Act was intended to regulate.

The Supreme Court reversed the lower court's ruling and upheld the federal law. Because Miller had fled and was not present to plead his case, the court heard only the government's side of the issue and did not hear a strong argument for permitting a citizen to maintain such a weapon. In the end, the Supreme Court denied Miller the right to carry a sawed-off shotgun, noting that no evidence had been presented as to the usefulness "at this time" of a sawed-off shotgun for military purposes. The court stated, "In the absence of any evidence tending to show that possession or use of a 'shotgun having a barrel of less than eighteen inches in length' at this time has some reasonable relationship to the preservation or efficiency of a well regulated militia, we cannot say that the Second Amendment guarantees the right to keep and bear such an instrument. Certainly it is not within judicial notice that this weapon is any part of the ordinary military equipment or that its use could contribute to the common defense."

Referring back to the debates of the Constitutional Convention and the discussion of the militia, the court observed that such deliberations showed "plainly enough that the Militia comprised all males physically capable of acting in concert for the common defense, 'A body of citizens enrolled for military discipline.' And further, that ordinarily when called for service these men were expected to appear bearing arms supplied by themselves and of the kind in common use at the time."

For the next seven decades *Miller* was the major Supreme Court ruling and precedent concerning gun control.

USING *MILLER* AS A PRECEDENT. The case of *United States v. Tot* (131 F.2d 261 [1942]) originated in the arrest of Frank Tot for stealing cigarettes from an interstate shipment. Tot had previously been convicted of a crime of violence. At the time of his arrest, a .32-caliber Colt automatic pistol was seized during a search of his home. The Third Circuit Court of Appeals did not accept Tot's argument that the Second Amendment prohibited the state of New Jersey from denying him the right to own a gun even if he was a convicted felon. Citing *Miller* as a precedent, the circuit court reasoned that "one could hardly argue seriously that a limitation upon the privilege of possessing weapons was unconstitutional when applied to a mental patient of the maniac type. The same would be true if the possessor were a child of immature years.... Congress has prohibited the receipt of weapons from interstate transactions by persons who have previously, by due process of law, been shown to be aggressors against society. Such a classification is entirely reasonable and does not infringe upon the preservation of the well regulated militia protected by the Second Amendment."

The circuit court noted that the Second Amendment, "unlike those providing for protection of free speech and freedom of religion, was not adopted with individual rights in mind, but as a protection for the States in the maintenance of their militia organizations against possible encroachments by the federal power."

In 1942 the First Circuit Court of Appeals cited *Miller* in *Cases v. United States* (131 F.2d 916) in an attempt to uphold the Federal Firearms Act of 1938. Jose Cases Velazquez had been convicted of a violent crime and, under the federal law, could not own a gun. The circuit court observed, "The Federal Firearms Act undoubtedly curtails to some extent the right of individuals to keep and bear arms.... [This] is not a right conferred upon the people by the federal constitution."

These rulings (*Tot* and *Cases*), perhaps more clearly than the *Miller* decision itself, established the precedent of interpreting the Second Amendment not as a guarantee of the individual's right to keep and bear arms but as a guarantee of that right so far as was necessary to ensure a well-regulated militia.

Miller was also cited by the Fifth Circuit Court of Appeals in *United States v. Emerson* (270 F.3d 203 [2001]). In this case Timothy Joe Emerson was charged with violating the Lautenberg Amendment to the Gun Control Act of 1968, which prohibits possession of a firearm by people under a domestic violence restraining order. Emerson's estranged wife had obtained such an order from a judge in 1998, after Emerson had acknowledged his mental instability. He was subsequently indicted for illegally possessing two 9mm pistols, a semiautomatic SKS assault rifle with bayonet, a semiautomatic M-14 assault rifle, and an M1 carbine. At his trial in district court, his lawyers argued that the case should be dismissed on the grounds that the federal ban on gun possession by those under a protective order for domestic violence violated the Second Amendment. The district court sided with Emerson and dismissed the charges, reasoning that the amendment to the 1968 law violates the Second Amendment because it allows a state court divorce proceeding to deprive a citizen of his or her right to keep and bear arms, even when that citizen has not been found guilty of anything.

In its ruling, the district court noted that it interpreted the Second Amendment as conferring individual rights on U.S. citizens. Nevertheless, U.S. Department of Justice prosecutors appealed the court's decision, stating that it directly conflicted with the long-established legal precedent (the collective rights interpretation) laid down by the Supreme Court in *Miller*.

When the Fifth Circuit Court of Appeals reversed the lower court decision and upheld the domestic violence gun ban against Emerson, gun control advocates viewed the decision as a victory for domestic violence victims and a safeguard for women across the country. Gun rights advocates also found something to praise in the decision, because it seemed to provide support for the argument that individuals are guaranteed the right under the U.S. Constitution to bear arms independent of the provision of a well-regulated militia. The decision of the court stated, "We conclude that *Miller* does not support the government's collective rights or sophisticated collective rights approach to the Second Amendment. Indeed, to the extent that *Miller* sheds light on the matter it cuts against the government's position."

Emerson was quickly seized on by gun rights advocates. In 2000 the attorney Gary Gorski filed a lawsuit in the Eastern District Court in California. The case, *Silveira v. Lockyer* (312 F.3d 1052 [2002]), sought to overturn California's ban on semiautomatic rifles on the basis of the individual right of a person to keep and bear arms under the Second Amendment. The *Silveira* lawsuit lost in the Eastern District Court and was appealed to the Ninth Circuit Court in 2002, which upheld the lower court's decision. The Ninth Circuit Court's written decision strongly rejected the reasoning behind the *Emerson* decision, stating "the debates of the founding era demonstrate that the second of the first ten amendments to the Constitution was included in order to preserve the efficacy of the state militias for the people's defense—not to ensure an individual right to possess weapons."

Silveira has been viewed by some observers as a significant setback for gun rights advocates.

United States v. Synnes: Gun Possession by a Convicted Felon

In *United States v. Synnes* (438 F.2d 764 [1971]), another case involving the possession of a firearm by a convicted felon, the Eighth Circuit Court of Appeals said this about the Second Amendment, "We see no conflict between [a law prohibiting the possession of guns by convicted criminals] and the Second Amendment since there is no showing that prohibiting possession of firearms by felons obstructs the maintenance of a 'well-regulated militia.'"

Most supporters of handgun possession have accepted the right of federal and state governments to deny weapons to former felons, drug abusers, and mentally disabled individuals.

United States v. Warin: Possession of a Machine Gun

In *United States v. Warin* (530 F.2d 103 [1976]), the defendant Francis J. Warin appealed his conviction for possessing an unlicensed 9mm prototype submachine gun. Warin, an engineer and designer of firearms, had built the firearm himself. Warin claimed before the Sixth Circuit Court of Appeals that he had not violated any law because, as a member of Ohio's "sedentary militia" (suggesting that he was prepared to serve in the state militia), the Second Amendment guaranteed his right to possess the weapon. The court upheld Warin's conviction, invoking both the *Miller* and *Tot* precedents in its reasoning that it is an "erroneous supposition that the Second Amendment is concerned with the rights of individuals rather than those of the States." This ruling was significant because it was another instance where the Second Amendment was interpreted to guarantee a collective state right to support a militia, rather than an individual's right to keep and bear arms.

Smith v. United States: Enhanced Penalties for "Use" of Firearms in a Drug Crime

The Supreme Court ruled in *Smith v. United States* (508 U.S. 223 [1993]) that the federal law authorizing stiffer penalties if the defendant "during and in relation to ... [a] drug trafficking crime, uses ... a firearm" applies not only to the use of firearms as weapons but also to firearms used as commerce, such as in a bartering or trading transaction. John Angus Smith and a companion traveled from Tennessee to Florida to buy cocaine, which they planned to resell for profit. During a drug transaction, an undercover agent posing as a pawnshop dealer examined Smith's MAC-10, a compact and lightweight firearm that can be equipped with a silencer and is popular among criminals. Smith told the agent he could have the gun in exchange for 2 ounces (56.7 g) of cocaine. The officer said he would try to get the drug and return in an hour. In the meantime Smith became suspicious and fled, and after a high-speed chase officers apprehended him.

A grand jury was convened to decide whether there was enough evidence to justify formal charges and a trial. Smith was charged with drug trafficking crimes and with knowingly using the MAC-10 in connection with a drug trafficking crime, among other offenses. Under 18 U.S.C. Section 924(c)(1)(B)(ii), a defendant who uses a firearm in such a way must be sentenced to five years' imprisonment, and if the firearm "is a machine gun or a destructive device, or is equipped with a firearm silencer or firearm muffler," as it was in this case, the sentence is 30 years. Smith was convicted on all counts.

On appeal Smith argued that the law applied only if the firearm was used as a weapon. The 11th Circuit Court of Appeals disagreed, ruling that the federal legislation did not require that the firearm be used as a weapon—"any use of 'the weapon to facilitate in any manner the commission of the offense' suffices." In a similar case, *United States v. Harris* (959 F.2d 246 [1992]), the U.S. Court of Appeals for the District of Columbia Circuit had arrived at the same conclusion. By contrast, the U.S. Court of Appeals for the Ninth Circuit held in *United States v. Phelps* (877 F.2d 28 [1989]) that trading a gun during a drug-related transaction was not "using" it within the meaning of the statute. To resolve the conflict among the different circuit courts, the Supreme Court heard Smith's appeal.

In a 6–3 decision, the court ruled in *Smith v. United States* that the "exchange of a gun for narcotics constitutes 'use' of a firearm 'during and in relation to ... [a] drug trafficking crime' within the meaning" of the federal statute. Delivering the opinion of the majority, Justice Sandra Day O'Connor (1930–) wrote that "when a word is not defined by statute, we normally construe it in accord with its ordinary or natural meaning." Definitions for the word *use* from various dictionaries and *Black's Law Dictionary* show the word to mean "convert to one's service; to employ; to avail oneself of; to utilize; to carry out a purpose or action by means of." In trying to exchange his MAC-10 for drugs, the defendant "used" or employed the gun as an item of trade to obtain drugs. The phrase "as a weapon" does not appear in the statute. O'Connor reasoned that if Congress had meant the narrow interpretation of "use" (as a weapon only), it would have worded the statute differently. Justice Antonin Scalia (1936–2016;) dissented from the majority's definition of "use," defining the normal usage of a gun as discharging, brandishing, or using as a weapon.

Bailey v. United States: New Interpretations of "Use"

In 1995 the Supreme Court narrowed the definition of the word *use* that had been established in *Smith*. In *Bailey v. United States* (516 U.S. 137), the court considered the criminal misdeeds of Roland J. Bailey and Candisha Robinson. Originally, Bailey and Robinson were defendants in separate, unrelated trials. After both were convicted, they filed a joint appeal to the Supreme Court, petitioning for clarification of the meaning of the word *use*.

Bailey's case began in 1988 with a routine traffic stop in the District of Columbia. When Bailey could not produce a driver's license, a police officer searched his car and found ammunition and just over 1 ounce (30 g) of cocaine. Another officer found a loaded pistol and more than $3,200 in small bills in the trunk. At his trial, Bailey was convicted of possession of cocaine with intent to deliver and using or carrying a firearm in connection with a drug offense. He appealed to the U.S. Court of Appeals for the District of Columbia Circuit.

Robinson's case originated in 1991, when an undercover police officer approached her to buy crack cocaine with a marked $20 bill. The officer noticed that she obtained the drugs from her one-bedroom apartment. Later, while executing a search warrant on Robinson's apartment, officers found a .22-caliber derringer, two rocks of crack cocaine, and the marked bill. Robinson was found guilty of cocaine distribution and, among other things, the use or carrying of a firearm during and in relation to a drug trafficking offense. She appealed to the U.S. Court of Appeals for the District of Columbia Circuit.

In Bailey's appeal, *United States v. Bailey* (995 F.2d 1113 [1993]), the defense argued there was no evidence that he had used the gun in connection with a drug offense. Robinson argued in her appeal, *United States v. Robinson* (997 F.2d 884 [1993]), that during the drug sale to the officer, the gun was unloaded and in a locked trunk and was not used in the commission of or in relation to a drug trafficking offense.

The court of appeals rejected Bailey's claim of insufficient evidence and held that he could be convicted for "using" a firearm if the jury could reasonably infer that the gun had assisted in the commission of a drug offense. In Robinson's case the court reversed her conviction for "using or carrying" because the presence of an unloaded gun in a locked trunk in a bedroom closet was not evidence of actual use. Because the decisions were contradictory, the court of appeals consolidated the two cases and reheard them. A majority of the judges then found that there was sufficient evidence to establish that each defendant had used a firearm in relation to a drug trafficking offense.

When Bailey and Robinson jointly appealed to the Supreme Court in 1995, the judges unanimously held that to establish "use," the government must show that the defendant actively employed a firearm so as to make it an "operative factor in relation to the predicate offense." This definition includes hiding a gun in a shirt or pants, threatening to use a gun, or actually using a gun during the commission of a drug crime. The court also found that Bailey's and Robinson's "use" convictions could not be supported because the evidence did not indicate that either defendant actively employed firearms during drug crimes.

In 1998 the Supreme Court put a much broader interpretation on the federal law, which mandates a five-year prison term for a person who "uses or carries" a gun "during and in relation to" a drug-trafficking crime. The three defendants in this case carried guns in the trunks of their cars. The court ruled in *Muscarello v. United States* (524 U.S. 125 [1998]) that having a gun in a car from which a person is dealing drugs fits the meaning of "carries" for purposes of the sentencing statutes.

United States v. Lopez: Possession of a Firearm near a School

The Gun-Free School Zones Act of 1990 made it unlawful for any individual to knowingly possess a firearm in a school zone, defined as within 1,000 feet (305 m) of school grounds, regardless of whether school was in session. Two federal appeals courts came to different conclusions about the constitutionality of the act.

FIFTH CIRCUIT COURT OF APPEALS: THE GUN-FREE SCHOOL ZONES ACT IS UNCONSTITUTIONAL. In March 1992 high school senior Alfonso Lopez Jr. carried a concealed revolver and five bullets into Edison High School in San Antonio, Texas. School officials caught him, and the student was subsequently charged with violating the Gun-Free School Zones Act. Lopez's attorneys moved to dismiss the changes because, they contended, the law was unconstitutional. The trial court did not accept their argument and convicted Lopez. The case then went to the U.S. Court of Appeals for the Fifth Circuit, which disagreed with the trial court, ruling in *United States v. Lopez* (2 F.3d 1342 [1993]) that Congress had exceeded the power granted to it under the commerce clause of the U.S. Constitution when it enacted the Gun-Free School Zones Act. The commerce clause gives Congress the power to regulate conduct that crosses state borders. According to the court, with few specific exceptions, "federal laws proscribing firearm possession require the government to prove a connection to commerce." Congress had made no attempt to link the Gun-Free School Zones Act to commerce in the debates before the law was enacted and in the law itself. The appeals court asserted:

> Both the management of education, and the general control of simple firearms possession by ordinary citizens, have traditionally been a state responsibility.... We are unwilling to ourselves simply assume that the concededly intrastate conduct of mere possession by any person of any firearm substantially affects interstate commerce, or the regulation thereof, whenever it occurs, or even most of the time that it occurs, within 1,000 feet of the grounds of any school, whether or not then in session. If Congress can thus bar firearms possession because of such a nexus [connection] to the grounds of any public or private school, and can do so without supportive findings or legislative history, on the theory that education affects commerce, then it could also similarly ban lead pencils, "sneakers," Game Boys, or slide rules.

Following this reasoning, the appeals court found the Gun-Free School Zones Act unconstitutional.

NINTH CIRCUIT COURT OF APPEALS: THE GUN-FREE SCHOOL ZONES ACT IS CONSTITUTIONAL. By contrast, the U.S. Court of Appeals for the Ninth Circuit ruled in *United States v. Edwards* (13 F.3d 291 [1993]) that the commerce clause of the U.S. Constitution did, indeed, give Congress the power to pass a law such as the Gun-Free School Zones Act. In 1991 Sacramento, California, police officers and school officials approached Ray Harold Edwards III and four other males at Grant Union High School. The officers discovered a .22-caliber rifle and a sawed-off rifle in the trunk of Edwards's car. One of the charges against Edwards was violation of the Gun-Free School Zones Act.

Edwards appealed, claiming the law violated the 10th Amendment because Congress did not have the authority under the commerce clause or any other delegated power to enact the Gun-Free School Zones Act. The 10th Amendment states that "the powers not delegated to the United States by the Constitution, nor prohibited by it to the States, are reserved to the States respectively, or to the people."

Disagreeing with *United States v. Lopez*, the court of appeals ruled that the Gun-Free School Zones Act "does not expressly require the Government to establish a nexus between the possession of a firearm in a school zone and interstate commerce.... It is unnecessary for Congress to make express findings that a particular activity or class of activities affects interstate commerce in order to exercise its legislative authority pursuant to the commerce clause."

THE U.S. SUPREME COURT: THE GUN-FREE SCHOOL ZONES ACT IS UNCONSTITUTIONAL. In 1995 the Supreme Court struck down the Gun-Free School Zones Act in *United States v. Lopez* (514 U.S. 549) on the grounds that Congress had overstepped its bounds because it had based the law on the commerce clause of the U.S. Constitution. The commerce clause empowers Congress to regulate interstate commerce, but Congress had failed to connect gun-free school zones with commerce. Chief Justice William H. Rehnquist (1924–2005) wrote that Congress had used the commerce clause as a general police power in a way that is generally retained by states. He also warned that the Gun-Free School Zones Act "is a criminal statute that by its terms has nothing to do with 'commerce' or any sort of economic enterprise, however broadly one might define those terms.... If we were to accept the Government's arguments, we are hard-pressed to posit any activity by an individual that Congress is without power to regulate."

Congress responded in 1996 by approving a slightly revised version of the Gun-Free School Zones Act in the form of amendments to the Department of Defense Appropriations Act of 1997. The amendments required prosecutors to prove an impact on interstate commerce as an element of the offense.

Printz v. United States: The Constitutionality of the Brady Law

Opponents of the Brady Handgun Violence Prevention Act of 1993 challenged its constitutionality soon after it passed. Under the Brady law, Congress had ordered local chief law enforcement officials nationwide to conduct background checks on prospective handgun purchasers who bought their guns through federally licensed dealers. Two sheriffs, Jay Printz of Ravalli County, Montana, and Richard Mack of Graham County, Arizona, charged that Congress exceeded its powers under the 10th Amendment of the U.S. Constitution, which defines the separation of powers—the relationship between the federal government and the sovereign powers of the individual states. They argued that the federal government had placed federal burdens on local police agencies with no federal compensation. Representing the federal government, the U.S. acting solicitor general Walter Dellinger (1941–) argued that the government had the right to require local agencies to carry out federal orders as long as those agencies were not forced to make policy.

In *Printz v. United States* (521 U.S. 898 [1997]), the Supreme Court struck down the Brady law provisions that required local chief law enforcement officials to conduct background checks on prospective handgun buyers and to accept the form on which that background check is based. The court declared that these provisions violated the 10th Amendment to the U.S. Constitution. Justice Scalia wrote, "The Federal Government may neither issue directives requiring the States to address particular problems, nor command the States' officers, or those of their political subdivisions, to administer or enforce a federal regulatory program."

The court unanimously upheld the Brady law's five-day waiting period for handgun purchases because the waiting period was directed at gun store owners and was not a federal mandate to state officials. Most chief law enforcement officers continued to conduct background checks voluntarily until the National Instant Criminal Background Check System, which was instituted by the Brady law, became effective in November 1998.

United States v. Stewart: Possession of Homemade Machine Guns

Robert Wilson Stewart Jr. was a convicted felon who sold parts kits to make Maadi-Griffin .50-caliber rifles. He advertised the kits on the Internet and in magazines. The Bureau of Alcohol, Tobacco, Firearms, and Explosives (ATF) began investigating Stewart when it realized that he had a prior conviction for the possession and transfer of a machine gun. During the investigation an ATF agent purchased parts kits from Stewart and determined that they could be used to make an unlawful firearm. After obtaining a warrant, the ATF searched Stewart's residence and discovered 31 firearms, including five machine guns that Stewart had machined and assembled. Stewart was convicted of being a felon in possession of a firearm, of unlawful possession of a machine gun, and of possessing several unregistered, homemade machine guns.

The case against Stewart had been based, in part, on an interpretation of the commerce clause of the U.S. Constitution that prohibits anyone except a licensed importer, manufacturer, or dealer of firearms to import, manufacture, or deal in firearms. In his appeal, Stewart claimed that Congress had exceeded its commerce clause power and violated the Second Amendment.

In *United States v. Stewart* (348 F.3d 1132 [2003]), the Ninth Circuit Court of Appeals overturned the lower court's ruling on violating the commerce clause, saying that Stewart did not have a substantial effect on interstate commerce. The court, however, affirmed his conviction for being a felon in possession of a firearm.

In 2005 the Department of Justice appealed the case to the Supreme Court. The high court would not hear the case but instructed the Ninth Circuit Court of Appeals to further consider the case in light of its recent ruling in *Gonzales v. Raich* (545 U.S. 1 [2005]). That case allowed Congress to use the commerce clause to ban the cultivation and possession of homegrown marijuana for personal medical use because, the court said, the marijuana could affect the supply and demand of the drug, thereby affecting interstate commerce. In June 2006 the Ninth Circuit Court of Appeals ruled that Congress has the power to regulate the sales of homemade machine guns because they can enter the interstate market and affect supply and demand. Thus, Congress had not exceeded its commerce clause power and had not violated the Second Amendment.

District of Columbia v. Heller: The Constitutionality of the District of Columbia's Handgun Ban

In 2008 the Supreme Court was once again confronted with making a decision that hinged on the Second Amendment and would issue its first decision interpreting the Second Amendment since the 1939 *Miller* decision. This time the interpretation was in the context of the District of Columbia's ban on the possession of handguns. The high court was expected to make a landmark ruling on the decades-old question of whether the Constitution's Second Amendment refers to an individual's right to gun ownership or strictly to militia service.

In 1976 the District of Columbia passed a law that banned the private possession of handguns and that required rifles and shotguns in the home to be outfitted with a trigger lock or kept unloaded and disassembled. Dick Anthony Heller was the named party in the suit, but the attorney Robert A. Levy (1941–) personally financed the lawsuit and worked on bringing it to the Supreme Court with the purpose of addressing the Second Amendment question. Attorneys for the District of Columbia contended that the Second Amendment does not give individuals the right to bear arms, whereas Levy and the other attorneys for Heller contended it does.

In 2007 a three-judge panel of the U.S. Court of Appeals for the District of Columbia struck down the gun control ordinance on Second Amendment grounds. Senior Judge Laurence H. Silberman (1935–) wrote for the 2–1 majority that the amendment provides an individual right just as other provisions of the Bill of Rights do. Handguns fall under the definition of "arms"; thus, the District of Columbia may not ban them.

The District of Columbia appealed the case to the Supreme Court. In March 2008 the high court heard arguments, and in June 2008 the court rendered its decision in *District of Columbia v. Heller*. The 5–4 landmark ruling interpreted the Second Amendment as protecting the individual's right to own a gun, thus supporting the 2007 decision of the U.S. Court of Appeals for the District of Columbia. The majority opinion, written by Justice Scalia, provided for gun control legislation by noting that "like most rights, the Second Amendment right is not unlimited. It is not a right to keep and carry any weapon whatsoever in any manner whatsoever and for whatever purpose." Justice John Paul Stevens (1920–), writing for the minority (dissenting) opinion, noted, "The Court's announcement of a new constitutional right to own and use firearms for private purposes upsets that settled understanding, but leaves for future cases the formidable task of defining the scope of permissible regulations."

McDonald v. City of Chicago: Must States Follow the Same Second Amendment Rules as the Federal Government?

The District of Columbia is under federal jurisdiction. As a result, the *Heller* ruling does not necessarily apply to state and local jurisdictions. One day after *Heller* was decided, Otis McDonald and three other Chicago residents filed suit in the U.S. District Court for the Northern District of Illinois in an attempt to determine whether states must follow the same Second Amendment rules as federal jurisdictions and to overturn a decades-long ban on handguns in the city of Chicago. In *McDonald v. City of Chicago* (No. 08 C 3645 [2008]), the district court ruled in favor of the city. The petitioners appealed the case to the U.S. Court of Appeals for the Seventh Circuit. In *National Rifle Association of America v. City of Chicago* (567 F.3d 856 [2009]), the appeals court upheld the decision of the district court, noting that the city's handgun ban did not violate the Constitution because the Supreme Court had not yet declared whether its decision in the *Heller* case established a fundamental right to guns applicable throughout the United States. The Supreme Court agreed to hear the case, *McDonald v. City of Chicago* (No. 08-1521), which was argued in March 2010. In June 2010 the high court returned its ruling. In a 5–4 decision, the court stated that the "Second Amendment right is fully applicable to the States."

After *Heller* and *McDonald*

In the years after the landmark *Heller* and *McDonald* rulings many gun rights advocates looked to other cases with the hope that the Supreme Court would clarify further dimensions of their Second Amendment rights. As of December 2016, the most prominent of these cases to have been in contention for the high court's consideration was *Drake v. Jerejian* (No. 12-1150), a case that challenged New Jersey's policy for issuing concealed weapons permits. Unlike shall-issue states, in which government officials are required to issue handgun carry permits to applicants except in extraordinary situations, New Jersey requires citizens to prove a justifiable need for a concealed-carry permit. Many gun rights advocates consider this requirement to be in conflict with the *Heller* ruling establishing the right to self-defense, given the stark contrast between the rights New Jersey gun owners enjoy in their home and the rights they enjoy outside of their home. In May 2014 the court refused to hear the case, which signaled that it was not willing to define a right to self-defense outside of the home.

STATE LAWS

Most state constitutions guarantee the right to bear arms, and this right has either been enacted or strengthened in more than a dozen states since 1970 (see the Appendix). Some states clearly tie this right to the militia, whereas other state constitutions and courts have ruled from the perspective of self-defense. States that have enacted laws to protect the right to use deadly force in self-defense are shown in Table 4.2. Two classic examples of cases involving self-defense are *Schubert v. DeBard* (398 N.E.2d 1339 [1980]) and *State v. Kessler* (289 Or. 359 [1980]).

Schubert v. DeBard: The Right to Possess a Handgun for Self-Defense Protected in Indiana

Joseph L. Schubert Jr. wanted a handgun to protect himself from his brother, who he believed was mailing him anonymous threats. Indiana law required that "a person desiring a license to carry a handgun shall apply to the chief of police or corresponding police officer"—in this case Robert L. DeBard, the superintendent of the Indiana State Police. The resulting investigation by DeBard's office found what it considered to be evidence that Schubert was mentally unstable and denied his request. When his application was denied, Schubert filed a petition for review.

Schubert's defense attorney accepted the conclusion of the police investigation that Schubert had some psychological problems, but he argued they were irrelevant to the matter at hand. Article 1, Section 32, of the Indiana constitution guarantees that "the people shall have a right to bear arms, for the defense of themselves and the State." Therefore, the attorney argued, when self-defense was properly indicated as the reason for desiring a firearms license, and the applicant was otherwise qualified, the license could not be withheld because an administrative official had subjectively determined that the applicant's need to defend himself was not justified. (At that point, Schubert had not been found to be mentally incompetent, which is an accepted reason to deny permission to carry a gun.)

The Third District Court of Appeals of the State of Indiana agreed with Schubert. After studying the debates surrounding the creation of the Indiana constitution in

TABLE 4.2

U.S. states with laws protecting the right to use deadly force in self-defense, 2016

Alabama	Nevada
Alaska	New Hampshire
Arizona	North Carolina
Arkansas	North Dakota
Florida	Oklahoma
Georgia	Oregon
Idaho	Pennsylvania
Indiana	South Carolina
Kansas	South Dakota
Kentucky	Tennessee
Louisiana	Texas
Maine	Utah
Maryland	Washington
Michigan	West Virginia
Mississippi	Wisconsin
Missouri	Wyoming
Montana	

SOURCE: Created by Erin Brown for Gale, © 2016.

1850, most of the appeals court judges concluded, "We think it clear that our constitution provides our citizenry the right to bear arms for their self-defense." If it were left to a police official to determine a "proper reason" for a person to claim self-defense, "it would supplant a right with a mere administrative privilege." Based on this conclusion, the court sent the case back to the lower court, asking it to determine if Schubert was mentally incompetent, which was an accepted basis for denying him the right to purchase a weapon.

Doe v. Portland Housing Authority: Maine State Law Preempts the Portland Public Housing Authority Provision against Gun Possession

Most state constitutions guarantee the right to bear arms, but state laws regulate their possession. The case of *Doe v. Portland Housing Authority* (656 A.2d 1200 [1995]) is an example of such regulation and its preemption (taking precedence) of a public housing authority provision.

A Maine couple identified as John and Jane Doe, who had lived in public housing since 1981, were threatened with eviction after the Portland Housing Authority (PHA) in Maine discovered guns in their apartment. John Doe was a U.S. Marine Corps veteran, a former firearms dealer, and a licensed hunter. Jane Doe, a target shooter, reported that she kept a handgun for self-protection when her husband worked late. The Does filed a petition to prevent the PHA from enforcing a provision in their lease that banned the possession of firearms.

The Does argued that a state law that regulates the possession of firearms preempted the lease. The Maine preemption statute declares, "The State intends to occupy and preempt the entire field of legislation concerning the regulation of firearms.... No political subdivision of the State, including, *but not limited to*, municipalities, counties, townships and village corporations, may adopt any...[law] concerning...firearms, components, ammunition or supplies."

The PHA claimed that the state law could not preempt its resolutions because the PHA is not a political subdivision listed in the statute. In 1995 the Maine Supreme Court ruled in *Doe v. Portland Housing Authority* that the PHA was indeed a political subdivision. The court also found that the state legislature intended to regulate uniformly the possession of firearms by all Maine residents whether they live in public housing or not. The case was appealed to the U.S. Supreme Court, but the court declined to hear it.

State v. Owenby: Mental Illness Limits the Right to Bear Firearms in Oregon

In 1991 the Circuit Court of Multnomah County, Oregon, ruled that Patrick Owenby, who suffered from mental illness and had carefully planned a murder, was a danger to himself and others. Because Owenby was unwilling, unable, or unlikely to seek voluntary treatment, the court had him committed to a psychiatric facility. It then ordered that Owenby be prohibited from purchasing or possessing firearms for a period of five years, in accordance with Oregon statutes. Owenby appealed the case.

The Oregon Court of Appeals ruled in *State v. Owenby* (111 Or. App. 270 [1992]) that the statute in question was a narrowly drawn and reasonable restriction on the right to bear arms, was not in violation of the state constitution, was supported by clear and convincing evidence, and did not violate the federal due process clause as expressed in the 14th Amendment: "Nor shall any State deprive any person of life, liberty, or property, without due process of law."

The court stated:

> The right to bear arms is not absolute. In the exercise of its police power, the legislature may enact reasonable regulations limiting the right and has done so....
>
> Given the nature of firearms ... the danger that the statute seeks to avert ... is a serious one. The restriction on the right of mentally ill persons to bear arms, on the other hand, is relatively minor. The statute is narrowly drawn and may be invoked only when it is shown that the prohibition is *necessary* "as a result of the mentally ill person's mental or psychological state," as demonstrated by past behavior that involves unlawful violence.

Benjamin v. Bailey: Ban on Semiautomatic Firearms in Connecticut Upheld

Connecticut—a state with deep economic ties to gun manufacturing and a strong pro-gun culture due to the popularity of hunting—made national headlines in 1993, when it narrowly passed a state law that banned the sale, possession, or transfer of 67 types of automatic and semiautomatic or burst-fire firearms. At that time, only two other states (California in 1989 and New Jersey in 1990) had passed assault weapons bans.

The National Rifle Association of America quickly mounted a challenge to the Connecticut law, recruiting a group of eight plaintiffs who were meant to represent a range of gun owners and others aggrieved by the law. The challenge was filed under one of the plaintiffs, DeForest H. Benjamin Jr., a small-town gunsmith and farmer. It charged that Connecticut's assault weapons ban violated the Second Amendment and was unenforceable and vague. John M. Bailey (1944–2003), the chief state's attorney, was the primary defendant.

In 1995 the Connecticut Supreme Court upheld the 1993 law in *Benjamin v. Bailey* (234 Conn. 455), ruling that the ban did not violate the state constitutional right to bear arms. The decision made Connecticut one of the first states to have an assault weapons ban pass legal challenge even though the right of self-defense was specified in its constitution. In April 2013, following the horrific massacre that took place in December 2012 at Sandy Hook Elementary School in Newtown, Connecticut, the state legislature expanded the assault weapons ban, adding 100 more weapons to the list of prohibited firearms.

State v. Wilchinski: Child Access Prevention in Connecticut Challenged

Florida was the first state to pass a child access prevention law (1989). Often referred to as the safe-storage law, it requires adults to either keep loaded guns in a place reasonably inaccessible to children or use a device to lock the gun. If a child (defined as anyone under the age of 16 years) obtains an improperly stored and loaded gun, the adult owner is held criminally liable. According to the Law Center to Prevent Gun Violence, in "Child Access Prevention Policy Summary" (http://smartgunlaws.org/gun-laws/policy-areas/consumer-child-safety/child-access-prevention/), as of 2016, 27 states and the District of Columbia had passed child access prevention laws.

Among these states is Connecticut, whose law was challenged by Joseph Wilchinski, a police officer employed by Central Connecticut State University. He was charged with criminal negligence and sentenced to three years' probation after his teenage son and another boy found a loaded revolver in Wilchinski's bedroom in 1993. The boy was accidentally shot by the Wilchinski boy and died two days later. Wilchinski appealed his conviction, claiming the law was unconstitutionally vague.

In 1997 the Connecticut Supreme Court upheld the law, declaring in *State v. Wilchinski* (242 Conn. 211) that the requirement to store firearms in a "securely locked box or other container or in a location which a reasonable person would believe to be secure" was sufficiently clear to inform Wilchinski of safe-storage practices.

Coalition of New Jersey Sportsmen v. Whitman: New Jersey's Assault Weapons Ban Challenged

A group of gun clubs and arms manufacturers sought to overturn New Jersey's 1999 ban on assault weapons on the grounds of vagueness, free speech, and equal protection. The U.S. District Court for the District of New Jersey rejected their challenge in March 1999. The court held in *Coalition of New Jersey Sportsmen v. Whitman* (44 F. Supp.2d 666) that the statute banning assault weapons is not vague because it addresses an understandable core of banned guns and "sufficiently puts on notice a person of ordinary intelligence that a weapon he or she possess[es], may be prohibited." In the ruling, the court further held that the statute's ban on specifically named weapons does not violate anyone's free speech, nor does the statute infringe on equal protection rights because "the rational link between public safety and a law proscribing possession of assault weapons is so obvious." The federal court's decision was affirmed in March 2001 by the U.S. Court of Appeals for the Third Circuit (No. 99-5296).

Oklahoma: Guns on Corporate Property Challenged

On November 1, 2004, amendments to the Oklahoma Firearms Act and the Oklahoma Self-Defense Act took effect, allowing guns in locked vehicles on corporate property in Oklahoma. The amendments were passed after 12 workers at an Oklahoma Weyerhaeuser paper mill were fired for violating a company ban on firearms in the company parking lot. Whirlpool Corporation, Williams Companies Inc., and ConocoPhillips filed a lawsuit against the state in federal court, arguing that the amendments were unconstitutional and prevented them from banning firearms in their parking lots to help ensure safe workplaces. A U.S. district court judge issued a temporary restraining order to prevent the amendments from going into effect until the courts made a final ruling.

In March 2005 the Oklahoma Court of Criminal Appeals ruled that the amendments were criminal in nature, rather than civil. This ruling was necessary to guide the U.S. district court judge in his determination. In October 2007 Terence C. Kern (1944–), the U.S. district judge, placed a permanent injunction against the Oklahoma amendments, ruling that they were in conflict with federal safety laws meant to protect employees at their jobs, primarily denoted in the Occupational Safety and Health Act of 1970, which created the Occupational Safety and Health Administration (OSHA). However, the issue remained in dispute, and in February 2009 the U.S. Court of Appeals for the 10th Circuit lifted the injunction in *Ramsey Winch Inc. v. Henry* (555 F.3d 1199). In its ruling, the court found that OSHA literature—including its website, guidelines, and citation history—did not indicate that employers should prohibit firearms from company parking lots and that the agency had explicitly deferred the regulation of workplace homicides to other federal, state, and local law enforcement agencies.

Numerous states followed Oklahoma's lead after 2009. In "Top Ten Issues Regarding Guns in the US Workplace" (ACC.com, March 8, 2016), the Association of Corporate Counsel notes that in 2016 more than 20 states had so-called parking lot laws, which explicitly required employers to allow employees to keep guns in their locked private vehicles on company property. Two additional states had similar laws, although the requirement was limited to state employers, such as government agencies and public (state-funded) schools, colleges, and universities.

Fiscal v. City and County of San Francisco: California State Law Preempts San Francisco Ordinance That Bans Handguns in the City

The 2008 California Supreme Court ruling in *Fiscal v. City and County of San Francisco* (158 Cal.App.4th 895) is an example of state law preempting a local ordinance. California state law regulates firearms within California, including their manufacture, distribution, sale, possession, and transfer. In November 2005 San Francisco voters passed Proposition H, a citywide ordinance that would ban the manufacture, distribution, sale, and transfer of firearms and ammunition within San Francisco, as well as prohibit San Francisco residents from possessing handguns within the city.

A few days after Proposition H was passed by city voters, the National Rifle Association of America and the Second Amendment Foundation, a nonprofit group that promotes the right to bear arms, filed suit to block the ordinance. In June 2006 Judge James L. Warren of the San Francisco Superior Court struck down the ordinance, stating that its key aspects were preempted by state law. He held that under California law local officials cannot ban the possession of firearms from law-abiding citizens.

The city appealed the superior court decision, and in January 2008 a three-judge panel of the California Court of Appeals unanimously ruled to uphold the lower court's decision. The city then appealed the decision to the California Supreme Court, which upheld the previous two decisions. The April 2008 decision by California's high court exhausted the city's possibilities to appeal the case further.

LOCAL RULINGS
Portland, Oregon: Possession of Guns Regulated

The city of Portland, Oregon, passed an ordinance prohibiting "any person on a public street or in a public place to carry a firearm upon his person, or in a vehicle under his control or in which he is an occupant, unless all ammunition has been removed from the chamber and from the cylinder, clip or magazine" (Portland Ordinance, No. 138210, Sec. No. 14.32.010). In 1982 Michael Boyce was convicted of violating this statute. He appealed, contending that the law violated Article 1, Section 27, of the Oregon constitution, which states, "The people shall have the right to bear arms for the defence of themselves, and the State, but the Military shall be kept in strict subordination to the civil power."

Boyce based his case on *State v. Kessler* (289 Or. 359 [1980]), in which the Oregon Supreme Court upheld the right to possess a billy club or any other type of small weapon for self-defense, and *State v. Blocker* (291 Or. 255 [1981]), in which the Oregon Supreme Court declared unconstitutional an Oregon law banning a number of weapons, including switchblades, billy clubs, and blackjacks. The court of appeals did not, however, see the similarities and upheld in *State v. Boyce* (61 Or. App. 662 [1983]) the lower court's conviction. The court of appeals observed that the statute in *Kessler* and *Blocker* forbids the "mere possession" of certain weapons and that was the characteristic that made it unconstitutional. The statute in this case regulates only the manner of possession, something both *Kessler* and *Blocker* recognized as permissible when the regulation was reasonable. As such, the city of Portland could regulate the use of weapons within its borders.

In fulfilling its obligation to protect the health, safety, and welfare of its citizens, a government body must sometimes pass legislation that touches on a right guaranteed by the state or federal constitution. Such an encroachment is permissible when the unrestricted exercise of the right poses a clear threat to the interests and welfare of the public in general, and the means chosen by the government body do not unreasonably interfere with the right.

The court of appeals agreed that individuals had a right to protect their property and themselves:

> When a threat to person or property arises in the victim's home or other private place, the ordinance will not interfere at all with the victim's defense capacity. It is true, on the other hand, that, when the threat arises in a public place, the fact that a person must have any ammunition separated from his firearm will hinder him to the extent that he is put to the trouble of loading the weapon. However, given the magnitude of the city's felt need to protect the public from an epidemic of random shootings, we think that the hindrance is permissible.

Renton, Washington: Guns Not Permitted Where Alcohol Is Served

The city of Renton, Washington, enacted Municipal Ordinance 3459, which states, "It is unlawful for anyone, on or in any premise in the City of Renton where alcoholic beverages are dispensed by the drink, to ... carry any rifle, shotgun or pistol, whether said person has a license or permit to carry said firearm or not, and whether said firearm is concealed or not."

Four residents, with the support of the Second Amendment Foundation, went to court seeking an injunction on the ordinance, claiming it violated state law and was unconstitutional. The Superior Court of King County upheld the city ordinance, so the Second Amendment Foundation took the case to the Court of Appeals of Washington.

Article 1, Section 24, of the Washington constitution states, "The right of the individual citizen to bear arms in defense of himself, or the state, shall not be impaired, but nothing in this section shall be construed as authorizing individuals or corporations to organize, maintain or employ an armed body of men."

The court indicated in *Second Amendment Foundation v. Renton* (35 Wn. App. 583 [1983]) that "it has long been recognized that the constitutional right to keep and bear arms is subject to reasonable regulation by the State under its police power." It also explained that simply because a right is guaranteed by either the state or federal constitution does not mean that it cannot be regulated. According to the court:

> The scope of permissible regulation must depend upon a balancing of the public benefit to be derived from the regulation against the degree to which it frustrates the purpose of the constitutional provision. The right to own and bear arms is only minimally reduced by limiting their possession in bars. The benefit to public safety by reducing the possibility of armed conflict while under the influence of alcohol outweighs the general right to bear arms in defense of self and state....
>
> On balance, the public's right to a limited and reasonable exercise of police power must prevail against the individual's right to bear arms in public places where liquor is served.

Furthermore, the court noted that the statutes "do not expressly state an unqualified right to be in possession of a firearm at any time or place." Had the city of Renton instituted "an absolute and unqualified local prohibition against possession of a pistol by the holder of a state permit," it would have conflicted with state law and Washington's constitution. This it did not do. Rather, the city had instituted a law "which is a limited prohibition reasonably related to particular places and necessary to protect the public safety, health, morals and general welfare."

Finally, the Court of Appeals of Washington noted that "while 36 states have constitutional provisions concerning the right to bear arms, in none is the right deemed absolute." Furthermore, "those states [Alabama, Indiana, Michigan, Oregon, and Wyoming] with constitutional provisions similar to ours have uniformly held the right subject to reasonable exercise of the police power." The city of Renton was within its rights when it passed the ordinance barring firearms from bars, and the court upheld the decision of the lower court.

Morton Grove, Illinois: Handguns Are Banned

As soon as Morton Grove, Illinois, a Chicago suburb, passed an ordinance banning handguns in 1981, handgun owners challenged the city in court. Article 1, Section 22, of the Illinois constitution provides that "subject only to the police power, the right of the individual citizen to keep and bear arms shall not be infringed." Handgun owners argued that the right to bear arms was protected by state and federal constitutions and further contended that if Morton Grove were allowed to pass such laws in contradiction to other towns and cities, a so-called patchwork quilt situation would result, in that handgun owners would never know if they were violating a law when traveling from town to town.

Morton Grove defended itself, claiming it was within its power to limit or ban the possession of handguns if city officials believed handgun possession was a threat to peace and stability. The city further claimed that its ordinance did not violate Section 22 of the Illinois constitution because it guaranteed the right to keep "some guns." The Morton Grove law did not ban all guns, only handguns.

LEGAL PROCEEDINGS. The cases of Victor Quilici, Robert Stengl, George Reichert, and Robert Metler were combined and brought to the Federal District Court of Northern Illinois in *Quilici v. Village of Morton Grove* (532 F. Supp. 1169 [1981]), in which the court upheld the town's right to ban handguns. The U.S. Court of Appeals for the Seventh Circuit also upheld in *Quilici v. Village of Morton Grove* (695 F.2d 261 [1982]) the findings of the district court, saying "the right to keep and bear handguns is not guaranteed by the second amendment." The case was appealed to the U.S. Supreme Court. The high court refused to hear the case, so the ruling of the lower court of appeals stood.

A NEW ROUTE. The Morton Grove handgun owners then went to the Circuit Court of Cook County for an injunction to prevent Morton Grove from instituting the ordinance that banned handguns. The county circuit court upheld the validity of the ordinance. The handgun owners next appealed to the Appellate Court of Illinois, First District, Third Division. In *Kalodimos v. Village of Morton Grove* (113 Ill. App.3d 488 [1983]), the court upheld the decisions of the lower courts. Although the court agreed with the handgun owners that "gun control legislation could vary from municipality to municipality, we find that the framers [of the Illinois constitution] envisioned this kind of local control."

The case was again appealed, and in October 1984 the Illinois Supreme Court upheld in *Kalodimos v. Village of Morton Grove* (103 Ill.2d 483) the lower courts' decisions. Agreeing with earlier observations, the state's highest court noted that "while the right to possess firearms for purposes of self-defense may be necessary to protect important personal liberties from encroachment by other individuals, it does not lie at the heart of the relationship between individuals and their government." Thus, Morton Grove needed only to have had a "rational" basis for instituting its ban on handguns. The Illinois Supreme Court concluded, "Because of the ease with which handguns can be concealed and handled, as compared with other types of weapons, a ban on handguns under the conditions set forth in the ordinance could rationally have been viewed by the village as a way of reducing the frequency of premeditated violent attacks as well as unplanned criminal shootings in the heat of passion or in overreaction to fears of assault, accidental shootings by children or by adults who are unaware that a handgun is loaded, or suicides.... [T]herefore, we conclude that the ordinance is a proper exercise of the police power authority."

Chicago: Limits on Handgun Possession

On March 19, 1982, the Chicago city council passed an ordinance prohibiting the registration of any handgun after April 10, 1982, the effective date of the ordinance, unless it was "validly registered to a current owner in the City of Chicago" before April 10, 1982 (*Municipal Code of the City of Chicago*, Chapter 8-20-050[c][1]). Jerome Sklar lived in neighboring Skokie, Illinois, when the law was passed. He owned a handgun and held a valid Illinois Firearms Identification Card. On April 15, 1982, after the ordinance had gone into effect, he moved to Chicago. He could not register the weapon and, therefore, was unable to bring it into the city.

Sklar went to court, claiming that the city of Chicago had violated the equal protection clause of the U.S. Constitution because he was unable to register the gun that he owned, whereas owners of firearms who resided in Chicago before the effective date of the ordinance had an opportunity to take advantage of the law's registration requirements. By this time, *Quilici v. Village of Morton Grove* had been decided by the Seventh Circuit Court of Appeals, a decision that applied to this judicial region. Therefore, the U.S. District Court for the Northern District of Illinois indicated in *Sklar v. Byrne* (556 F. Supp. 736 [1983]) that because of the *Quilici* ruling it "conclude[d] that the Chicago firearms ordinance does not infringe on a constitutionally protected right." The court indicated that the city of Chicago had legitimately and rationally used its police power to promote the health and safety of its citizens. Sklar's argument that Chicago could have chosen better ways to protect its citizens from the negative effects of firearms was irrelevant.

The court also concluded that the ordinance did not violate the equal protection clause of the U.S. Constitution by limiting new registrations instead of banning handguns altogether. The city was under no legal requirement to take an all-or-nothing approach to limiting handguns.

Sklar appealed the district court's decision to the U.S. Court of Appeals for the Seventh Circuit, the same court that had ruled on *Quilici v. Village of Morton Grove*. In his appeal, Sklar claimed that his constitutionally guaranteed right to travel had been violated because he could not move into Chicago without giving up his gun. In *Sklar v. Byrne* (727 F.2d 633 [1984]), the court upheld the lower court's decision, citing the precedent established in *Quilici*. The court did not believe that a fundamental constitutional issue was involved. Therefore, the city of Chicago had a right to institute local regulations as long as it did not go overboard. The court stated:

> The Chicago handgun ordinance as a whole promotes legitimate government goals. The city council set forth its purposes in the preamble to the ordinance. The council found that handguns and other firearms play a major role in crimes and accidental deaths and injuries, and that the "convenient availability" of firearms and ammunition contributed to deaths and injuries in Chicago. The council therefore enacted the ordinance to restrict the availability of firearms and thereby to prevent some deaths and injuries among Chicago citizens.

> The city's primary goals are thus classic examples of the city's police power to protect the health and safety of its citizens.

Sklar argued that it was irrational and "inconsistent with the overall purposes of the ordinance" to allow some people to have handguns and others not to, and not to classify gun owners on the basis of their ability to handle handguns safely. In dismissing his claim, the court stated, "that argument essentially asks this court to second-guess the judgment of the city council. The Constitution does not require the city council to act with a single purpose or to be entirely consistent. Indeed, the council is a political body for the accommodation of many conflicting interests.... The Constitution does not require the city council to enact the perfect law. The council may proceed step by step, 'adopting regulations that only partially ameliorate a perceived evil and deferring complete elimination of the evil to future regulations.'"

In both *Quilici* and *Sklar*, the courts were not saying that handgun control is or is not a good decision for any local authority to make. They did not see the possession of a handgun as a fundamental right protected by either the federal or the state constitution. The courts stated that a town or city, under the police powers granted it by American tradition and the Illinois constitution, has the right to decide and implement such an ordinance for its own people. A local ordinance does not have to be consistent, as long as the city council can prove that it thought out its decision rationally.

As noted earlier, the U.S. Supreme Court held in *McDonald v. City of Chicago* that the Second Amendment right of individuals to keep and bear arms for self-defense may not be preempted by state or local laws. The high court found that the restrictions on gun ownership in Chicago and neighboring Oak Park, Illinois, were too restrictive because they denied citizens the right to legally possess a gun in their own home for self-defense. The court left open the possibility, however, that some state and local gun control measures could withstand constitutional scrutiny applied in the case. Writing for the majority, Justice Samuel A. Alito Jr. (1950–) noted that "incorporation does not imperil every law regulating firearms." Following this ruling, Chicago and other municipalities throughout the country began revising their gun control policies.

West Hollywood, California: Saturday Night Specials Banned

In 1998 the California Supreme Court let stand a ruling by the California Court of Appeals on inexpensive handguns known as Saturday Night Specials. The court of appeals had upheld a municipal ban by the city of West Hollywood on the sale of this type of weapon. In its opinion, the court rejected the gun lobby's claim that California state law preempted the ordinance; furthermore, the court found that the ban did not violate the principles of equal protection or due process. Gun rights advocates maintained that singling out inexpensive weapons denies poor people an affordable means of self-defense.

Seattle, Washington: State Law Overrides City Gun Ban

In October 2009 the Second Amendment Foundation and several other gun rights groups sued the city of Seattle, claiming that the city had no right to ban guns in certain places, such as parks and community centers. Judge Catherine Shaffer of the King County Superior Court ruled in February 2010 that Washington State law did not allow Seattle to regulate the possession of firearms. The city of Seattle appealed the ruling to the U.S. District Court on Second Amendment grounds, citing *Heller*. In March 2010 Judge Marsha J. Pechman (1951–) of the U.S. District Court for the Western District of Washington upheld the lower court ruling and stated that the *Heller* ruling applied only to federal jurisdictions. However, the U.S. Supreme Court ruling in *McDonald v. City of Chicago* in June 2010 clarified that Second Amendment rights also apply to the states. The article "Justices Extend Gun Owner Rights Nationwide; McKenna Statement Inside" (KHQ.com, June 30, 2010) quotes Rob McKenna (1962–), the attorney general of Washington State, as saying that he was "gratified that the U.S. Supreme Court has affirmed that the Second Amendment may not be infringed by state and local governments" and that "the right to bear arms shouldn't be infringed just because a person crosses state or city lines."

RESPONSIBILITY FOR HANDGUN DEATHS

The cases presented thus far on the federal, state, and local levels focus primarily on the right to bear arms. The following cases probe the responsibilities and liabilities associated with the use of those arms. Some victims of the use of certain weapons have tried to place that responsibility and liability on the manufacturers of the weapons, whereas others fault the people who made the weapons available to criminals. Each of the decisions presented here was based on state laws that differ greatly.

California: Gun Dealers Can Be Held Liable for Gun Violations of Others

Nineteen-year-old Jeff Randa had mentioned to a gun dealer many times that he wanted to buy a handgun and ammunition. The dealer told the youth that he could not buy a gun until he was 21 years old. Randa asked if his grandmother could purchase the weapon. The dealer replied that if she were a qualified buyer she could, but that the dealer could not sell her the weapon "just so she could give the gun to her grandson."

Subsequently, Randa's grandmother came into the store with him and purchased the handgun the youth wanted. Twelve days later, Randa took the gun to a party. Bryan Hoosier, who was also at the party, told Randa to point the gun and shoot. Randa did so, killing Hoosier, and was later convicted of voluntary manslaughter.

Hoosier's father sued the gun dealer for negligence, accusing the dealer of knowing that the gun would be given to a minor after being sold to an adult. The dealer argued that he could not be liable and that the state laws imposed criminal penalties only on violators.

The California Court of Appeals ruled in *Hoosier v. Randa* (17 Cal. Rptr. 2d 518 [1993]) that the dealer was indeed liable for injuries. The state gun control laws were passed not only to establish criminal penalties but also to protect the public. If a dealer violated the law, he also violated his responsibility of care owed to the public. Consequently, any person harmed by such a violation may sue the violator.

Ohio: Gun Show Promoters Must Provide Adequate Security against Juvenile Gun Theft

During a 1992 gun show that was promoted by Niles Gun Show Inc., four youths under the age of 18 years stole several handguns. The corporation from which the vendors rented space had no policy that required the vendors to protect their wares from being stolen, although it had an unenforced policy barring minors from entering the show.

After leaving the show, the youths also stole a car. While driving around in the car, the juveniles confronted two men, Greg L. Pavlides and Thomas E. Snedeker. One of the boys, Edward A. Tilley III, shot Pavlides in the chest and Snedeker in the head with one of the stolen guns. Tilley was arrested, charged, and convicted of two counts of attempted murder and one count of unauthorized use of a motor vehicle.

Pavlides and Snedeker survived their injuries and sued Niles Gun Show for negligence for not protecting them and the rest of the public from criminal acts by third parties who stole weapons that had not been properly secured. The trial court dismissed the case, stating that the promoters had no such responsibility, but the Ohio Court of Appeals reversed the lower court's decision, sending the case back to be tried. The court ruled that the promoters of gun shows have a duty to provide adequate security to protect the public from criminal acts that might occur if guns are stolen. The court explained that gun show operators should require vendors to secure their firearms and make a reasonable effort to bar minors from stealing or purchasing weapons and ammunition. The court further stated that it is "common knowledge" that minors possessing guns can create dangerous situations, and consequently gun show promoters should be aware that minors stealing guns might use them in criminal activity.

Florida: Store Responsible in Criminal Act for Selling Ammunition to Juveniles

In 1991 a Florida Wal-Mart store employee sold ammunition to two teenagers without asking about age or requesting identification, which is a violation of federal law. Several hours later the teenagers used the ammunition in a robbery of an auto parts store, during which they fatally shot Billy Wayne Coker. Coker's wife filed suit against Wal-Mart.

Although Wal-Mart acknowledged that the sale was illegal, it argued that the perpetrators' intervening act of murder was not foreseeable and, therefore, the illegal sale was not the legal cause of Coker's death. The court agreed with this argument and dismissed the case. However, the Florida Court of Appeals ruled that an ammunition vendor's illegal sale could be the legal cause of an injury or death caused by the buyer's intentional or criminal act. In July 1998 the Florida Supreme Court upheld in *Wal-Mart Stores, Inc. v. Coker* (1998 Fla. Lexis 861) the $2.6 million verdict against Wal-Mart for negligence in selling handgun ammunition to underaged buyers.

GUN INDUSTRY AND LIABILITY FOR GUNSHOT INJURIES

Maryland: *Kelley v. R.G. Industries Inc.*

In 1981 Olen J. Kelley was injured when he was shot in the chest during an armed robbery of the grocery store where he worked. The gun used was a Rohm revolver handgun model RG-38S that was designed and marketed by Rohm Gesellschaft, a German corporation. The handgun was assembled and initially sold by R.G. Industries Inc., a Miami-based subsidiary of the German corporation. Kelley and his wife filed suits against Rohm Gesellschaft and R.G. Industries in the Circuit Court of Montgomery County, Maryland.

Two counts charged that the handgun was "abnormally dangerous" and "defective in its marketing, promotion, distribution, and design." A third count charged negligence. The case revolved around whether the gun in question was a Saturday Night Special, which was banned from import by the ATF. The Federal District Court of Baltimore, where the case was first brought, asked the state court for a ruling on whether the manufacturer could be held liable under Maryland law.

The Maryland Court of Appeals ruled in *Kelley v. R.G. Industries Inc.* (304 Md. 124 [1985]) that the manufacturer and marketers could not be held strictly liable because handguns are "abnormally dangerous products" and their manufacturing and marketing are "abnormally dangerous activit[ies]." In its decision, the court noted, "Contrary to Kelley's argument, a handgun is not defective merely because it is capable of being used during criminal activity to inflict harm. A consumer would expect a handgun to be dangerous, by its very nature, and to have the capacity to fire a bullet with deadly force."

The court of appeals also stated that Kelley confused a product's normal function, which may be dangerous by its very nature, with a defect in its design and function. Kelley had cited as an example that a car is dangerous if it is used to run down pedestrians. The injury that results is from the nature of the product—the ability to be propelled to great speeds at great force. Nevertheless, if the gas tank of the car leaked in such a way as to cause an explosion in the event of a rear-end collision, then the design of the product would be defective, and the manufacturer would be liable. The court concluded that to impose "strict liability upon the manufacturers or marketers of handguns for gunshot injuries resulting from the misuse of handguns by others, would be contrary to Maryland public policy."

The Maryland court's opinion differed on Saturday Night Specials, which it defined as guns "characterized by short barrels, light weight, easy concealability, low cost, use of cheap quality materials, poor manufacture, inaccuracy and unreliability." The court considered these guns "largely unfit for any of the recognized legitimate uses sanctioned by the Maryland gun control legislation. They are too inaccurate, unreliable and poorly made for use by law enforcement personnel, sportsmen, homeowners or businessmen.... The chief 'value' a Saturday Night Special handgun has is in criminal activity, because of its easy concealability and low price."

The court determined that manufacturers and marketers are liable because they should know that this type of gun is made primarily for criminal activity. Judge John C. Eldridge quoted an R.G. Industries salesperson as telling a prospective handgun marketer, "If your store is anywhere near a ghetto area, these ought to sell real well. This is most assuredly a ghetto gun." The salesperson allegedly went on to say that although the gun sold well, it was virtually useless, and that he would be afraid to fire it.

The court of appeals did not rule on whether the gun in question fell within the category of Saturday Night Specials but referred that decision to the U.S. District Court. The court of appeals, however, did indicate that strong evidence had been presented that the gun fit many of the qualifications; if it were found to be a Saturday Night Special, liability against both the manufacturer and marketer could be imposed. This decision applied only in Maryland, and the Maryland legislature soon passed a law overriding it. Few courts have accepted this interpretation.

New Mexico: *Armijo v. Ex Cam Inc.*

Dolores Armijo's brother, Steven Armijo, shot and killed James Salusberry, Dolores's husband, in front of Dolores and her daughter. He then tried to shoot them, but the gun jammed. Dolores, claiming the gun used was a Saturday Night Special, sued Ex Cam Inc., the importer and distributor of the weapon.

The suit was based on four theories: strict product liability (the product was defective and unreasonably dangerous; therefore, the manufacturer was responsible for the actions of the product), "ultra-hazardous activity" liability (a gun is a dangerous product and the manufacturer is accountable for the results of its use), negligence liability (the manufacturer did not show reasonable care while marketing a product that carried some degree of risk that it might be used to commit a crime), and a narrow form of strict product liability for Saturday Night Specials put forth in *Kelley v. R.G. Industries Inc.*

The U.S. District Court in New Mexico did not believe that any court in New Mexico would ever recognize any of these theories as the basis of a court case under New Mexico law. In *Armijo v. Ex Cam Inc.* (656 F. Supp. 771 [1987]), the court said: "It would be evident to any potential consumer that a gun could be used as a murder weapon. So could a knife, an axe, a bow and arrows, a length of chain. The mere fact that a product is capable of being misused to criminal ends does not render the product defective. [Based on New Mexico law, such a case] would not result in liability for a manufacturer of guns, as guns are commonly distributed and the dangers ... are so obvious as to not require any manufacturers' warnings."

The court showed little respect for *Kelley v. R.G. Industries Inc.*, indicating that it went against common law in the state of New Mexico; therefore, it would not be considered. Furthermore, the court concluded that "all firearms are capable of being used for criminal activity. Merely to impose liability upon the manufacturers of the cheapest types of handguns will not avoid that basic fact. Instead, claims against gun manufacturers will have the anomalous [unusual] result that only persons shot with cheap guns will be able to recover, while those shot with expensive guns, admitted by the *Kelley* court to be more accurate and therefore deadlier, would take nothing."

District of Columbia: The Federal Government Settles the Manufacturer Liability Issue

In 1998, after the tobacco industry was found to be responsible for lung cancer deaths caused by smoking cigarettes, many cities and counties across the United States began filing lawsuits against gun manufacturers and dealers. At issue was the gun distribution system, which was thought to allow guns to pass too easily to criminals and youth. However, Fox Butterfield reports in "Gun Industry Is Gaining Immunity from Suits" (NYTimes.com, September 1, 2002) that by 2002, 30 states had passed laws granting immunity to the gun industry from these civil lawsuits. In a countermeasure, California passed a bill repealing such an immunity law in its state.

In 2005 Congress settled this issue with the passage of the Protection of Lawful Commerce in Arms Act. President George W. Bush (1946–) signed the bill into law in October 2005. This act prohibits liability actions (charging legal responsibility) against firearms and ammunition manufacturers and sellers for unlawful misuse of their products. An exception to the act allows petitioners to sue firearms manufacturers and dealers if they knowingly violate state or federal statutes. The act also prohibits the sale of a handgun unless the purchaser is provided with a secure gun storage or safety device, and it provides for particular sentences when armor-piercing ammunition is used in certain crimes.

According to David Stout, in "Justices Decline New York Gun Suit" (NYTimes.com, March 9, 2009), since the passage of the Protection of Lawful Commerce in Arms Act, many city and state officials have sued gun manufacturers without success, but some litigants have won suits against gun dealers.

Williams v. Beemiller: Gun Industry May Be Liable When Engaged in Illegal Trafficking

In *Williams v. Beemiller, Inc.* (100 A.D.3d 143 [2012]), the New York Court of Appeals, Fourth Judicial Department, reversed a lower court's dismissal of a case in which a shooting victim brought suit against the gun manufacturer, distributor, and dealer who supplied the weapon used to injure him. *Williams v. Beemiller* stemmed from a drive-by shooting that occurred in Buffalo, New York, in 2003. In a case of mistaken identity, teenager Daniel Williams was shot by gang member Cornell Caldwell with a 9mm Hi-Point handgun that had been acquired by James Nigel Bostic in a straw purchase, with his girlfriend as the buyer of record. Bostic had purchased about 250 guns in Ohio that he sold illegally in Buffalo; 140 of the guns were purchased from one dealer, Charles Brown, and 87 guns were exchanged in the transaction that included the one that was used to shoot Williams. Caldwell was apprehended, tried, found guilty, and sent to prison for the crime.

Williams, who survived the shooting, filed a suit against the manufacturer of the gun, the distributor, and Brown, who he claimed should have recognized that Bostic was involved in criminal activity from the number and type of weapons that he purchased. Brown claimed that Bostic told him he was planning to open a gun shop. In 2011 the defendants had sought and received a dismissal of the case based on the Protection of Lawful Commerce in Arms Act; however, the New York State Court of Appeals overturned that dismissal in October 2012. Justice Erin Peradotto of the Appellate Division, Fourth Department, stated in the ruling, "Although the complaint does not specify the statutes allegedly violated [by the defendants], it sufficiently alleges facts supporting a finding that defendants knowingly violated federal gun laws." The ruling allowed Williams's case to move forward, therefore, holding open the possibility that the gun industry may be criminally liable when engaged in illegal sales. As of December 2016, this case had not been resolved.

CHAPTER 5
FIREARMS AND CRIME

Many of the statistics on the frequency and ways in which guns are used to commit crimes come from the Federal Bureau of Investigation (FBI) of the U.S. Department of Justice, which collects a range of U.S. crime statistics. One of the primary sources of FBI statistics is the Uniform Crime Reports (UCR) program, on which the annual publication *Crime in the United States* is based. *Crime in the United States* and other UCR data releases are derived from police investigation reports and arrests; crimes that are not reported to the police are not included. As of December 2016, the most recent final report from the UCR program was *Crime in the United States, 2015* (September 2016, https://ucr.fbi.gov/crime-in-the-u.s/2015/crime-in-the-u.s.-2015). The FBI's Bureau of Justice Statistics (BJS) also regularly releases publications based on its National Crime Victimization Survey (NCVS). The NCVS, which consists of a survey of a representative sample of U.S. households, seeks to measure rates and levels of victimization for nonfatal violent and property crimes. It is different from the UCR program in two important ways. First, it attempts to capture both crimes that are reported to police and those that are not. Second, it cannot capture information about homicides because it consists of surveys of living people. As such, UCR and NCVS statistics are not identical, but they are often broadly consistent with each other. They complement one another in key ways and thus contribute to a comprehensive, albeit incomplete, understanding of criminal activity and incidents in the United States.

In *Crime in the United States, 2015*, the FBI reports that an estimated 1,197,704 violent crimes were reported to law enforcement agencies in 2015, an increase of 3.9% over the previous year. The five-year period between 2011 and 2015 saw significant fluctuation in the number of violent crimes reported to law enforcement. (See Figure 5.1.) The 2015 level of violent crime was 0.7% lower than in 2011 and 16.5% lower than in 2006, and the overall rate of violent crime in the United States was 372.6 per 100,000 residents.

The FBI divides violent crime into four categories: murder and nonnegligent manslaughter, forcible rape, robbery, and aggravated assault. The category of murder and nonnegligent manslaughter is defined as "the willful (nonnegligent) killing of one human being by another." Prior to 2013 forcible rape was defined as "the carnal knowledge of a female forcibly and against her will." In 2013, however, the UCR program redefined this term. Rape is now defined as "penetration, no matter how slight, of the vagina or anus with any body part or object, or oral penetration by a sex organ of another person, without the consent of the victim," and includes "attempts or assaults to commit rape" but excludes "statutory rape and incest." Although some UCR statistics were reported under this new definition of rape, others were reported under the older definition. The 2015 data provide figures for the current (revised) definition of rape as well as for the old (or "legacy") definition, maintaining the latter to facilitate comparison with previous years. Robbery is defined as "the taking or attempting to take anything of value from the care, custody, or control of a person or persons by force or threat of force or violence and/or by putting the victim in fear." Aggravated assault is defined as "an unlawful attack by one person upon another for the purpose of inflicting severe or aggravated bodily injury." The FBI includes in this category, "Attempted aggravated assault that involves the display of—or threat to use—a gun, knife, or other weapon," and if aggravated assault occurs in combination with larceny-theft (a property crime), the bureau classifies the offense as robbery. Together, these four categories of crime represent those in which firearms are typically used. Firearms were used in 71.5% of U.S. murders, 40.8% of robberies, and 24.2% of aggravated assaults. Forcible rape reports to law enforcement are not classified according to weapons used.

In assessing the distribution of violent crime in the United States in 2015, the FBI divides cities into six groups: Group I, cities with populations of 250,000 and over, which had a total 2015 population of 59.1 million;

FIGURE 5.1

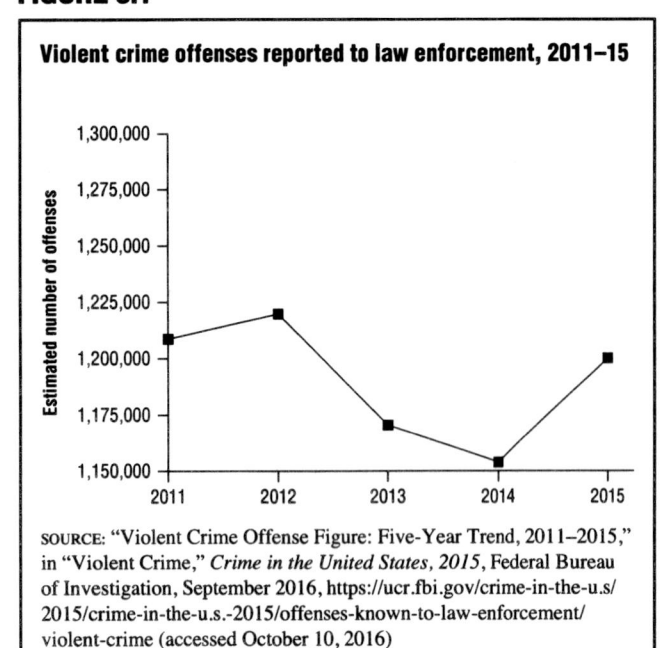

Violent crime offenses reported to law enforcement, 2011–15

SOURCE: "Violent Crime Offense Figure: Five-Year Trend, 2011–2015," in "Violent Crime," *Crime in the United States, 2015*, Federal Bureau of Investigation, September 2016, https://ucr.fbi.gov/crime-in-the-u.s/2015/crime-in-the-u.s.-2015/offenses-known-to-law-enforcement/violent-crime (accessed October 10, 2016)

Group II, cities with populations of 100,000 to 249,999, which had a total population of 32.2 million; Group III, cities with populations of 50,000 to 99,999, which had a total population of 33.5 million; Group IV, cities with populations of 25,000 to 49,999, which had a total population of 29.8 million; Group V, cities with populations of 10,000 to 24,999, which had a total population of 28.9 million; and Group VI, cities with populations under 10,000, which had a total population of 23.3 million. (See Table 5.1.) Group I cities, which made up 19% of the total U.S. population, accounted for 38% of the total number of violent crimes, including 6,146 murders (41% of all U.S. murders), 26,947 forcible rapes (28%), 154,760 robberies (49%), and 248,913 aggravated assaults (34%).

HOMICIDE AND MURDER
Murders, Weapons, and Circumstances

In *Homicide Trends in the United States, 1980–2008* (November 2011, http://www.bjs.gov/content/pub/pdf/htus8008.pdf), Alexia Cooper and Erica L. Smith of the BJS note that the homicide victimization rate (the number of people killed by another per 100,000 population) was relatively low during the 1950s and the early 1960s, under five homicides per 100,000 people in most years. By the mid-1960s, however, the rate rose dramatically and continued to rise through the mid-1970s, reaching nearly 10 homicides per 100,000 people. Following a slight dip, it peaked in 1980 at 10.2 homicides per 100,000 people. After a decline to 7.9 homicides per 100,000 people in 1984, the homicide rate rose again during the late 1980s and early 1990s to 9.8 homicides per 100,000 people in 1991. The homicide rate fell throughout the 1990s. In 2000 it reached the lowest level since 1966, at 5.5 homicides per 100,000 people, and remained stable through 2008. Smith and Cooper indicate in *Homicide in the U.S. Known to Law Enforcement, 2011* (December 2013, http://www.bjs.gov/content/pub/pdf/hus11.pdf), the most recent BJS report on this topic as of December 2016, that after 2008 the homicide rate declined further, reaching 4.7 per 100,000 people in 2011, the lowest rate since 1963 and one that represented a 54% decline from the 1980 peak.

DEMOGRAPHICS OF HOMICIDE. The vast majority of murderers and murder victims are men. Smith and Cooper note that between 1992 and 2011 the homicide rate for males and females declined almost equally (by 50% for males and 49% for females), but over that whole period the homicide rate for males remained, on average, 3.6 times higher than the rate for females. In 2011 the homicide rate for males was 7.4 per 100,000, and there were 11,370 male murder victims. (See Table 5.2.) By comparison, the homicide rate for females was 2 per 100,000, and there were 3,240 female murder victims.

According to Smith and Cooper, over the 10-year period between 2002 and 2011 the homicide rate for African Americans was, on average, 6.3 times higher than the homicide rate for whites. During this period, however, homicides among both groups declined at similar rates (19% for African Americans and 17% for whites). In 2011 the homicide rate for African Americans was 17.3 per 100,000, compared with a rate of 2.8 per 100,000 for whites. (See Table 5.2.)

Historically, young adult males between the ages of 20 and 29 years are far more likely than any other age group to be the victims of homicide. Of the 13,455 total homicide victims in the United States in 2015, 3,835 (28.5%) were men aged 20 to 29 years. (See Table 5.3.) Among all male homicide victims (10,608), young men in this age bracket accounted for 36.2%. Young women between the ages of 20 and 29 were also more likely than younger or older women to be homicide victims, although their numbers were much lower. Of the 13,455 total homicide victims in the United States in 2015, 667 were women in this age bracket. These victims accounted for just 5% of all homicide victims in 2015, but they amounted to nearly a quarter (23.7%) of the total number of female victims (2,818).

Especially in the 20- to 29-year-old age groups, homicide victims also tend to be disproportionately African American. Of the 4,502 total homicide victims aged 20 to 29 years in 2015, 64.1% (2,886) were African American. (See Table 5.3.) By comparison, 1,481 (32.9%) of the victims in this age bracket were white and 735 (16.3%) were Hispanic. In all, African Americans were more likely than any other race or ethnicity to be the victims of homicide, with a total of 7,039 (52.3%) deaths in 2015. Still, in some age brackets whites were more likely than people of other races to be victims. For example, of the

TABLE 5.1
Crime trends, by population group, 2014–15

Population group		Violent crime	Murder and nonnegligent manslaughter	Rape (revised definition)[a]	Rape (legacy definition)[b]	Robbery	Aggravated assault	Property crime	Burglary	Larceny-theft	Motor vehicle theft	Arson	Number of agencies	2015 estimated population
Total all agencies:	2014	1,118,918	13,594	89,098	7,115	310,511	698,600	7,730,805	1,613,258	5,464,398	653,149	42,914	15,545	305,159,115
	2015	1,160,664	15,192	94,717	7,586	315,660	727,509	7,549,676	1,489,820	5,383,743	676,113	41,376		
	Percent change	+3.7	+11.8	+6.3	+6.6	+1.7	+4.1	−2.3	−7.7	−1.5	+3.5	−3.6		
Total cities	2014	894,588	10,513	66,134	4,782	272,092	541,067	6,103,377	1,176,651	4,404,971	521,755	32,603	11,122	206,728,529
	2015	929,348	11,862	70,759	5,123	276,831	564,773	5,998,043	1,092,856	4,362,498	542,689	31,557		
	Percent change	+3.9	+12.8	+7.0	+7.1	+1.7	+4.4	−1.7	−7.1	−1.0	+4.0	−3.2		
Group I (250,000 and over)	2014	420,517	5,370	24,607	1,040	150,505	238,995	2,006,899	404,912	1,361,161	240,826	11,730	80	59,097,749
	2015	437,895	6,146	26,947	1,129	154,760	248,913	1,987,100	378,045	1,365,521	243,534	11,931		
	Percent change	+4.1	+14.5	+9.5	+8.6	+2.8	+4.1	−1.0	−6.6	+0.3	+1.1	+1.7		
1,000,000 and over (Group I subset)	2014	177,366	2,017	10,367	—	67,346	97,636	737,431	139,640	504,988	92,803	4,149	11	27,404,679
	2015	187,350	2,231	12,147	—	69,301	103,671	726,998	131,803	501,699	93,496	3,921		
	Percent change	+5.6	+10.6	+17.2	—	+2.9	+6.2	−1.4	−5.6	−0.7	+0.7	−5.5		
500,000 to 999,999 (Group I subset)	2014	132,955	1,719	7,755	612	43,570	79,299	675,699	138,006	460,506	77,187	3,184	22	15,610,663
	2015	135,732	2,066	7,919	620	46,025	79,102	668,848	129,169	463,493	76,186	3,480		
	Percent change	+2.1	+20.2	+2.1	+1.3	+5.6	−0.2	−1.0	−6.4	+0.6	−1.3	+9.3		
250,000 to 499,999 (Group I subset)	2014	110,196	1,634	6,485	428	39,589	62,060	593,769	127,266	395,667	70,836	4,397	47	16,082,407
	2015	114,813	1,849	6,881	509	39,434	66,140	591,254	117,073	400,329	73,852	4,530		
	Percent change	+4.2	+13.2	+6.1	+18.9	−0.4	+6.6	−0.4	−8.0	+1.2	+4.3	+3.0		
Group II (100,000 to 249,999)	2014	143,201	1,797	11,010	678	44,227	85,489	1,056,023	213,004	741,802	101,217	5,356	216	32,227,725
	2015	150,861	1,999	11,875	789	44,821	91,377	1,054,055	198,842	746,268	108,945	4,999		
	Percent change	+5.3	+11.2	+7.9	+16.4	+1.3	+6.9	−0.2	−6.6	+0.6	+7.6	−6.7		
Group III (50,000 to 99,999)	2014	108,643	1,116	8,527	641	30,670	67,689	896,590	169,982	656,450	70,158	4,358	481	33,490,047
	2015	111,117	1,243	8,772	729	30,638	69,735	891,484	157,891	657,274	76,319	4,142		
	Percent change	+2.3	+11.4	+2.9	+13.7	−0.1	+3.0	−0.6	−7.1	+0.1	+8.8	−5.0		
Group IV (25,000 to 49,999)	2014	83,487	915	7,803	763	21,487	52,519	772,777	141,808	585,795	45,174	3,728	861	29,777,604
	2015	86,591	1,013	8,325	841	21,376	55,036	753,094	131,479	574,041	47,574	3,476		
	Percent change	+3.7	+10.7	+6.7	+10.2	−0.5	+4.8	−2.5	−7.3	−2.0	+5.3	−6.8		
Group V (10,000 to 24,999)	2014	74,277	731	7,206	793	15,606	49,941	731,328	134,348	560,627	36,353	3,183	1,805	28,863,408
	2015	76,808	834	7,693	767	15,685	51,829	704,558	122,779	544,020	37,759	3,044		
	Percent change	+3.4	+14.1	+6.8	−3.3	+0.5	+3.8	−3.7	−8.6	−3.0	+3.9	−4.4		
Group VI (under 10,000)	2014	64,463	584	6,981	867	9,597	46,434	639,760	112,597	499,136	28,027	4,248	7,679	23,271,996
	2015	66,076	627	7,147	868	9,551	47,883	607,752	103,820	475,374	28,558	3,965		
	Percent change	+2.5	+7.4	+2.4	+0.1	−0.5	+3.1	−5.0	−7.8	−4.8	+1.9	−6.7		
Metropolitan counties	2014	179,958	2,340	16,116	1,802	35,727	123,973	1,300,465	329,479	861,902	109,084	7,564	1,946	73,347,881
	2015	185,888	2,476	17,206	1,842	36,195	128,169	1,247,681	298,050	838,438	111,193	7,379		
	Percent change	+3.3	+5.8	+6.8	+2.2	+1.3	+3.4	−4.1	−9.5	−2.7	+1.9	−2.4		
Nonmetropolitan counties[c]	2014	44,372	741	6,848	531	2,692	33,560	326,963	107,128	197,525	22,310	2,747	2,477	25,082,705
	2015	45,428	854	6,752	621	2,634	34,567	303,952	98,914	182,807	22,231	2,440		
	Percent change	+2.4	+15.2	−1.4	+16.9	−2.2	+3.0	−7.0	−7.7	−7.5	−0.4	−11.2		
Suburban areas[d]	2014	313,352	3,670	28,816	3,001	67,667	210,198	2,680,468	566,379	1,928,061	186,028	14,074	8,480	132,726,573
	2015	323,844	3,900	30,758	3,048	68,209	217,929	2,582,715	514,887	1,875,964	191,864	13,388		
	Percent change	+3.3	+6.3	+6.7	+1.6	+0.8	+3.7	−3.6	−9.1	−2.7	+3.1	−4.9		

[a]The figures shown in this column for the offense of rape were reported using the revised Uniform Crime Reporting (UCR) definition of rape.
[b]The figures shown in this column for the offense of rape were reported using the legacy UCR definition of rape.
[c]Includes state police agencies that report aggregately for the entire state.
[d]Suburban areas include law enforcement agencies in cities with less than 50,000 inhabitants and county law enforcement agencies that are within a Metropolitan Statistical Area. Suburban areas exclude all metropolitan agencies associated with a principal city. The agencies associated with suburban areas also appear in other groups within this table.

Note: No agencies over 1,000,000 in population submitted rape data using the legacy UCR definition in 2015; therefore, the UCR Program could not provide a 2-year comparison for this agency group size.

SOURCE: "Crime Trends by Population Group, 2014–2015," in "Violent Crime," *Crime in the United States, 2015*, Federal Bureau of Investigation, September 2016, https://ucr.fbi.gov/crime-in-the-u.s.-2015/tables/table-12 (accessed October 10, 2016)

TABLE 5.2

Number and rate of homicides, by victim demographic characteristics, 2002–11

		Sex		Race			Age						
Year	All victims	Male	Female	White	Black/African American	Other*	11 or younger	12–17	18–24	25–34	35–49	50–64	65 or older
Number of homicides													
2002	16,230	12,475	3,755	7,805	7,990	435	790	825	4,345	4,305	3,875	1,365	720
2003	16,530	12,835	3,690	7,985	8,080	465	790	785	4,510	4,375	3,890	1,425	750
2004	16,150	12,600	3,550	7,980	7,755	415	765	860	4,165	4,295	3,800	1,510	755
2005	16,740	13,180	3,560	8,050	8,255	435	745	925	4,405	4,480	3,880	1,560	740
2006	17,310	13,655	3,655	8,135	8,710	460	780	1,035	4,615	4,545	3,920	1,755	660
2007	17,130	13,460	3,665	8,110	8,610	405	800	1,035	4,485	4,570	3,875	1,680	675
2008	16,465	12,900	3,565	8,020	8,070	375	810	955	4,065	4,445	3,745	1,705	745
2009	15,400	11,880	3,520	7,485	7,495	425	730	840	3,840	3,975	3,590	1,685	745
2010	14,720	11,410	3,315	6,885	7,450	385	710	770	3,655	4,025	3,200	1,700	665
2011	14,610	11,370	3,240	6,830	7,380	400	740	665	3,680	3,850	3,260	1,710	700
Rate per 100,000 U.S. residents													
2002	5.6	8.8	2.6	3.3	21.2	2.7	1.6	3.3	15.2	10.9	5.9	3.0	2.0
2003	5.7	9.0	2.5	3.4	21.2	2.9	1.6	3.1	15.6	11.1	5.9	3.0	2.1
2004	5.5	8.7	2.4	3.4	20.1	2.5	1.6	3.4	14.2	10.9	5.8	3.1	2.1
2005	5.7	9.1	2.4	3.4	21.2	2.5	1.5	3.6	14.9	11.3	5.9	3.1	2.0
2006	5.8	9.3	2.4	3.4	22.1	2.6	1.6	4.1	15.5	11.4	6.0	3.4	1.8
2007	5.7	9.1	2.4	3.3	21.5	2.2	1.6	4.1	15.0	11.4	5.9	3.1	1.8
2008	5.4	8.6	2.3	3.3	19.9	2.0	1.6	3.8	13.4	10.9	5.8	3.1	1.9
2009	5.0	7.8	2.3	3.0	18.3	2.2	1.5	3.4	12.5	9.6	5.6	3.0	1.9
2010	4.8	7.5	2.1	2.8	17.7	1.8	1.4	3.0	11.9	9.8	5.0	2.9	1.6
2011	4.7	7.4	2.0	2.8	17.3	1.8	1.5	2.7	11.9	9.2	5.2	2.8	1.7

*Includes persons identified as American Indian, Alaska Native, Asian, Native Hawaiian, or other Pacific Islander.
Note: Data may not sum to total due to rounding. Counts rounded to the nearest 5. Homicide rates by Hispanic or Latino origin were not calculated due to missing data on ethnicity.

SOURCE: Erica L. Smith and Alexia Cooper, "Table 1. Number and Rate of Homicides in the U.S., by Victim Demographic Characteristics, 2002–2011," in *Homicide in the U.S. Known to Law Enforcement, 2011*, U.S. Department of Justice, Bureau of Justice Statistics, December 2013, http://www.bjs.gov/content/pub/pdf/hus11.pdf (accessed September 8, 2016)

---

580 homicide victims aged 55 to 59 years, 363 (62.6%) were white, compared with 183 (31.6%) who were African American.

FIREARMS USED IN HOMICIDES. As noted earlier, firearms are the most common weapon used to commit murder. Table 5.4 shows murder circumstances by weapon in 2015. Of the 13,455 total murder victims in the United States in 2015, 9,616 (71.5%) were killed by firearms. This was more than six times the number of people (1,544) who were killed with knives or cutting instruments, the next most common weapon. Handguns were the most common type of firearm used to commit murder. Of the 9,616 murders committed with firearms in 2015, 6,447 (67%) were committed with handguns. Among the 2,014 murders committed in combination with another type of felony, 1,458 (72.3%) were committed with firearms, including 467 of 595 (78.5%) robberies and 389 of 468 (83.1%) narcotic drug offenses. Meanwhile, firearms accounted for 3,941 of 5,958 (66.1%) murders committed in circumstances that did not involve additional types of felonies, including 175 of 188 (93%) gangland killings and 578 of 604 (95.7%) juvenile gang killings. Among murders committed as a result of arguments, including arguments over money or property (which resulted in 184 murders) and other arguments (2,941), firearms were the murder weapon in 62.4% of cases reported to law enforcement agencies.

Table 5.5 shows murder by state and type of weapon in 2015, among cases in which supplemental homicide data were available. As a result, the total number of murders for each state is not necessarily comprehensive but rather represents only those for which supplemental data about weapon type were available. Among such homicide cases, firearms were the most likely type of weapon used, and handguns were the most common type of firearm used. For example, firearms accounted for 1,275 of 1,861 (68.5%) murders in California and 906 of 1,276 (71%) murders in Texas, the states with the highest homicide totals.

Murder often occurs at the hands of a family member, friend, romantic partner, or acquaintance. As Table 5.6 shows, the relationship of the victim to the murderer was unknown in 6,432 of 13,455 (47.8%) homicide cases. In 2,801 (20.8%) cases the victim and murderer were acquaintances. In 1,721 (12.8%) cases the victim was either the husband, wife, mother, father, son, daughter, brother, sister, or other family member of the perpetrator. Among all murders of family members, wives were victims in 509 cases, or 29.6% of the total. Friend (365 cases) and girlfriend (496) were among the other most common relationship types among murder victims.

As the previous data indicate, women who are murdered are especially likely to have a preexisting relationship with

TABLE 5.3

Homicide victims by age, sex, race, and ethnicity, 2015

		Sex			Race				Ethnicity		
Age	Total	Male	Female	Unknown	White	Black or African American	Other[a]	Unknown	Hispanic or Latino	Not Hispanic or Latino	Unknown
Total	13,455	10,608	2,818	29	5,854	7,039	366	196	2,028	7,971	2,224
Percent distribution[b]	100.0	78.8	20.9	0.2	43.5	52.3	2.7	1.5	16.6	65.2	18.2
Under 18[c]	1,093	761	329	3	504	544	26	19	193	611	189
Under 22[c]	2,624	2,135	487	2	954	1,598	44	28	488	1,484	443
18 and older[c]	12,228	9,773	2,448	7	5,296	6,451	339	142	1,823	7,305	1,974
Infant (under 1)	168	104	63	1	89	65	8	6	24	104	28
1 to 4	260	153	106	1	130	117	6	7	32	160	58
5 to 8	91	49	42	0	58	30	1	2	12	57	16
9 to 12	56	35	20	1	30	22	3	1	9	30	6
13 to 16	282	218	64	0	110	167	4	1	62	141	49
17 to 19	996	870	126	0	325	647	16	8	191	546	159
20 to 24	2,431	2,102	329	0	764	1,596	45	26	414	1,398	368
25 to 29	2,071	1,733	338	0	717	1,290	48	16	321	1,201	331
30 to 34	1,647	1,340	307	0	650	927	49	21	266	974	254
35 to 39	1,263	1,021	241	1	539	666	45	13	200	728	223
40 to 44	925	701	222	2	467	428	23	7	150	545	158
45 to 49	781	586	194	1	399	330	33	19	118	485	115
50 to 54	737	568	169	0	428	282	17	10	84	468	120
55 to 59	580	425	154	1	363	183	27	7	57	389	102
60 to 64	360	250	109	1	226	114	12	8	36	231	54
65 to 69	235	168	67	0	166	59	6	4	14	168	41
70 to 74	159	93	65	1	113	33	9	4	11	100	33
75 and over	279	118	161	0	226	39	13	1	15	191	48
Unknown	134	74	41	19	54	44	1	35	12	55	61

[a]Includes American Indian or Alaska Native, Asian, and Native Hawaiian or other Pacific Islander.
[b]Because of rounding, the percentages may not add to 100.0.
[c]Does not include unknown ages.

SOURCE: "Expanded Homicide Data Table 2. Murder Victims by Age, Sex, Race, and Ethnicity, 2015," in "Violent Crime," *Crime in the United States, 2015*, Federal Bureau of Investigation, September 2016, https://ucr.fbi.gov/crime-in-the-u.s/2015/crime-in-the-u.s.-2015/tables/expanded_homicide_data_table_2_murder_victims_by_age_sex_and_race_2015.xls (accessed October 10, 2016)

those who kill them. In *When Men Murder Women: An Analysis of 2014 Homicide Data* (September 2016, http://www.vpc.org/studies/wmmw2016.pdf), the Violence Policy Center, a national nonprofit educational foundation that conducts research on violence in the United States, analyzes FBI data on homicides in which a single female is the victim of a single male offender. The center finds that in 2014, among homicides in which the relationship between victim and offender could be determined, 93% of female victims (1,388 out of 1,495) were murdered by a male they knew. Moreover, the number of females murdered by a male they knew (1,388) was 13 times higher than those who were killed by a male stranger (107 victims). Of these 1,388 victims, 63% (870) were wives or intimate acquaintances of their killer, and 239 of these victims were shot and killed by their husband or intimate acquaintance during the course of an argument. More women were killed with firearms (54%) than with any other type of weapon in 2014. Of the homicides committed with firearms, 69% involved handguns.

The Criminal Advantages of Guns

A gun offers a criminal several advantages over other weapons. A gun offender can keep a greater physical distance from the victim to ensure his or her own safety and increase the chances of escaping. A gun also allows the offender to maintain psychological distance, keeping the confrontation more impersonal and minimizing the emotional involvement, a factor that some people claim leads to more killings than would otherwise occur. Control over potential victims can be easier to maintain with a gun; victims are less likely to run from a gun-carrying offender than from an offender who brandishes other types of weapons, such as knives, for fear of being shot from a distance.

Finally, multiple-victim homicides are particularly dependent on access to firearms. The higher the number of homicide victims, the greater the likelihood that a firearm was the weapon used. In 2011, 67.1% of all homicides involved a firearm, and the percentage of single-victim homicides involving a firearm was 66.5%. (See Table 5.7.) By comparison, 77.3% of two-victim homicides, 82.3% of three-victim homicides, and 90.8% of homicides in which there were four or more victims involved firearms. As these statistics show, mass murders are almost always mass shootings.

Mass Shootings in the United States

Firearms, unlike other commonly available weapons, make it possible to kill a large number of people in a

TABLE 5.4

Homicide circumstances, by weapon, 2015

Circumstances	Total murder victims	Total firearms	Handguns	Rifles	Shotguns	Other guns or type not stated	Knives or cutting instruments	Blunt objects (clubs, hammers, etc.)	Personal weapons (hands, fists, feet, etc.)	Poison	Pushed or thrown out window	Explosives	Fire	Narcotics	Drowning	Strangulation	Asphyxiation	Other
Total	13,455	9,616	6,447	252	269	2,648	1,544	437	623	7	1	1	82	70	14	96	120	844
Felony type total:	2,014	1,458	1,042	36	40	340	178	84	72	1	0	0	34	40	2	16	14	115
Rape	12	2	2	0	0	0	1	2	4	0	0	0	0	0	0	1	0	2
Robbery	595	467	366	5	15	81	42	27	26	0	0	0	0	0	0	7	5	21
Burglary	102	66	33	7	4	22	13	8	2	0	0	0	3	0	0	0	2	8
Larceny-theft	16	12	8	0	0	4	1	0	1	0	0	0	0	0	0	0	1	1
Motor vehicle theft	41	19	9	1	1	8	10	2	2	0	0	0	0	0	0	0	1	7
Arson	19	0	0	0	0	0	0	0	0	0	0	0	14	0	0	0	0	4
Prostitution and commercialized vice	6	3	3	0	0	0	1	1	0	0	0	0	0	0	0	2	0	1
Other sex offenses	15	6	5	0	1	0	1	1	2	1	0	0	0	0	0	0	0	3
Narcotic drug laws	468	389	306	7	2	74	19	7	4	0	0	0	0	35	0	2	0	13
Gambling	5	4	3	0	0	1	1	0	0	0	0	0	0	0	0	0	0	0
Other-not specified	735	490	307	16	17	150	88	36	31	0	0	0	17	5	2	6	5	55
Suspected felony type	117	84	56	1	4	23	9	3	2	0	0	0	0	0	0	2	4	13
Other than felony type total:	5,958	3,941	2,884	140	162	755	913	203	410	4	0	0	11	20	9	47	72	328
Romantic triangle	106	76	54	3	7	12	19	4	4	0	0	0	0	0	0	1	1	1
Child killed by babysitter	36	1	0	0	0	1	0	4	19	0	0	0	0	0	1	0	1	10
Brawl due to influence of alcohol	112	41	27	2	2	10	31	9	14	0	0	0	1	0	0	0	0	15
Brawl due to influence of narcotics	75	41	26	2	1	12	8	1	6	0	0	0	0	9	0	0	1	9
Argument over money or property	184	127	95	10	5	17	31	12	3	2	0	0	7	1	0	5	1	5
Other arguments	2,941	1,822	1,325	50	99	348	633	111	186	2	0	0	1	7	5	29	29	116
Gangland killings	188	175	146	5	0	24	8	1	1	0	0	0	1	0	0	0	0	1
Juvenile gang killings	604	578	465	6	3	104	20	1	2	0	0	0	0	0	0	0	0	3
Institutional killings	24	0	0	0	0	1	2	2	13	0	0	0	0	0	0	2	3	1
Sniper attack	5	4	1	2	1	0	0	1	0	0	0	0	0	0	0	0	0	0
Other-not specified	1,683	1,075	745	60	44	226	161	57	162	2	0	0	2	10	3	10	34	167
Unknown	5,366	4,133	2,465	75	63	1,530	444	147	139	2	1	1	37	10	3	31	30	388

SOURCE: "Expanded Homicide Data Table 11. Murder Circumstances by Weapon, 2015," in "Violent Crime," *Crime in the United States, 2015*, Federal Bureau of Investigation, September 2016, https://ucr.fbi.gov/crime-in-the-u.s/2015/crime-in-the-u.s.-2015/tables/expanded_homicide_data_table_11_murder_circumstances_by_weapon_2015.xls (accessed October 10, 2016)

TABLE 5.5

Homicides, by state and weapon type, 2015

State	Total murders[a]	Total firearms	Handguns	Rifles	Shotguns	Firearms (type unknown)	Knives or cutting instruments	Other weapons	Hands, fists, feet, etc.[b]
Alabama[c]	3	3	1	0	1	1	0	0	0
Alaska	57	39	12	2	1	24	7	8	3
Arizona	278	171	128	4	3	36	42	55	10
Arkansas	164	110	51	10	4	45	18	30	6
California	1,861	1,275	855	34	33	353	263	233	90
Colorado	176	115	65	12	6	32	25	19	17
Connecticut	107	73	29	0	2	42	16	9	9
Delaware	63	52	26	0	0	26	6	3	2
District of Columbia	162	121	65	1	0	55	28	8	5
Georgia	565	464	394	16	10	44	38	60	3
Hawaii	19	4	1	1	0	2	6	2	7
Idaho	30	24	17	4	0	3	3	3	0
Illinois[c]	497	440	431	2	1	6	29	25	3
Indiana	272	209	147	9	5	48	22	29	12
Iowa	72	49	31	0	2	16	10	5	8
Kansas	125	91	51	4	4	32	9	14	11
Kentucky	209	141	91	1	9	40	30	26	12
Louisiana	474	379	207	7	9	156	46	43	6
Maine	23	16	15	0	0	1	3	1	3
Maryland	372	279	266	3	3	7	44	33	16
Massachusetts	126	81	33	1	0	47	27	11	7
Michigan	571	389	148	10	10	221	52	103	27
Minnesota	133	79	47	6	9	17	21	23	10
Mississippi	159	126	102	3	5	16	11	14	8
Missouri	499	418	233	12	11	162	26	40	15
Montana	36	18	13	0	0	5	8	3	7
Nebraska	61	43	41	1	1	0	9	8	1
Nevada	177	113	17	1	1	94	25	30	9
New Hampshire	14	8	4	0	0	4	6	0	0
New Jersey	353	255	200	3	0	52	38	41	19
New Mexico	94	56	14	1	2	39	18	19	1
New York	611	383	331	1	5	46	104	104	20
North Carolina	506	353	235	10	21	87	41	77	35
North Dakota	17	9	5	0	0	4	3	1	4
Ohio	480	316	184	12	7	113	39	103	22
Oklahoma	233	149	113	8	9	19	33	32	19
Oregon	71	34	17	1	5	11	12	22	3
Pennsylvania	651	497	386	10	11	90	65	67	22
Rhode Island	27	10	3	0	1	6	10	6	1
South Carolina	394	312	201	7	10	94	34	34	14
South Dakota	27	12	5	1	5	1	8	2	5
Tennessee	402	297	204	13	7	73	40	51	14
Texas	1,276	906	610	19	36	241	146	135	89
Utah	54	34	26	2	0	6	5	6	9
Vermont	10	8	2	4	0	2	0	1	1
Virginia	383	275	148	6	7	114	51	43	14
Washington	209	141	84	3	5	49	18	36	14
West Virginia	57	30	19	1	1	9	6	17	4
Wisconsin	238	170	122	5	4	39	37	25	6
Wyoming	16	10	4	0	3	3	2	3	1
Guam	6	3	1	0	0	2	1	2	0
U.S. Virgin Islands	35	26	12	1	0	13	3	6	0

[a]Total number of murders for which supplemental homicide data were received.
[b]Pushed is included in hands, fists, feet, etc.
[c]Limited supplemental homicide data were received.

SOURCE: "Table 20. Murder by State, Types of Weapons, 2015," in "Violent Crime," *Crime in the United States, 2015*, Federal Bureau of Investigation, September 2016, https://ucr.fbi.gov/crime-in-the-u.s/2015/crime-in-the-u.s.-2015/tables/table-20 (accessed October 10, 2016)

matter of seconds or minutes. Although the perpetrators are usually apprehended or commit suicide, there is inevitably a lag time after the perpetrators begin shooting but before they are subdued. Christopher Ingraham explains in "We Have Three Different Definitions of 'Mass Shooting,' and We Probably Need More" (Washington Post.com, February 26, 2016) that the exact definition of a mass shooting varies among the organizations and agencies that study them, but the general understanding is that a mass shooting involves four or more gunshot victims (wounded or killed).

Table 5.8 provides an overview of notable multiple or mass shootings that occurred in the United States between 1949 and 2008. The first mass shooting of its kind occurred in Camden, New Jersey, in 1949, when 28-year-old Howard Unruh (1921–2009), an unemployed World War II (1939–1945) veteran, killed 13 people and wounded three while walking through his neighborhood

TABLE 5.6

Homicide circumstances, by relationship of victim to offender, 2015

Circumstances	Total murder victims	Husband	Wife	Mother	Father	Son	Daughter	Brother	Sister	Other family	Acquaintance	Friend	Boyfriend	Girlfriend	Neighbor	Employee	Employer	Stranger	Unknown
Total	13,455	113	509	125	131	255	162	108	32	286	2,801	365	152	496	95	8	10	1,375	6,432
Felony type total:	2,014	3	37	8	12	28	18	7	9	36	533	68	9	34	18	5	1	388	800
Rape	12	0	0	0	0	1	0	0	0	0	6	0	0	2	1	0	0	0	2
Robbery	595	0	0	1	0	0	0	0	0	3	128	20	1	0	2	3	1	211	225
Burglary	102	0	0	0	1	1	0	0	0	7	30	1	0	2	4	1	0	24	31
Larceny-theft	16	0	0	0	0	0	0	0	0	0	6	2	0	1	0	0	0	5	2
Motor vehicle theft	41	0	0	0	0	0	0	0	0	4	11	0	0	3	0	0	0	13	9
Arson	19	0	2	0	0	1	0	0	1	1	2	0	0	0	0	0	0	6	6
Prostitution and commercialized vice	6	0	0	0	0	0	0	0	0	0	1	0	1	0	0	0	0	1	3
Other sex offenses	15	0	1	0	0	0	0	0	0	0	5	0	1	0	1	0	0	3	3
Narcotic drug laws	468	0	0	0	0	1	1	0	0	4	174	22	0	4	0	0	0	28	234
Gambling	5	0	0	0	0	0	0	0	0	0	5	0	0	0	0	0	0	0	0
Other-not specified	735	3	34	7	10	24	17	7	8	17	165	23	6	21	10	1	0	97	285
Suspected felony type	117	4	7	1	0	7	3	0	1	4	19	2	0	6	0	0	0	8	55
Other than felony type total:	5,958	84	367	80	92	171	110	88	19	176	1,714	229	120	355	59	2	7	588	1,697
Romantic triangle	106	3	14	2	0	0	0	0	0	1	52	4	3	9	0	0	0	8	10
Child killed by babysitter	36	0	0	0	0	2	1	0	0	3	30	1	0	0	1	0	0	0	1
Brawl due to influence of alcohol	112	0	2	0	2	1	1	11	0	4	37	8	1	1	4	1	0	20	19
Brawl due to influence of narcotics	75	0	2	2	1	5	2	1	0	1	28	8	3	3	0	0	0	7	14
Argument over money or property	184	2	5	3	2	4	0	4	1	10	100	9	3	3	5	0	0	11	22
Other arguments	2,941	64	242	45	64	35	20	62	5	100	940	150	97	276	34	1	4	281	521
Gangland killings	188	1	0	0	0	1	0	0	0	1	36	6	0	0	0	0	0	28	115
Juvenile gang killings	604	0	0	0	0	0	0	0	0	0	105	3	0	0	0	0	0	56	440
Institutional killings	24	0	1	0	0	0	0	0	0	0	18	0	0	0	0	0	0	3	2
Sniper attack	5	0	0	0	0	0	0	0	0	0	1	0	0	0	0	0	0	0	3
Other-not specified	1,683	14	100	28	23	125	87	10	13	56	367	40	15	63	15	0	3	174	550
Unknown	5,366	22	98	36	27	49	31	13	3	70	535	66	23	101	18	1	2	391	3,880

Note: The relationship categories of husband and wife include both common-law and ex-spouses. The categories of mother, father, sister, brother, son, and daughter include stepparents, stepchildren, and stepsiblings. The category of acquaintance includes homosexual relationships and the composite category of other known to victim.

SOURCE: "Expanded Homicide Data Table 10. Murder Circumstances by Relationship, 2015," in "Violent Crime," Crime in the United States, 2015, Federal Bureau of Investigation, September 2016, https://ucr.fbi.gov/crime-in-the-u.s/2015/crime-in-the-u.s.-2015/tables/expanded_homicide_data_table_10_murder_circumstances_by_relationship_2015.xls (accessed October 10, 2016)

TABLE 5.7

Homicides, by number of victims and weapon type, 2011

			Percent of homicides involving a—			
				Firearm		
Number of homicide victims	Number of homicide incidents	Total	Any firearm	Handgun	Other firearm*	Other weapon
Total	13,750	100%	67.1	49.4	17.7	32.9
1 victim	13,050	100%	66.5	49.3	17.2	33.5
2 victims	565	100%	77.3	52.0	25.3	22.7
3 victims	110	100%	82.3	47.1	35.2	17.7
4 or more victims	25	100%	90.8	44.2	46.6	9.2

*Includes rifles, shotguns, and firearms of unspecified type, including automatic weapons.
Note: Due to limitations of the data, the incident count presented above may not accurately reflect the total number of unique homicide incidents in the United States.

SOURCE: Erica L. Smith and Alexia Cooper, "Table 5. Homicides in the U.S., by the Number of Victims Killed and Weapon Type, 2011," in *Homicide in the U.S. Known to Law Enforcement, 2011*, U.S. Department of Justice, Bureau of Justice Statistics, December 2013, http://www.bjs.gov/content/pub/pdf/hus11.pdf (accessed September 8, 2016)

and firing on his neighbors with a handgun. The first mass shooting in U.S. history at a school occurred in 1966 on the campus of the University of Texas, Austin, when 25-year-old Charles Joseph Whitman (1941–1966) barricaded himself inside the observation deck of the campus clock tower, shooting at random victims below for more than an hour and a half. By the time police subdued the shooter, 16 people were dead and another 32 were wounded. Four decades later the American consciousness was forever altered by the 1999 incident at Columbine High School in Littleton, Colorado, in which Dylan Klebold (1982–1999) and Eric Harris (1981–1999) killed 13 people and wounded 24 others before killing themselves. This was the deadliest school shooting in U.S. history until the 2007 massacre at Virginia Polytechnic Institute and State University (Virginia Tech), in which Seung-Hui Cho (1984–2007) used two handguns to kill 32 students and faculty members and wound 17 others before fatally shooting himself. Other notable mass shootings have occurred in the workplace, including the 1986 shooting at the Edmond, Oklahoma, post office, in which Patrick Henry Sherrill (1941–1986) killed 14 coworkers and wounded six before committing suicide. Mass shootings have also occurred in public places such as restaurants. For example, in 1991 George Hennard (1956–1991) opened fire inside a crowded Luby's Cafeteria in Killeen, Texas, killing 23 and wounding 27 more before fatally shooting himself.

Even while mass shootings had become increasingly common in the United States, the incidents that occurred between 2009 and 2016 drove home the grim reality that these sudden massacres could take place anywhere—including a military base, a movie theater, and even a place of worship—at any time. (See Table 5.9.) Still, perhaps no previous incident provoked such an outpouring of public horror and grief as the 2012 shooting at Sandy Hook Elementary School in Newtown, Connecticut, in which 20-year-old Adam Lanza (1992?–2012) killed 20 first graders and six school staff members before committing suicide. As subsequent chapters in this book reflect, the Sandy Hook shooting became a watershed moment in the national debate about gun control versus gun rights. Meanwhile, the 2007 Virginia Tech massacre remained the deadliest mass shooting by a lone gunman in U.S. history until 2016, when 29-year-old Omar Mateen (1986–2016) shot more than twice as many victims—killing 49 people and wounding 53—at the Pulse night club in Orlando, Florida, leaving the American public once again stunned and searching for answers about how to understand these deadly attacks and prevent them in the future.

Table 5.8 and Table 5.9 are by no means comprehensive lists of mass shootings in the United States, but they do reflect the deadliest attacks as well as their approximate relative distribution over time. Prior to the mid-1990s, incidents of this type were relatively rare, but their rate of frequency has since accelerated. According to Mark Follman, Gavin Aronsen, and Deanna Pan, in "A Guide to Mass Shootings in America" (MotherJones.com, September 24, 2016), since 1982 there have been at least 84 mass shootings in 34 states, 47 of which occurred between 2006 and 2016. The researchers further note that more than three-quarters of the guns used in these attacks were acquired legally.

ACTIVE SHOOTER INCIDENTS

Among law enforcement (who tend to be the first to respond to these mass shootings while they are in progress), these events are often referred to as "active shooter" incidents. In September 2014 the FBI released *A Study of Active Shooter Incidents in the United States between 2000 to 2013* (https://www.fbi.gov/file-repository/active-shooter-study-2000-2013-1.pdf), the first major study of its kind. As the FBI explains, U.S. government agencies—including the White House, the Department of Justice/FBI, the U.S. Department of Education, and the U.S. Department of Homeland Security/Federal Emergency Management Agency—define the term *active shooter* as "an individual actively engaged in killing or

TABLE 5.8

Notable mass shootings, 1949–2008

Date	Location	Shooter	Profile	Incident	Duration, Termination	Casualties	Weapons	Gun acquisition
Feb-14-08	Northern Illinois University; DeKalb, Illinois	• Name: Steven Phillip Kazmierczak • Age: 27 • Nationality: American • Race: white	Attempted suicide multiple times in adolescence, diagnosed with psychiatric problems. At 21, Kazmierczak was discharged from the army when it was discovered that he had falsified his application to hide his mental health history. He had reportedly stopped taking his medication in the weeks before the shooting.	Gunman burst into auditorium lecture hall and opened fire on a class with approximately 120 students in attendance.	~6 minutes; gunman shot and killed himself	5 killed, 21 wounded	• Remington 12-gauge shotgun • Glock 19 semiautomatic pistol • Sig Sauer 9 mm semiautomatic pistol • HiPoint 380 caliber semiautomatic pistol	Legally purchased from a federally licensed firearms dealer; at least one background check was performed.
Apr-16-07	Virginia Polytechnic Institute and State University (Virginia Tech)	• Name: Seung-Hui Cho • Age: 23 • Nationality: South Korean; permanent U.S. resident since 1984 • Race: Asian	A Virginia Tech senior majoring in English with a long history of depression. Cho's academic writings incuded a paper describing fantasies about the 1999 Columbine shooting. In 2005 a Virginia judge deemed him "an imminent danger to himself because of mental illness," ordering him to undergo outpatient treatment, but not committing him to a mental health facility.	Gunman shot and killed two students in a dormitory, then returned to his room to change out of bloody clothes; later that morning, Cho entered an academic building and began firing on students and faculty in classrooms on the second floor.	~10 minutes (2nd indicent only); gunman shot and killed himself	32 killed (in both incidents), 17 wounded	• Walther P32 semiautomatic pistol • Glock 19 semiautomatic pistol	Legally purchased from federally licensed firearms dealers; at least one background check was performed.
Apr-20-99	Columbine High School, Littleton, Colorado	• Name: Dylan Klebold • Age: 17 • Nationality: American • Race: white • Name: Eric Harris • Age: 18 • Nationality: American • Race: white	The shooters' diaries, video recordings, and other personal documents revealed nihilism, alienation, and contempt for their classmates. Both were preoccupied with guns, explosives, historical murders, and Nazis. Planned for more than a year, their attack took place on Adolf Hitler's birthday.	After homemade explosives planted inside the school failed to detonate, gunmen exited their cars, entered the campus and began shooting at students outside. Gunmen entered the school throwing pipe bombs and shooting at random; entered the library where they shot most of their victims. Leaving the library, gunmen roamed the school, occasionally firing at law enforcement outside.	~40 minutes; gunmen shot and killed themselves	13 killed, 24 wounded	• Savage Stevens 311-D12-gauge shotgun (sawed off) • Intratec TEC-DC9 assault pistol • Savage Springfield 67H pump-action 12-gauge shotgun • Hi-Point 9 mm Carbine semiautomatic rifle	Illegally purchased: Three firearms were acquired through a 'straw purchaser,' who bought the firearms from an unlicensed dealer at a gun show; the fourth firearm was purchased in an informal cash transaction with an acquaintance who knew the buyers were under age.
May-21-98	Thurston High School; Springfield, Oregon	• Name: Kip Kinkel • Age: 15 • Nationality: American • Race: white	A Thurston student who had been arrested and suspended the day before the shooting when a gun was found in his locker. Kinkel had a history of emotional and disciplinary problems, and a known interest in guns and explosives.	Gunman shot and killed both of his parents the day before the shooting. The next morning, he drove to school and shot one student in the hall before entering the cafeteria, where he opened fire on those inside.	~5 minutes; gunman subdued by students while he was trying to reload	4 killed (including the gunman's parents), 25 wounded	• Glock 9 mm semiautomatic pistol • Ruger .22 semiautomatic rifle • Ruger .22 semiautomatic pistol	Two of the guns Kinkel used in the shooting belong had been purchased for him by his father, and the third belonged to his father.
Oct-16-91	Luby's Cafeteria; Killeen, Texas	• Name: George Hennard • Age: 35 • Nationality: American • Race: white	Described by family and friends as paranoid and mentally troubled; vocalized misogynistic feelings. After the shooting, it emerged that Hennard had a history of stalking women.	Gunman crashed his vehicle into the crowded family restaurant, stepped out and opened fire on those inside, choosing women as his preferred victims.	~15 minutes; gunman shot and killed himself	23 killed, 27 wounded	• Ruger P-89 9 mm semiautomatic pistol • Glock 9 mm semiautomatic pistol	Legally purchased.

TABLE 5.8
Notable mass shootings, 1949–2008 [CONTINUED]

Date	Location	Shooter	Profile	Incident	Duration, Termination	Casualties	Weapons	Gun acquisition
Aug-20-86	U.S. Post Office, Edmond, Oklahoma	• Name: Patrick Henry Sherrill • Age: 44 • Nationality: American • Race: white	A loner and ex-marine sharp shooter; known locally as "Crazy Pat" due to his hostile attitude. Sherrill had no record of criminal activity or mental illness. The day before the shooting, he had been reprimanded by a supervisor.	Gunman entered the post office dressed in his employee uniform, carrying three firearms and more than 300 rounds of ammunition; opened fire on those inside.	~10 minutes; gunman shot and killed himself	14 killed, 6 injured	• Two .45 semiautomatic pistols • .22 semiautomatic pistol	Two of the guns were checked out from the National Guard armory, where Sherrill was a member of the marksmanship team; the third was legally purchased.
Jul-18-84	McDonald's restaurant; San Ysidro, California	• Name: James Huberty • Age: 41 • Nationality: American • Race: white	An unemployed father of two, with a history of domestic violence. Huberty had had openly expressed his hatred for "Mexicans, children, and America."	Gunman remarked to his wife that he was going "hunting humans" before he entered the restaurant and opened fire on those inside.	~77 minutes, gunman killed by police sniper	21 killed (including five children), 19 and injured	• Browning P-35 Hi-Power 9 mm pistol • Winchester 1200 12-gauge shotgun pump-action • Israeli Military Industries 9 mm Model A Carbine (Uzi)	Legally purchased.
Aug-1-66	University of Texas; Austin, Texas	• Name: Charles Joseph Whitman • Age: 25 • Nationality: American • Race: white	A former marine and engineering student at the university; an accomplished marksman whose father, a gun collector, had taught him to shoot. In the days and weeks before the shooting, Whitman had complained of headaches and an altered mental state. A suicide note confessed to having irrational thoughts and requested that an autopsy be performed on his body. The autopsy reveal a brain tumor that was was pressing against a region of the brain that regulates emotions.	Gunman killed his wife and mother in their Austin homes before barricading himself inside the 28th-floor observation deck of the campus clock tower with seven firearms and a footlocker full of weapons, equipment, food, and over 700 rounds of ammunition.	~96 minutes; gunman killed by police who stormed the tower.	16 killed, 32 injured	• Remington .35 pump rifle • Remington 6 mm bolt-action rifle • .30 M-1 Carbine • .25 pistol • Luger 9 mm pistol • Smith & Wesson .357 magnum pistol • Sawed-off 12-gauge shotgun	All of Whitman's guns were legal and easy to acquire in 1966
Sep-6-49	Camden, New Jersey	• Name: Howard Unruh • Age: 28 • Nationality: American • Race: white	An unemployed World War II veteran who was then living with his mother. Known to be reclusive and angry, Unruh kept a journal of grievances and perceived slights, including names of people he wished to retaliate against. The shooter was subsequently diagnosed as paranoid schizophrenic. He confessed to the killings but was judged mentally incompetent and was never tried.	Gunman walked through his neighborhood firing randomly on his neighbors.	~12 minutes; gunman subdued by police after ammunition ran out	13 killed, 3 injured	• Luger P08 .9 mm—one of the first semi-automatic pistols	Legally purchased.

Note: Casualty figures do not include the death of the shooter. In cases where the shooter was killed or took his own life, this fact is indicated under Duration, Termination.

SOURCE: Created by Erin Brown for Gale, © 2016.

TABLE 5.9

Notable mass shootings, 2009–2016

Date	Location	Shooter	Profile	Incident	Duration, Termination	Casualties	Weapons	Gun acquisition
Jun-12-16	Pulse nightclub; Orlando, Florida	• Name: Omar Mateen • Age: 29 • Nationality: American citizen, Afghan descent • Race: Pashtun (ethnicity)	Born in the U.S. to Afghan immigrant parents, Mateen had a history of belligerent behavior. Teased for being chubby as a child, he became a serious body builder and steroid user. His first marriage ended after only a few months, due to his alleged domestic abuse. Mateen had been twice interviewed by the FBI in terror-related investigations, but he was not under surveillance at the time of the attack. Questions about Mateen's character and his motives centered on the degree to which he was or was not a radicalized Muslim and also to what extent he was homophobic, or possibly a closeted homosexual himself.	Gunman opened fire inside a crowded nightclub, spraying bullets in all directions. Police responded within minutes, and the gunman barricaded himself inside a bathroom with several hostages. A lengthy standoff ensued, during which the gunman professed terrorist sympathies to a 911 dispatcher. As SWAT officers attempted to breach a bathroom wall with explosives, the gunman ran out and began firing on police. A gunfight ensued, during which an estimated 150 shots were fired, before the gunman was finally subdued.	~3 hours; gunman was fatally shot by police	49 killed, 53 wounded	• Sig Sauer MCX .223-caliber semi-automatic rifle • Glock 179 mm semi-automatic pistol	Legally purchased from a federally licensed firearms dealer.
Dec-2-15	Inland Regional Center, San Bernadino, California	• Name: Syed Rizwan Farook • Age: 28 • Nationality: American • Race: Middle Eastern descent • Name: Tashfeen Malik • Age: 29 • Nationality: Pakistani; permanent U.S. resident • Race: Middle Eastern descent	Farook and Malik met through an online dating service. Both were devout Muslims. Malik moved from Saudi Arabia to the U.S. on a K-1 ("fiance") visa in July 2014 and the couple married a few weeks later. FBI investigators reported that both had been radicalized for some time before the attack. They had stockpiled weapons and ammunition in their home and had practiced their shooting skills at several target ranges across the Los Angeles region. The FBI found no evidence that the couple belonged to a larger terror network, although there were indications that they might have been inspired by foreign terrorist organizations.	The couple left their 6-month-old daughter with her grandmother and went to the facility where the San Bernardino County Department of Public Health (for which Farook worked) was having a holiday party in a conference room. Wearing masks and body armor, the attackers stormed the party and began shooting. They reloaded several times, firing more than 100 rounds in a span of minutes before fleeing in a black SUV. Police apprehended them hours later.	~2–3 minutes; both killed in a gunfight with police	14 killed, 22 wounded	• Smith & Wesson .223-caliber M&P15 semi-automatic rifle • DPMS Model A15 .223-caliber semi-automatic rifle • Llama 9 mm semi-automatic pistol • Springfield Armory 9 mm semi-automatic pistol	Rifles illegally acquired through a "straw purchaser"; pistols legally purchased by Farook
Jun-17-15	Emanuel African Methodist Episcopal Church, Charleston, South Carolina	• Name: Dylann Roof • Age: 21 • Nationality: American • Race: white	A directionless high school drop out, Roof was described by those who knew him as quiet and unsocial. In the months leading up to the attack, he was arrested twice—once on a misdemeanor drug charge and again for trespassing. Police investigators subsequently found racist hate videos among his belongings, and his online activity revealed a fascination with historical racism—including South African Apartheid and the American Confederacy. The attack was classified as a hate crime.	Gunman attended an evening bible study session at the church. After sitting with the group for about an hour, he stood up and began shooting. According to witnesses, the gunman made racist remarks towards his victims and reloaded multiple times before he fled in a black sedan. He was apprehended and arrested by police the following day, about 245 miles north of Charleston.	~6 minutes; gunman fled the scene	9 killed, 1 wounded	• Glock 41 Gen 4 .45-caliber semi-automatic pistol	Legally purchased from a federally licensed firearms dealer.
Sep-16-13	Washington Navy Yard, Washington, D.C.	• Name: Aaron Alexis • Age: 34 • Nationality: American • Race: African American	A former Navy reservist, Alexis was working for The Experts, a Hewlett-Packard defense department IT subcontractor. He had been involved in two prior gun-related incidents. He had sought medical treatment for insomnia but had not received psychiatric treatment. After the shooting, writings found scratched into the barrel of Alexis's gun and on his electronic devices revealed signs of mental illness, including his belief that he was being controlled by extremely low-frequency electromagnetic waves.	Gunman used his security clearance to enter Navy Yard Building 197, where he had been working as an IT contractor. He went to a men's bathroom carrying a backpack and emerged with a sawed-off shotgun. He proceeded to the fourth floor and began firing down into a crowded first-floor atrium. Police arrived after about seven minutes and pursued the gunman throughout the building, as he continued firing on victims at random.	~30 minutes; gunman was fatally shot by police	12 killed, 3 wounded	• Remington 870 12-gauge shotgun, sawed off	Legally purchased from a federally licensed firearms dealer.

TABLE 5.9
Notable mass shootings, 2009–2016 [CONTINUED]

Date	Location	Shooter	Profile	Incident	Duration, Termination	Casualties	Weapons	Gun acquisition
Dec-14-12	Sandy Hook Elementary School, Newtown, Connecticut	• Name: Adam Lanza • Age: 20 • Nationality: American • Race: white	Lanza had displayed warning signs of severe developmental, and psychiatric problems since early childhood, but had never received proper treatment. For three months leading up to the attack, he stayed locked in his bedroom, where he had covered the windows with black plastic. He communicated only online, especially with a community of people interested in mass murder. A 114-page report released by the Connecticut Office of the Child Advocate almost two years after the shooting indicated that Lanza likely suffered from anxiety disorder, obsessive-compulsive disorder, and suicidal disorder. The report also noted that he had unfettered access to the stockpile of firearms his mother kept in the home.	Gunman shot and killed his mother in her home, then drove her car, loaded with weapons and ammunition, to Sandy Hook Elementary School. He shot his way into the locked school and opened fire, concentrating his assault in two first-grade classrooms. More than 154 shots were fired before the gunman committed suicide, as law enforcement approached.	~5 minutes; gunman shot and killed himself	26 killed, 1 wounded	• Bushmaster XM15 .223-caliber semi-automatic rifle • Glock 20SF 10 mm semi-automatic pistol • SIG Sauer P226 9 mm semi-automatic pistol	Legally purchased by the gunman's mother.
Jul-20-12	Century 16 movie theatre, Aurora, Colorado	• Name: James Holmes • Age: 24 • Nationality: American • Race: white	Intellectually gifted, with a B.A. in neuroscience, Holmes was in the process of withdrawing from a PhD program at the University of Colorado at Denver. Acquaintances described him as quiet, shy, and reclusive, but not menacing. A university psychiatrist who had treated Holmes prior to the attack described him in courtroom testimony as an anxious, antisocial oddball who talked of his daily and increasing homicidal thoughts.	Wearing ballistic body armor and a gas mask, the gunman entered the crowded movie theater through an emergency exit. He tossed two gas canisters into the crowd and opened fire on the audience. When his AR-15-type assault rifle jammed, he continued shooting with a shotgun and a handgun. The gunman had also booby-trapped his apartment with explosives, which were defused by police.	~2 minutes; gunman apprehended by police in theater parking lot	12 killed, 58 wounded	• Remington 870 12-gauge pump action shotgun • Smith & Wesson model M&P15T semi-automatic rifle • 2 Glock 45-caliber semi-automatic pistols	Purchased legally from federally licensed firearms dealers.
Nov-5-09	Fort Hood Army Base, near Killeen, Texas	• Name: Nidal Hasan • Age: 39 • Nationality: American • Race: white	A U.S. Army major and psychiatrist of Muslim faith. In 2013 Hasan expressed his motive for the first time, telling a judge he had sought to protect the leadership of the Islamic Emirate of Afghanistan, the Taliban—thereby confirming that his attack was an act of terrorism.	Dressed in his military uniform, the gunman entered the Soldier Readiness Processing Center where an estimated 300 soldiers were awaiting medical care before deployment. Standing on a desk, he shouted "Allahu Akbar" ("God is great") before opening fire, unleashing more than 100 rounds. Exiting the building, the gunman engaged in gun fire with civilian police, who eventually felled him.	~4 minutes; gunman subdued by civilian police	13, killed, 38 wounded	• FN Herstal 5.7 semi-automatic pistol	Legally purchased from a federally licensed firearms dealer.
Apr-3-09	American Civic Association, Binghamton, New York	• Name: Jiverly Wong • Age: 41 • Nationality: Naturalized U.S. citizen, born in Vietnam • Race: Asian	A naturalized U.S. citizen since 1995, Wong was unemployed and frustrated over his lack of English proficiency. On the day of the shooting, he mailed a package to a Syracuse TV station containing personal documentation and a letter describing his sense of being persecuted by police.	After barricading the rear door of the American Civic Association building with his vehicle, the gunman entered through the front door and immediately shot two receptionists. Proceeding to an English class where he had recently been a student, he shot everyone in the room multiple times. The gunman fired at least 98 shots in less than two minutes. When he heard sirens, he turned the gun on himself.	~2 minutes; gunman shot and killed himself	13 killed, 4 wounded	• Beretta 92FS Vertec Inox 9mm semi-automatic pistol • Beretta Px4 Storm .45 caliber semi-automatic pistol	Legally purchased and licensed.

Note: Casualty figures do not include the death of the shooter. In cases where the shooter was killed or took his own life, this fact is indicated under Duration, Termination.

SOURCE: Created by Erin Brown for Gale, © 2016.

attempting to kill people in a confined and populated area." Implicitly, this definition assumes that the perpetrator's actions involve the use of firearms. The FBI also makes an important distinction about the nature of these events: "Unlike a defined crime, such as a murder or mass killing, the active aspect inherently implies that both law enforcement personnel and citizens have the potential to affect the outcome of the event based upon their responses."

Figure 5.2 shows a steady increase in the frequency of active shooter incidents per year. According to the FBI, between 2000 and 2006 the average number of active shooter events was 6.4. Between 2007 and 2013 the average had more than doubled to 16.4 incidents per year. Moreover, Jack Date, Pierre Thomas, and Mike Levine report in "Active Shooter Incidents Continue to Rise, New FBI Data Show" (ABCNews.com, June 15, 2016) that FBI data indicated 20 such incidents per year in 2014 and 2015—more than for any two-year period in the past 16 years, and nearly six times the total number of incidents that occurred in 2000 and 2001.

Besides their increasing frequency, active shooter incidents claimed an increasing number of casualties between 2000 and 2013. (See Figure 5.3.) The FBI notes that of the total 1,043 casualties (486 killed and 557 wounded) that occurred in 160 mass shooter incidents between 2000 and 2013, 247 casualties occurred between 2000 to 2006, compared with 796 casualties (more than triple) between 2007 to 2013. The FBI notes that if the shooter died or was wounded as a result of the incident, that individual was not included in the casualty count.

The FBI also analyzes active shooter incidents according to the type of location where they took place. As Figure 5.4 illustrates, of the 160 active shooter incidents that occurred between 2000 to 2013, 73 (45.6%) occurred in areas of commerce (or businesses), including those open to pedestrian traffic (such as a grocery store), closed to pedestrian traffic (such as an office), and malls. Educational environments, including schools and institutions of higher learning (colleges and universities), were the second most likely place for an active shooter incident to occur, accounting for 39 (24.4%) of the incidents between 2000 to 2013. In all, seven out of 10 active shooter incidents took place in one of these two location categories.

POLICE DEATHS AND INJURIES

In *Law Enforcement Officers Killed and Assaulted, 2014* (2015, https://ucr.fbi.gov/leoka/2014), the FBI indicates that 51 law enforcement officers were killed in the line of duty in 2014, 46 of them with firearms. (See Table 5.10.) Handguns were used in 33 of the killings, rifles in 10, and shotguns in three. The South was the U.S. region in which law enforcement officers were the most likely to be killed with a firearm while on duty, accounting for 15 of the 46 total deaths in 2014. The West had the second-highest total, with 12 officers shot and killed in the line of duty.

The National Law Enforcement Officers Memorial Fund (NLEOMF), a private nonprofit organization that tracks police fatalities, notes in "Causes of Law Enforcement Deaths" (July 18, 2016, http://www.nleomf.org/facts/officer-fatalities-data/causes.html) that 521 federal, state, and local law enforcement officers had been killed

FIGURE 5.2

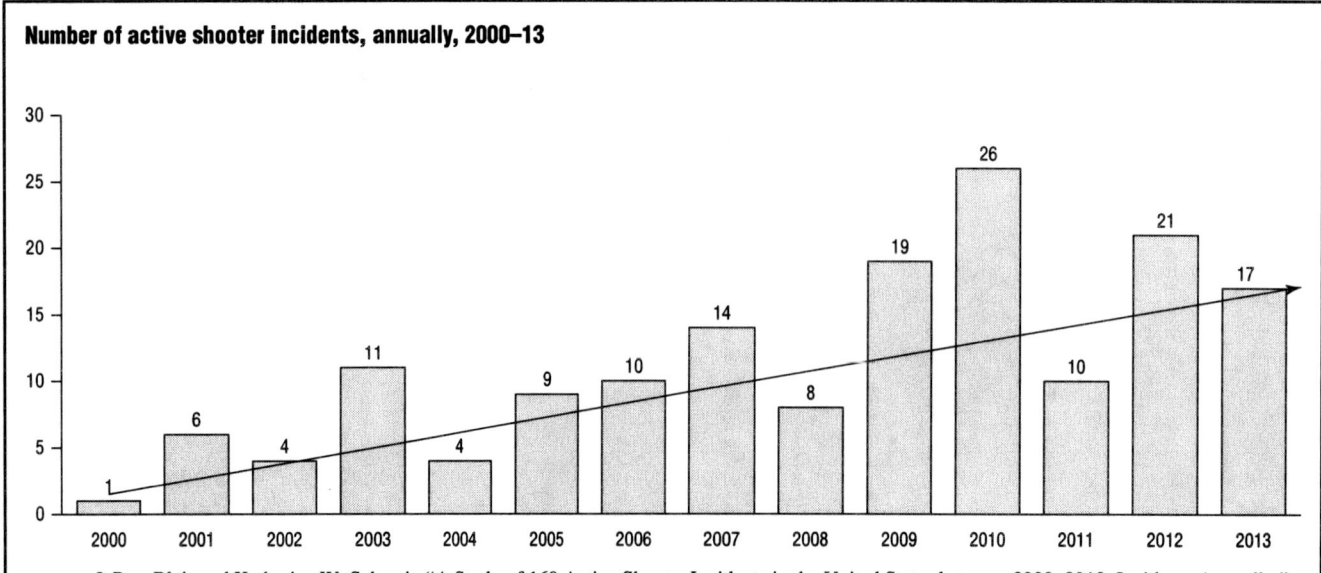

Number of active shooter incidents, annually, 2000–13

Year	Incidents
2000	1
2001	6
2002	4
2003	11
2004	4
2005	9
2006	10
2007	14
2008	8
2009	19
2010	26
2011	10
2012	21
2013	17

SOURCE: J. Pete Blair and Katherine W. Schweit, "A Study of 160 Active Shooter Incidents in the United States between 2000–2013: Incidents Annually," in *A Study of Active Shooter Incidents, 2000–2013*, Texas State University and U.S. Department of Justice, Federal Bureau of Investigation, 2014, https://www.fbi.gov/file-repository/active-shooter-study-2000-2013-1.pdf (accessed September 1, 2016)

FIGURE 5.3

Number of casualties (killed or wounded) by active shooters, annually, 2000–13

Year	Casualties
2000	7
2001	43
2002	29
2003	51
2004	20
2005	51
2006	46
2007	126
2008	63
2009	143
2010	86
2011	84
2012	208
2013	86

SOURCE: J. Pete Blair and Katherine W. Schweit, "A Study of 160 Active Shooter Incidents in the United States between 2000–2013: Annual Totals of 1,043 Casualties," in *A Study of Active Shooter Incidents, 2000–2013*, Texas State University and U.S. Department of Justice, Federal Bureau of Investigation, 2014, https://www.fbi.gov/file-repository/active-shooter-study-2000-2013-1.pdf (accessed September 1, 2016)

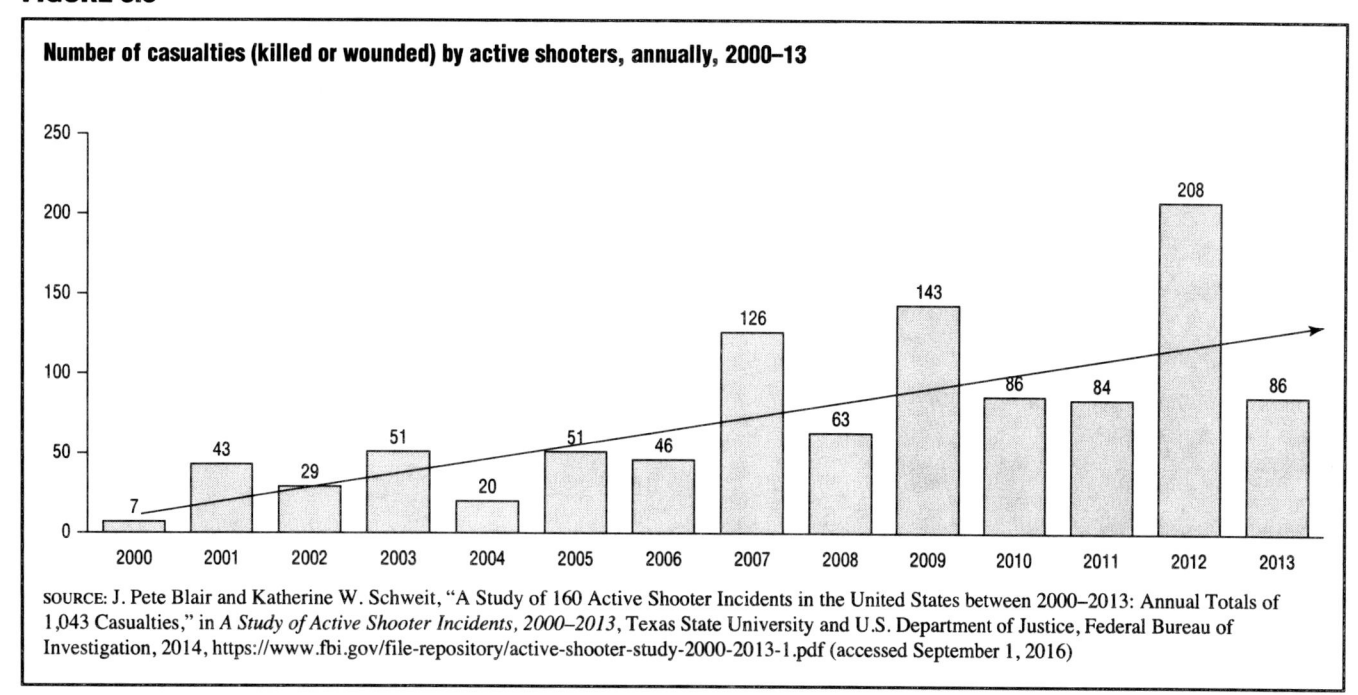

in firearm-related incidents between 2006 and 2015. Shootings were the most common cause of death for law enforcement officers, accounting for 36.2% of the total 1,439 law enforcement deaths over the course of that decade. Auto crashes, accounting for 408 deaths (28.4%), were the second-leading cause of death for law enforcement officers.

The NLEOMF indicates that the number of officer fatalities held stable between 2014 (122 officers died) and 2015 (123 officers died). Among these fatalities, the number of officers killed by firearms decreased 16%, from 49 to 41. However, during the first half of 2016 this decline saw a dramatic reversal. According to the NLEOMF, in "2016 Mid-year Law Enforcement Officer Fatalities Report" (July 26, 2016, http://www.nleomf.org/assets/pdfs/reports/2016-Mid-Year-Officer-Fatalities-Report.pdf), of the 67 law enforcement officers killed in the line of duty during the first half of the year, 32 were killed in firearm-related incidents—a 78% increase over the same period in 2015, when 18 officers were fatally shot. Of particular concern was the fact that of the 32 fatal officer shootings, 14 occurred in ambush-style attacks carried out on unsuspecting officers—a 300% increase over the three officers killed in ambushes during the first half of 2015.

Two ambush attacks on police made national headlines in July 2016. Both occurred in an atmosphere of acute racial tension following the high-profile police killings of two African American men: Alton Sterling (1979–2016) in Baton Rouge, Louisiana, on July 5, and Philando Castile (1983–2016) in a suburb of Minneapolis–St. Paul, Minnesota, on July 6. The first ambush occurred at the conclusion of a peaceful protest march in Dallas, Texas, on July 7. A lone gunman with a semi-automatic rifle began firing on police officers from an elevated position, killing five and injuring nine more. A second ambush occurred in Baton Rouge on July 17, when another lone gunman, also armed with a semiautomatic rifle, opened fire on police officers in a parking lot, killing three and wounding three others. Both of the attackers had served in the U.S. military and were skilled shooters. Prior to the attacks, both men had been outspoken about their intentions to seek retaliation for police killings of African American men.

The ambushes in Dallas and Baton Rouge contributed to the dramatic data in the NLEOMF 2016 midyear report and fueled the public perception that civilian attacks against police were on the rise, in spite of an overall trend to the contrary. In "Number of Law Enforcement Officers Fatally Shot this Year up Significantly after Ambush Attacks, Report Says" (Washington Post.com, July 27, 2016), Mark Berman notes, "The new numbers, released with the attacks in Texas and Louisiana still fresh in the public's mind, come at a time when overall line-of-duty deaths have fallen significantly. On average, fewer officers have been fatally shot in each of the past four decades, overall line-of-duty deaths have decreased over that span, and deadly assaults are on a steady decline."

Table 5.11 details the circumstances under which federal, state, and local police officers were killed in the line of duty between 2005 and 2014. Of the 505 officers killed over the course of the decade, 95 (18.8%) were killed in arrest situations and 93 (18.4%)

FIGURE 5.4

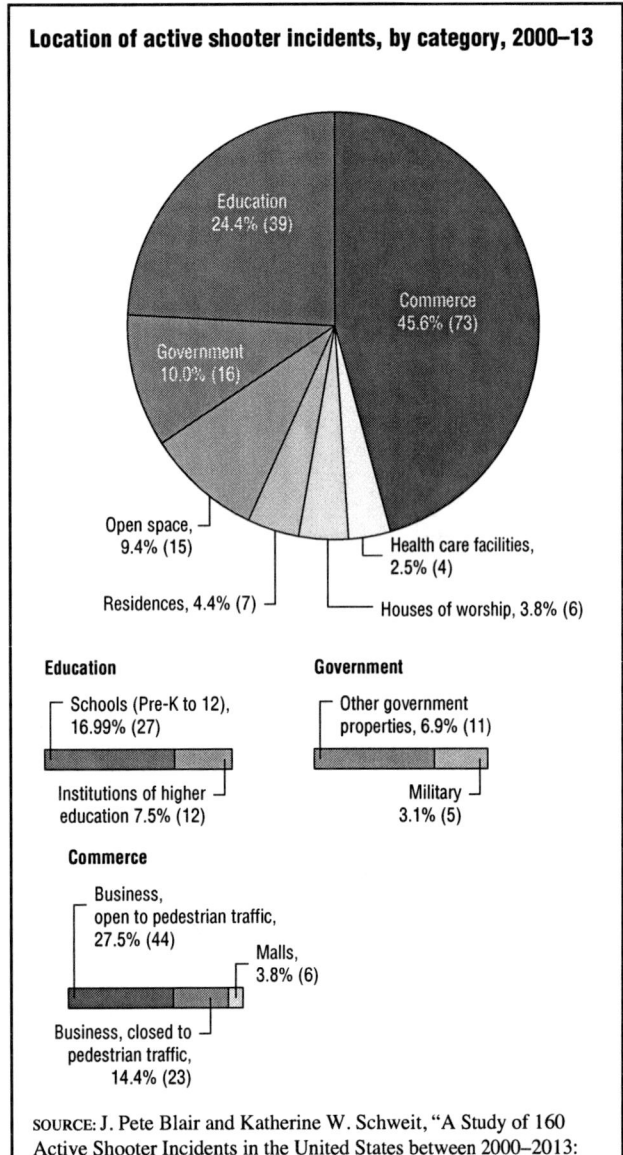

Location of active shooter incidents, by category, 2000–13

- Education 24.4% (39)
- Commerce 45.6% (73)
- Government 10.0% (16)
- Open space, 9.4% (15)
- Health care facilities, 2.5% (4)
- Residences, 4.4% (7)
- Houses of worship, 3.8% (6)

Education
- Schools (Pre-K to 12), 16.99% (27)
- Institutions of higher education 7.5% (12)

Government
- Other government properties, 6.9% (11)
- Military 3.1% (5)

Commerce
- Business, open to pedestrian traffic, 27.5% (44)
- Malls, 3.8% (6)
- Business, closed to pedestrian traffic, 14.4% (23)

SOURCE: J. Pete Blair and Katherine W. Schweit, "A Study of 160 Active Shooter Incidents in the United States between 2000–2013: Location Categories," in *A Study of Active Shooter Incidents, 2000–2013*, Texas State University and U.S. Department of Justice, Federal Bureau of Investigation, 2014, https://www.fbi.gov/file-repository/active-shooter-study-2000-2013-1.pdf (accessed September 1, 2016)

were killed during traffic stops. Another 68 (13.4%) officers were killed while investigating suspicious people or circumstances, 62 (12.3%) while responding to disturbance calls, and 57 (11.3%) in unprovoked attacks. Relatively few officers (36, or 7.1%) were killed in ambush situations.

Similar numbers of law enforcement officers are also killed accidentally in the line of duty. As Table 5.12 shows, 599 officers lost their lives on duty as a result of accidents between 2005 and 2014, a majority of them (348, or 58.1%) in automobile accidents. Accidental shootings accounted for relatively few officer deaths (29, or 4.8%) over the course of the decade.

Officers Assaulted

Table 5.13 shows the number of assaults on federal officers between 2010 and 2014. During this period there were 8,617 assaults on federal law enforcement officers. Firearms were not the most common type of weapon used in assaults; they were used in 447 incidents, which accounted for only 5.2% of the total. By comparison, personal weapons (e.g., fists) were used in 3,019 (35%) cases. Of the 77 firearm assaults on federal officers reported in 2014, none resulted in death, 10 resulted in injuries, and 50 in no injuries.

In 2014, 48,315 state and local law enforcement officers were assaulted. (See Table 5.14.) Again, firearms were the type of weapons used in only a small number of cases, accounting for 1,950 (4%) of the total. Personal weapons such as fists and feet accounted for the overwhelming majority of these assaults (38,611, or 79.9%).

NONFATAL VIOLENT CRIMES INVOLVING FIREARMS

The incidence of nonfatal firearm violence, like that of the firearm homicide rate and the homicide rate in general, peaked during the mid-1990s and then fell markedly thereafter. Among people aged 12 years and older,

TABLE 5.10

Law enforcement officers killed, by type of weapon and region, 2014

Region	Total	Total firearms	Handgun	Rifle	Shotgun	Type of firearm not reported	Knife or other cutting instrument	Bomb	Blunt instrument	Personal weapons	Vehicle	Other
Number of victim officers	51	46	33	10	3	0	0	0	0	1	4	0
Northeast	8	7	5	2	0	0	0	0	0	0	1	0
Midwest	8	8	7	1	0	0	0	0	0	0	0	0
South	17	15	11	2	2	0	0	0	0	1	1	0
West	14	12	6	5	1	0	0	0	0	0	2	0
Puerto Rico and other outlying areas	4	4	4	0	0	0	0	0	0	0	0	0

SOURCE: "Table 31. Law Enforcement Officers Feloniously Killed, Region by Type of Weapon, 2014," in *Law Enforcement Officers Killed and Assaulted, 2014*, Federal Bureau of Investigation, 2015, https://ucr.fbi.gov/leoka/2014/tables/table_31_leos_fk_region_by_type_of_weapon_2014.xls (accessed September 8, 2016)

TABLE 5.11

Law enforcement officers killed, by circumstance at scene of incident, 2005–14

Circumstance	Total	2005	2006	2007	2008	2009	2010	2011	2012	2013	2014
Number of victim officers											
Total	505	55	48	58	41	48	56	72	49	27	51
Disturbance call											
Total	62	7	8	5	1	6	6	10	4	4	11
Disturbance (bar fight, person with firearm, etc.)	39	2	6	3	1	4	2	5	3	3	10
Domestic disturbance (family quarrel, etc.)	23	5	2	2	0	2	4	5	1	1	1
Arrest situation											
Total	95	8	12	17	9	8	14	10	7	6	4
Burglary in progress/pursuing burglary suspect	9	1	0	1	2	1	3	0	1	0	0
Robbery in progress/pursuing robbery suspect	39	4	6	7	1	3	6	6	3	3	0
Drug-related matter	8	0	2	1	1	0	1	0	2	0	1
Attempting other arrest	39	3	4	8	5	4	4	4	1	3	3
Civil disorder (mass disobedience, riot, etc.)											
Total	0	0	0	0	0	0	0	0	0	0	0
Handling, transporting, custody of prisoner											
Total	11	1	1	1	1	2	1	1	3	0	0
Investigating suspicious person/circumstance											
Total	68	7	6	4	7	4	8	12	8	5	7
Ambush (entrapment/premeditation)											
Total	36	4	1	9	1	6	2	2	3	1	7
Unprovoked attack											
Total	57	4	9	7	5	9	11	6	1	4	1
Investigative activity (surveillance, search, interview, etc.)											
Total	25	4	0	1	2	0	2	4	6	1	5
Handling person with mental illness											
Total	9	2	1	0	0	0	0	0	3	0	3
Traffic pursuit/stop											
Total	93	15	8	11	8	8	9	14	9	2	9
Felony vehicle stop	33	5	0	5	5	2	3	7	5	0	1
Traffic violation stop	60	10	8	6	3	6	6	7	4	2	8
Tactical situation (barricaded offender, hostage taking, high-risk entry, etc.)											
Total	49	3	2	3	7	5	3	13	5	4	4

SOURCE: "Table 21. Law Enforcement Officers Feloniously Killed, Circumstance at Scene of Incident, 2005–2014," in *Law Enforcement Officers Killed and Assaulted, 2014*, Federal Bureau of Investigation, 2015, https://ucr.fbi.gov/leoka/2014/tables/table_21_leos_fk_circumstance_at_scene_of_incident_2005-2014.xls (accessed September 8, 2016)

---

nonfatal firearm violence declined from 725.3 incidents per 100,000 in 1993 to 174.8 incidents per 100,000 in 2014. (See Figure 5.5.)

The firearm violence rate (the incidence of firearm violence per each 1,000 people aged 12 years and older) declined slightly between 2005 and 2014: after peaking at 2.5 per 1,000 in 2006, it fell to a low of 1.3 in 2013 before climbing back up to 1.7 in 2014. (See Table 5.15.) In 2014, 81.9% of all serious violent crimes involving a firearm were reported to police. This was a significant increase over 2010, when only 50.9% of firearm crime incidents were reported to police.

Michael Planty and Jennifer L. Truman of the BJS indicate in *Firearm Violence, 1993–2011* (May 2013, http://www.bjs.gov/content/pub/pdf/fv9311.pdf), the most recent BJS report on this topic as of December 2016, that although the number of violent incidents involving firearms has fallen drastically since the mid-1990s, handguns have consistently accounted for the overwhelming majority of firearm violence during this period. Handguns accounted for between 84.2% (in 1997) and 92.6% (2010) of all nonfatal crimes involving firearms, and no clear upward or downward trend in this percentage is evident. Notably, the raw number of nonfatal firearm crimes for both handguns and other types of firearms has fallen dramatically since 1994, whereas among homicides committed with other types of firearms it has remained relatively constant.

Characteristics of Nonfatal Firearm Violence Victims

Planty and Truman explain that nonfatal firearm violence rates differ by sex, but the rates for males and females converged between 1994 and 2011 as violence declined dramatically among males. The rate of nonfatal

TABLE 5.12

Law enforcement officers accidentally killed, by circumstance at scene of incident, 2005–14

Circumstance	Total	2005	2006	2007	2008	2009	2010	2011	2012	2013	2014
Number of victim officers											
Total	599	67	66	83	68	48	72	53	48	49	45
Automobile accident											
Total	348	39	38	49	39	34	45	30	23	23	28
Motorcycle accident											
Total	54	4	8	6	6	3	7	4	6	4	6
Aircraft accident											
Total	18	2	3	3	2	1	2	1	3	1	0
Struck by vehicle											
Total	97	11	13	12	13	7	11	5	10	9	6
Traffic stop, roadblock, etc.	31	5	4	7	1	3	4	3	2	1	1
Directing traffic, assisting motorist, etc.	66	6	9	5	12	4	7	2	8	8	5
Accidental shooting											
Total	29	4	4	4	2	2	3	4	2	2	2
Crossfire, mistaken for subject, firearm mishap	22	2	3	4	2	2	1	3	2	2	1
Training session	3	1	0	0	0	0	1	0	0	0	1
Self-inflicted, cleaning mishap (not apparent or confirmed suicide)	4	1	1	0	0	0	1	1	0	0	0
Drowning											
Total	11	2	0	2	1	0	0	3	0	2	1
Fall											
Total	14	3	0	1	0	0	1	2	3	4	0
Other accidental											
Total	28	2	0	6	5	1	3	4	1	4	2

SOURCE: "Table 67. Law Enforcement Officers Accidentally Killed, Circumstance at Scene of Incident, 2005–2014," in *Law Enforcement Officers Killed and Assaulted, 2014*, Federal Bureau of Investigation, 2015, https://ucr.fbi.gov/leoka/2014/tables/table_67_leos_ak_circumstance_at_scene_of_incident_2005-2014.xls (accessed September 8, 2016)

---

firearm violence for males declined from 10.1 per 1,000 to 1.9 per 1,000 during this time, for a decline of 81%. Over the same period the rate of nonfatal firearm violence for females declined from 4.7 per 1,000 to 1.6 per 1,000, for a decline of 67%.

Rates of nonfatal firearm violence differed dramatically by race and Hispanic origin in 1994 but the rates for different demographic groups converged over the course of the following 17 years. (See Figure 5.6.) The rates of nonfatal firearm violence for non-Hispanic African Americans declined from more than 16 per 1,000 in 1994 to 2.8 per 1,000 in 2011, while the rates for Hispanics declined from almost 13 per 1,000 to 2.2 per 1,000. Meanwhile, the rates for non-Hispanic whites fell from just over 5 per 1,000 to 1.4 per 1,000. Thus, although the rates for non-Hispanic African Americans and Hispanics were consistently higher than those for non-Hispanic whites, all three groups saw their nonfatal firearm violence rates fall at comparable levels: 83% for both non-Hispanic African Americans and Hispanics and 74% for non-Hispanic whites.

Planty and Truman indicate that although the rates of nonfatal firearm violence victimization declined for all age groups between 1993 and 2011, the likelihood of being a victim of nonfatal firearm violence remained greatest between the ages of 18 and 24 years. In 2011 young adults aged 18 to 24 years were the victims of nonfatal firearm violence at a rate of 5.2 per 1,000, compared with much lower rates for other age brackets: 1.4 per 1,000 for children aged 12 to 17 years, 2.2 for those aged 25 to 34 years, 1.4 for those aged 35 to 49 years, and 0.7 for those aged 50 years and older.

Of the 2.2 million victims of nonfatal firearm violence between 2007 and 2011, 77% escaped the incidents in which they were involved without physical injury. Twenty-three percent received some sort of injury, including 6.7% whose injuries were classified as serious, 16.1% whose injuries were classified as minor, and 0.2% who suffered rape without being injured by the firearms themselves.

Robbery and Aggravated Assault

Historical data indicate that firearms are consistently used in the overwhelming majority of homicides. According to the FBI, in *Crime in the United States, 2015*, firearms were used in 71.5% of all U.S. homicides in 2015. By comparison, firearms are not as prevalent in other types of violent victimization, such as robbery and aggravated assault.

TABLE 5.13

Assaults on federal officers, by extent of injury and type of weapon, 2010–14

Year	Extent of injury	Total	Firearm	Knife or other cutting instrument	Bomb	Blunt instrument	Personal weapons	Vehicle	Other
Number of victim officers	Total	8,617	447	122	32	420	3,019	505	4,072
2010	Total	1,886	86	17	0	16	387	76	1,304
	Killed	1	1	0	0	0	0	0	0
	Injured	351	12	0	0	3	117	18	201
	Not injured	1,534	73	17	0	13	270	58	1,103
2011	Total	1,689	92	18	15	19	663	102	780
	Killed[a]	3	3	0	0	0	0	0	0
	Injured	254	7	4	0	6	147	20	70
	Not injured	1,432	82	14	15	13	516	82	710
2012	Total	1,858	104	33	16	38	733	104	830
	Killed	1	1	0	0	0	0	0	0
	Injured	206	7	3	0	5	125	7	59
	Not injured	1,096	71	30	1	22	414	47	511
	Extent of injury not reported[b]	555	25	0	15	11	194	50	260
2013[c]	Total	1,774	88	27	1	198	626	121	713
	Killed[d]	4	2	1	0	0	0	0	1
	Injured	292	10	4	0	10	198	17	53
	Not injured	1,478	76	22	1	188	428	104	659
2014	Total	1,410	77	27	0	149	610	102	445
	Killed	0	0	0	0	0	0	0	0
	Injured	170	10	3	0	4	110	12	31
	Not injured	867	50	24	0	15	324	52	402
	Extent of injury not reported[e]	373	17	0	0	130	176	38	12

[a]For 2011, detailed incident data were not reported by the U.S. Immigration and Customs Enforcement for a killed victim officer; therefore, 1 of these deaths was not represented in the feloniously killed section of this publication.
[b]For 2012, extent of injury data were not reported by the U.S. Customs and Border Protection, Office of Border Patrol.
[c]For 2013, data were not reported by the U.S. Immigration and Customs Enforcement.
[d]For 2013, detailed incident data were not reported by the U.S. Customs and Border Protection, Office of Border Patrol, for 2 killed victim officers, by the Bureau of Indian Affairs for 1 killed victim officer, and by the U.S. Drug Enforcement Administration for 1 killed victim officer; therefore, these 4 deaths were not represented in the feloniously killed section of this publication.
[e]For 2014, extent of injury data were not reported by the U.S. Customs and Border Protection, Office of Border Patrol.

SOURCE: "Table 132. Federal Law Enforcement Officers Killed and Assaulted, Extent of Injury of Victim Officer by Type of Weapon, 2010–2014," in *Law Enforcement Officers Killed and Assaulted, 2014*, Federal Bureau of Investigation, 2015, https://ucr.fbi.gov/leoka/2014/tables/table_132_fed_leos_k_and_a_extent_of_injury_of_victim_officer_by_type_of_weapon_2010-2014.xls (accessed September 8, 2016)

TABLE 5.14

Law enforcement officers assaulted, by circumstance at scene of incident and type of weapon, 2014

Circumstance	Total	Percent distribution	Firearm Total	Firearm Percent distribution	Knife or other cutting instrument Total	Knife or other cutting instrument Percent distribution	Other dangerous weapon Total	Other dangerous weapon Percent distribution	Personal weapons Total	Personal weapons Percent distribution
Number of victim officers	48,315	100.0	1,950	4.0	951	2.0	6,803	14.1	38,611	79.9
Disturbance call	14,901	100.0	574	3.9	404	2.7	1,444	9.7	12,479	83.7
Burglary in progress/pursuing burglary suspect	697	100.0	40	5.7	22	3.2	117	16.8	518	74.3
Robbery in progress/pursuing robbery suspect	419	100.0	78	18.6	30	7.2	73	17.4	238	56.8
Attempting other arrest	7,343	100.0	228	3.1	106	1.4	773	10.5	6,236	84.9
Civil disorder (mass disobedience, riot, etc.)	713	100.0	25	3.5	9	1.3	190	26.6	489	68.6
Handling, transporting, custody of prisoner	6,303	100.0	20	0.3	31	0.5	673	10.7	5,579	88.5
Investigating suspicious person/circumstance	4,486	100.0	270	6.0	92	2.1	627	14.0	3,497	78.0
Ambush situation	167	100.0	44	26.3	4	2.4	21	12.6	98	58.7
Handling person with mental illness	1,499	100.0	92	6.1	85	5.7	172	11.5	1,150	76.7
Traffic pursuit/stop	4,022	100.0	270	6.7	34	0.8	1,333	33.1	2,385	59.3
All other	7,765	100.0	309	4.0	134	1.7	1,380	17.8	5,942	76.5

Note: Because of rounding, the percentages may not add to 100.0.

SOURCE: "Table 79. Law Enforcement Officers Assaulted, Circumstance at Scene of Incident by Type of Weapon and Percent Distribution, 2014," in *Law Enforcement Officers Killed and Assaulted, 2014*, Federal Bureau of Investigation, 2015, https://ucr.fbi.gov/leoka/2014/tables/table_79_leos_asltd_circumstance_at_scene_of_incident_by_type_of_weapon_and_percent_distribution_2014.xls (accessed September 8, 2016)

Table 5.16 shows the prevalence of different kinds of weapons used among the robberies and aggravated assaults reported to law enforcement in 2014 and 2015 (in cases for which weapon information was known). In 2015, 123,358 (40.9%) robberies of the total (301,567) reported to police involved firearms. Strong-arm incidents, which involved physical force but no other external weapon, accounted for a slightly higher proportion of robberies (127,053). Among aggravated assaults reported to law enforcement agencies in 2015, those involving firearms accounted for 170,941, or 24.4% of the total (701,593). This was higher than the number of aggravated assaults committed with knives or other cutting instruments (127,189) and lower than the number committed with hands, fists, feet, or other body parts (183,451) and with weapons other than firearms, knives, or body parts (220,012).

There were regional differences, as well, in the percentage distribution of firearm robberies and aggravated assaults among all robberies and aggravated assaults in 2015. Robberies involving firearms accounted for 51.4% of total robberies in the South, 46.5% in the Midwest, 29.6% in the West, and 29.2% in the Northeast. (See Table 5.17.) Similarly, firearm aggravated assaults represented a higher proportion of total aggravated assaults in the South (29.1%) and Midwest (27%) than in the West (19.5%) or Northeast (14.4%). (See Table 5.18.)

Justifiable Homicide

The FBI also compiles data on the category of justifiable homicide, which it defines in *Crime in the United States, 2015* as "the killing of a felon by a peace officer in the line of duty" or as "the killing of a felon, during the commission of a felony, by a private citizen."

Table 5.19 shows the data for justifiable homicides committed by law enforcement officers between 2011 and 2015. Almost all such incidents involved firearms. In both 2011 and 2012 a firearm was used in all but three (401 of 404 and 423 of 426, respectively) cases in which a felon was killed by a law enforcement officer in the line of duty. Firearms were used in 467 (all but four) cases in 2013, 451 (all but two) cases in 2014, and 441 (all but one) cases in 2015. Handguns were the most frequently used type of firearm in cases of justifiable homicide by law enforcement officers. In 2015 out of 441 cases in which a firearm was used, 302 (68.5%) involved a handgun.

Table 5.20 shows the data for justifiable homicides committed by private citizens between 2011 and 2015. Firearms were used in the overwhelming majority of cases, but their use in justifiable homicides by citizens was not as universal as in justifiable homicides by law enforcement officers. Firearms were used in 209 of 270 (77.4%) cases of justifiable homicide by private citizens in 2011 and in 268 of 328 (81.7%) cases in 2015. As with cases involving law enforcement officers, handguns were the most frequently used type of firearm in cases of justifiable homicide by private citizens. Out of 268 cases

FIGURE 5.5

Nonfatal firearm violence, 1993–2014

[Nonfatal violent firearm crime victimizations per 100,000 people ages 12 and older]

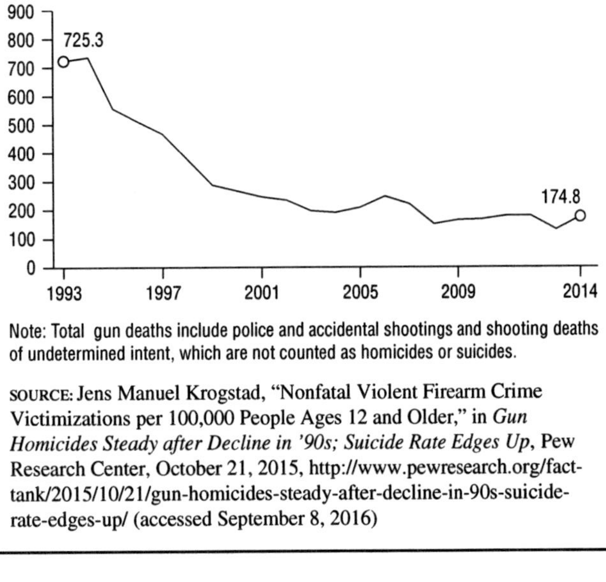

Note: Total gun deaths include police and accidental shootings and shooting deaths of undetermined intent, which are not counted as homicides or suicides.

SOURCE: Jens Manuel Krogstad, "Nonfatal Violent Firearm Crime Victimizations per 100,000 People Ages 12 and Older," in *Gun Homicides Steady after Decline in '90s; Suicide Rate Edges Up*, Pew Research Center, October 21, 2015, http://www.pewresearch.org/fact-tank/2015/10/21/gun-homicides-steady-after-decline-in-90s-suicide-rate-edges-up/ (accessed September 8, 2016)

TABLE 5.15

Firearm violence, 2005–14

	2005	2006	2007	2008	2009	2010	2011	2012	2013	2014[a]
Firearm incidents	446,370	552,040	448,410	331,620	383,390	378,800	415,160	427,700	290,620	414,700
Firearm victimizations	503,530	614,410	554,780	371,290	410,110	415,000	467,930	460,720	332,950	466,110
Rate of firearm victimizations[b]	2.1	2.5	2.2	1.5	1.6	1.6	1.8	1.8	1.3	1.7
Percent of firearm victimizations reported to police	72.3%	71.0%	52.9%	70.8%	61.5%	50.9%	73.5%	66.4%	75.3%	81.9%

[a]Comparison year.
[b]Per 1,000 persons age 12 or older.
Note: Includes violent incidents and victimizations in which the offender had, showed, or used a firearm.

SOURCE: Jennifer L. Truman and Lynn Langton, "Table 2. Firearm Victimizations, 2005–2014," in *Criminal Victimization, 2014*, U.S. Department of Justice, Bureau of Justice Statistics, August 2015, http://www.bjs.gov/content/pub/pdf/cv14.pdf (accessed September 8, 2016)

FIGURE 5.6

Nonfatal firearm violence, by race and Hispanic origin, 1994–2011

*Excludes persons of Hispanic or Latino origin.
Note: Data based on 2-year rolling averages beginning in 1993.

SOURCE: Michael Planty and Jennifer L. Truman, "Figure 7. Nonfatal Firearm Violence, by Race and Hispanic Origin, 1994–2011," in *Firearm Violence, 1993–2011*, U.S. Department of Justice, Bureau of Justice Statistics, May 2013, http://www.bjs.gov/content/pub/pdf/fv9311.pdf (accessed October 14, 2016)

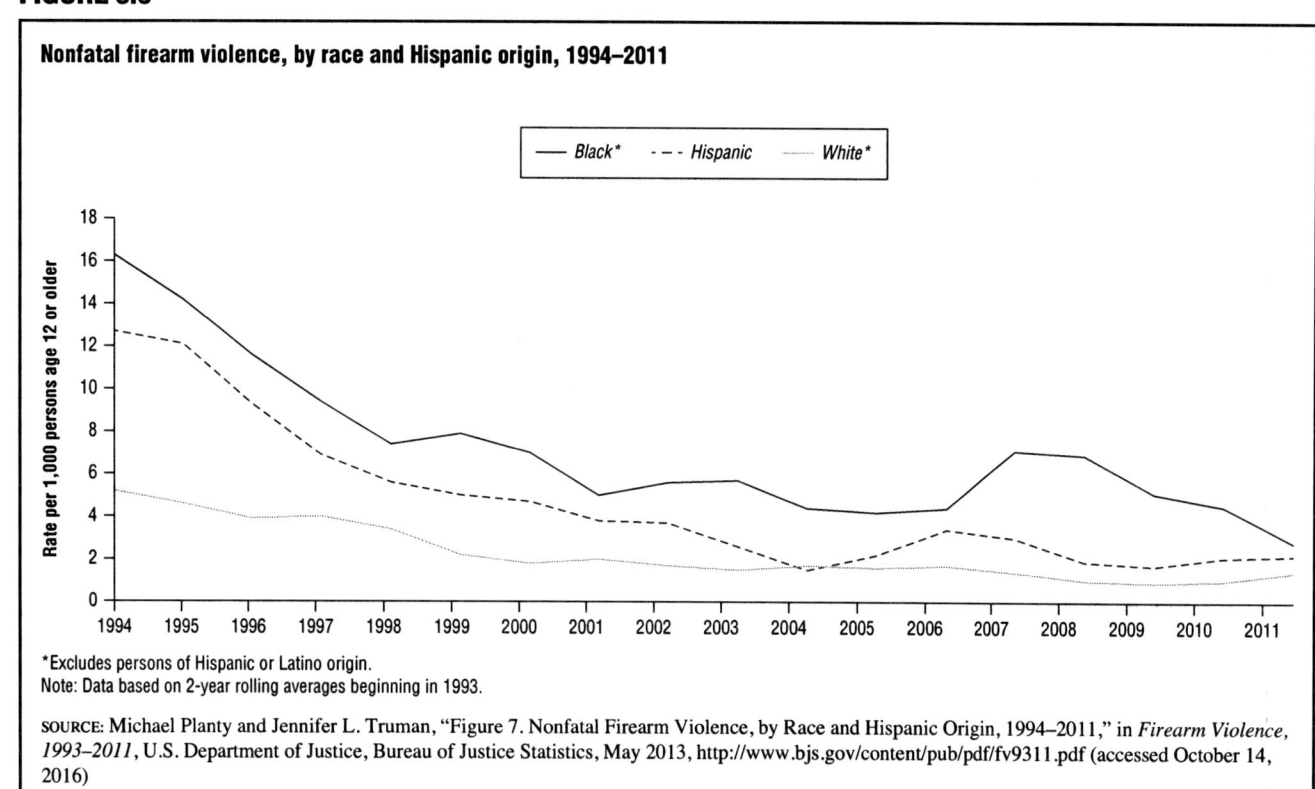

TABLE 5.16

Robberies and aggravated assaults, by type of weapon used, 2014–15

	Robbery				Aggravated assault			
Population group	Firearm	Knife or cutting instrument	Other weapon	Strong-arm	Firearm	Knife or cutting instrument	Other weapon	Hands, fists, feet, etc.
Total all agencies:								
2014	119,583	23,486	26,181	126,903	151,957	126,457	215,342	179,850
2015	123,358	23,759	27,397	127,053	170,941	127,189	220,012	183,451
Percent change	+3.2	+1.2	+4.6	+0.1	+12.5	+0.6	+2.2	+2.0

SOURCE: Adapted from "Table 15. Crime Trends, Additional Information about Selected Offenses by Population Group, 2014–2015," in *Crime in the United States, 2015*, Federal Bureau of Investigation, September 2016, https://ucr.fbi.gov/crime-in-the-u.s/2015/crime-in-the-u.s.-2015/tables/table-15 (accessed October 10, 2016)

in which a firearm was used in 2015, 215 (80.2%) involved a handgun.

WHERE DO CRIMINALS GET FIREARMS?

Many criminals obtain their guns legally. For example, James Eagan Holmes (1987–), accused of killing 12 people and wounding 58 others in the Aurora, Colorado, movie theater shooting in July 2012, was armed with an AR-15 assault rifle, a Remington 12-gauge shotgun, and two Glock handguns at the time of the incident. His four weapons had been purchased at shops in the Denver area, and he bought a supply of ammunition online. In Colorado no permit or waiting period is required for gun purchases. However, firearms dealers are required by law to maintain records on all sales, including personal information about the purchaser, the serial number and identifying information about the gun, and the date and terms of the sale. Speaking at a press conference covered by ABC News on July 20, 2012 (http://abcnews.go.com/US/colorado-movie-theater-shooting-suspect-bought-guns-6000/story?id=16817842), the Aurora chief of police Dan Oates said of Holmes, "All the ammunition he possessed, he possessed legally, all the weapons he possessed, he possessed legally, all the clips he possessed, he possessed legally."

Guns can also be obtained illegally in a variety of ways, primarily by stealing them or by illegally buying or

TABLE 5.17

Robbery, by type of weapon used, percentage distribution by region, 2015

[Percent distribution by region, 2015]

Region	Total all weapons*	Armed Firearms	Armed Knives or cutting instruments	Armed Other weapons	Strong-arm
Total	100	40.8	7.9	9.1	42.2
Northeast	100	29.2	10.3	8.8	51.7
Midwest	100	46.5	5.6	8.7	39.2
South	100	51.4	6.4	8.2	33.9
West	100	29.6	9.8	10.7	49.9

*Because of rounding, the percentages may not add to 100.0.

SOURCE: "Robbery Table 3. Robbery, Types of Weapons Used, Percent Distribution by Region, 2015," in *Crime in the United States, 2015*, Federal Bureau of Investigation, September 2016, https://ucr.fbi.gov/crime-in-the-u.s/2015/crime-in-the-u.s.-2015/tables/robbery_table_3_robbery_types_of_weapons_used_percent_distribution_by_region_2015.xls (accessed October 10, 2016)

TABLE 5.18

Aggravated assault, by type of weapon used, percentage distribution by region, 2015

Region	Total all weapons*	Firearms	Knives or cutting instruments	Other weapons (clubs, blunt objects, etc.)	Personal weapons (hands, fists, feet, etc.)
Total	100	24.2	18.1	31.4	26.3
Northeast	100	14.4	22.0	31.2	32.4
Midwest	100	27.0	15.8	27.6	29.6
South	100	29.1	17.8	32.0	21.1
West	100	19.5	17.7	32.9	29.9

*Because of rounding, the percentages may not add to 100.0.

SOURCE: "Aggravated Assault Table. Aggravated Assault, Types of Weapons Used, Percent Distribution by Region, 2015," in *Crime in the United States, 2015*, Federal Bureau of Investigation, September 2016, https://ucr.fbi.gov/crime-in-the-u.s/2015/crime-in-the-u.s.-2015/tables/aggravated_assault_table_aggravated_assault_types_of_weapons_used_percent_distribution_by_region_2015.xls (accessed October 10, 2016)

TABLE 5.19

Justifiable homicides committed by law enforcement officers, by weapon used, 2011–15

Year	Total*	Total firearms	Handguns	Rifles	Shotguns	Firearms, type not stated	Knives or cutting instruments	Other dangerous weapons	Personal weapons
2011	404	401	305	36	11	49	2	0	1
2012	426	423	339	38	7	39	0	3	0
2013	471	467	334	46	9	78	0	3	1
2014	453	451	331	45	4	71	1	1	0
2015	442	441	302	42	7	90	0	1	0

*The killing of a felon by a law enforcement officer in the line of duty.

SOURCE: "Expanded Homicide Data Table 14. Justifiable Homicide, by Weapon, Law Enforcement, 2011–2015," in *Crime in the United States, 2015*, Federal Bureau of Investigation, September 2016, https://ucr.fbi.gov/crime-in-the-u.s/2015/crime-in-the-u.s.-2015/tables/expanded_homicide_data_table_14_justifiable_homicide_by_weapon_law_enforcement_2011-2015.xls (accessed October 10, 2016)

trading them on the black market (a market where products are bought and sold illegally). However, there are few up-to-date studies of the relationship between criminality and gun ownership. This is partly a result of the Tiahrt Amendment (discussed in greater detail in Chapter 2), which began preventing the dissemination of certain statistics on the sales of multiple handguns and on firearms tracing statistics (gun trace statistics) in 2003. Additionally, federal funding for studies that might be used to support increased gun control has been staunchly opposed by the National Rifle Association of America and other gun rights advocates. In *Access Denied: How the Gun Lobby Is Depriving Police, Policy Makers, and the Public of the Data We Need to Prevent Gun Violence*

TABLE 5.20

Justifiable homicides committed by private citizens, by weapon used, 2011–15

Year	Total*	Total firearms	Handguns	Rifles	Shotguns	Firearms, type not stated	Knives or cutting instruments	Other dangerous weapons	Personal weapons
2011	270	209	156	13	11	29	49	9	3
2012	315	263	198	20	15	30	35	6	11
2013	286	227	174	6	11	36	35	13	11
2014	277	231	181	10	11	29	35	2	9
2015	328	268	215	8	13	32	37	9	14

*The killing of a felon, during the commission of a felony, by a private citizen.

SOURCE: "Expanded Homicide Data Table 15. Justifiable Homicide, by Weapon, Private Citizen, 2011–2015," in *Crime in the United States, 2015*, Federal Bureau of Investigation, September 2016, https://ucr.fbi.gov/crime-in-the-u.s/2015/crime-in-the-u.s.-2015/tables/expanded_homicide_data_table_15_justifiable_homicide_by_weapon_private_citizen_2011-2015.xls (accessed October 10, 2016)

(January 2013, http://3gbwir1ummda16xrhf4do9d21bsx.wpengine.netdna-cdn.com/wp-content/uploads/2014/04/AccessDenied.pdf), Mayors against Illegal Guns (MAIG), a gun control advocacy group, explains that pressure from the gun rights lobby has led the National Institute of Justice (NIJ), a primary research branch of the Department of Justice, to defund much of the firearms research that was once a top priority. The MAIG notes that although the NIJ funded 32 studies related to firearms between 1993 and 1999, "it has not funded a single public study on firearms during the Obama Administration."

Thus, the most reliable studies on the topic date from the 1990s and the first decade of the 21st century. The most comprehensive study conducted by the government on where criminals get their guns was released in 2001 and was based on interviews with 18,000 state prison inmates. Caroline Wolf Harlow of the BJS indicates in *Firearm Use by Offenders* (November 2001, http://bjs.ojp.usdoj.gov/content/pub/pdf/fuo.pdf) that 13.9% of those who carried a firearm during the offense for which they were serving time in 1997 bought their gun from a retail store, pawn shop, flea market, or gun show. This figure was down from 20.8% in 1991, when the previous survey was conducted. Another 39.6% acquired their firearms from family or friends, up from 33.8% in 1991. The remaining 39.2% acquired their firearms "on the street" from an illegal source, down from 40.8% in 1991.

The MAIG attempted to fill the research gap by hiring an investigative services firm to determine where criminals get their guns. The results of this investigation were published in *Inside Straw Purchasing: How Criminals Get Guns Illegally* (April 2008). The coalition presents findings on straw purchasing, a situation in which a person who would be denied a gun purchase, such as a convicted felon or an underaged buyer, has another person (the straw purchaser) fill out the paperwork and obtain the gun for him or her. Sometimes a straw purchaser is used when a person simply does not want a gun purchase listed in his or her name.

The MAIG explains that straw purchasing occurred frequently in so-called easy stores. Straw purchasers often paid for their purchases with both money and drugs and bought several guns at one time. Gun dealers appear to be the key in this illegal activity: some encourage straw purchases and actually coach straw purchasers, whereas others discourage the practice by asking many probing questions until the straw purchaser leaves the store. These dealers also train their employees how to spot straw purchasers and thwart them.

Many criminals get their guns from the black market. Sometimes, these guns are legally purchased in states with less restrictive gun laws and are then transported to states with strong gun laws, a phenomenon known as gunrunning or gun trafficking (buying, moving, and selling guns illegally). In *Trace the Guns: The Link between Gun Laws and Interstate Gun Trafficking* (September 2010, http://everytownresearch.org/documents/2015/04/trace-the-guns.pdf), the MAIG analyzes data related to guns recovered from crime scenes in 2009 and gun laws in the states in which the guns were purchased, finding that states with weak gun laws were more likely to serve as a source state in supplying guns to criminals. The coalition reports that the Bureau of Alcohol, Tobacco, Firearms, and Explosives (ATF) was able to identify the source state of 145,321 (61%) out of 238,107 guns that were recovered at U.S. crime scenes in 2009. Of this number, 70% (102,067) of the guns were used in crimes in the same state in which they were purchased, and 30% (43,254) of the traced guns were recovered in a different state. Nearly half (20,996 firearms, or 48.5%) of the traced guns that were used in crimes in other states in 2009 originated in just 10 states: Georgia (2,781), Florida (2,640), Virginia (2,557), Texas (2,240), Indiana (2,011), Ohio (1,806), Pennsylvania (1,777), North Carolina (1,775), California (1,772), and Arizona (1,637). In 2009 the states that had the highest rates of guns recovered at crime scenes in other states (i.e., the number of guns that were used in crimes in other states per 100,000 inhabitants) were: Mississippi (50.3), West Virginia

(46.8), Kentucky (34.9), Alaska (33.4), Alabama (33.2), South Carolina (33), Virginia (32.4), Indiana (31.3), Nevada (30.6), and Georgia (28.3). All these states had rates that were more than twice the national average of 14.1 per 100,000 inhabitants.

In *Trace the Guns*, the MAIG also analyzes "time-to-crime" (TTC) data (the time between the initial sale of a gun and the time it is recovered at a crime scene) to determine the likelihood that the gun was acquired illegally before being used in a crime. Firearms that have a TTC of less than two years are considered most likely to have been trafficked illegally. The national average TTC in 2009 was 10.8 years. The MAIG reports that the 10 states with the highest gun export rates also had a higher proportion of guns recovered in other states in less than two years from the initial retail sale.

In addition, the MAIG finds that states that had enacted strong gun laws were less likely to serve as a source for interstate trafficking of firearms in 2009. When used together, state regulations that require the prosecution of straw purchasers or those who falsify purchase information, stipulate background checks for handgun sales at gun shows, require purchase permits, authorize local officials to approve or deny concealed carry permits, deny guns to people convicted of violent misdemeanors, require owners to report missing guns to law enforcement, support local firearms regulations, and require state oversight of gun dealers seemed to offer some effectiveness in limiting illegal trafficking. The MAIG finds that the states that had not enacted these laws were twice as likely to serve as a source state for guns used in crimes in other states and for guns with a short TTC.

WEAPONS OFFENSES

Weapons offenses are violations of statutes or regulations that control deadly weapons, which include firearms and their ammunition, silencers, explosives, and certain knives. All 50 states, many cities and towns, and the federal government have laws concerning deadly weapons, including restrictions on their possession, carrying, use, sales, manufacturing, importing, and exporting.

Table 5.21 shows arrest trends for crimes committed in 2006 and 2015, based on reports from 9,581 law enforcement agencies. Total arrests across all age groups decreased 22.3% during this 10-year period, and arrests for weapons offenses across all age groups decreased 24.2%. Arrests of people under the age of 18 years fell 54.8% during this period, and arrests of those under the age of 18 years for weapons offenses fell 58.1%. By comparison, arrests of those aged 18 years and older fell 16.7%, and arrests of those aged 18 years and older on weapons offenses fell 13.4%.

Table 5.22 shows arrests by race and age for crimes committed in 2015. (The data in Table 5.22, which were drawn from a larger number of law enforcement agencies [12,692], differ slightly from the data in Table 5.21.) White offenders accounted for 69.7% of total arrests in 2015, and African American offenders accounted for 26.6%. Whites accounted for 57.7% of weapons offenses and African Americans accounted for 40%. In both categories of arrests, as well as in many specific categories, African Americans were overrepresented. According to the U.S. Census Bureau (July 2015, https://www.census.gov/quickfacts/table/PST045215/00), African Americans accounted for approximately 42.7 million (13.3%) of the United States' total population of 321.4 million in 2015. Brad Heath reports in "Racial Gap in U.S. Arrest Rates: 'Staggering Disparity'" (USAToday.com, November 19, 2014) that FBI data reveal that in almost every U.S. city and for almost every type of crime, African Americans are more likely to be arrested (often by a factor of 10) than people of other races. Heath explains, "Those disparities are easier to measure than they are to explain. They could be a reflection of biased policing; they could just as easily be a byproduct of the vast economic and educational gaps that persist across much of the USA—factors closely tied to crime rates." In the second decade of the 21st century, there is a growing national awareness of the need to understand and address the causes of these striking racial disparities in U.S. arrest rates.

TABLE 5.21

Arrest trends, by offense and age, 2006–15

[9,581 agencies; 2015 estimated population 199,921,204; 2006 estimated population 186,371,331]

	Total all ages			Under 18 years of age			18 years of age and over		
Offense charged	2006	2015	Percent change	2006	2015	Percent change	2006	2015	Percent change
Total[a]	8,676,456	6,739,363	−22.3	1,280,195	578,538	−54.8	7,396,261	6,160,825	−16.7
Murder and nonnegligent manslaughter	7,104	6,201	−12.7	642	421	−34.4	6,462	5,780	−10.6
Rape[b]	14,120	13,945	—	2,111	2,239	—	12,009	11,706	—
Robbery	68,437	54,003	−21.1	18,201	9,753	−46.4	50,236	44,250	−11.9
Aggravated assault	271,740	231,828	−14.7	35,984	17,717	−50.8	235,756	214,111	−9.2
Burglary	188,122	136,465	−27.5	51,953	22,056	−57.5	136,169	114,409	−16.0
Larceny-theft	679,290	753,665	+10.9	180,623	101,898	−43.6	498,667	651,767	+30.7
Motor vehicle theft	72,650	46,463	−36.0	17,651	7,547	−57.2	54,999	38,916	−29.2
Arson	10,445	5,737	−45.1	5,223	1,811	−65.3	5,222	3,926	−24.8
Violent crime[c]	361,401	305,977	−15.3	56,938	30,130	−47.1	304,463	275,847	−9.4
Property crime[c]	950,507	942,330	−0.9	255,450	133,312	−47.8	695,057	809,018	+16.4
Other assaults	792,178	678,537	−14.3	152,396	83,689	−45.1	639,782	594,848	−7.0
Forgery and counterfeiting	68,319	34,911	−48.9	2,288	632	−72.4	66,031	34,279	−48.1
Fraud	181,863	86,484	−52.4	5,090	2,776	−45.5	176,773	83,708	−52.6
Embezzlement	13,008	9,923	−23.7	949	391	−58.8	12,059	9,532	−21.0
Stolen property; buying, receiving, possessing	79,077	57,918	−26.8	14,032	6,600	−53.0	65,045	51,318	−21.1
Vandalism	189,867	124,058	−34.7	76,202	27,793	−63.5	113,665	96,265	−15.3
Weapons; carrying, possessing, etc.	114,104	86,443	−24.2	27,707	11,614	−58.1	86,397	74,829	−13.4
Prostitution and commercialized vice	33,402	17,663	−47.1	733	269	−63.3	32,669	17,394	−46.8
Sex offenses (except rape and prostitution)	50,656	32,356	−36.1	9,948	5,643	−43.3	40,708	26,713	−34.4
Drug abuse violations	1,078,156	928,122	−13.9	113,132	63,035	−44.3	965,024	865,087	−10.4
Gambling	2,807	1,657	−41.0	292	133	−54.5	2,515	1,524	−39.4
Offenses against the family and children	79,568	58,377	−26.6	3,348	2,208	−34.1	76,220	56,169	−26.3
Driving under the influence	925,818	675,960	−27.0	12,947	4,294	−66.8	912,871	671,666	−26.4
Liquor laws	408,511	174,230	−57.3	94,729	29,530	−68.8	313,782	144,700	−53.9
Drunkenness	368,271	251,424	−31.7	11,093	3,528	−68.2	357,178	247,896	−30.6
Disorderly conduct	422,187	240,723	−43.0	129,948	45,659	−64.9	292,239	195,064	−33.3
Vagrancy	18,106	15,017	−17.1	3,475	598	−82.8	14,631	14,419	−1.4
All other offenses (except traffic)	2,477,981	1,997,799	−19.4	248,829	107,250	−56.9	2,229,152	1,890,549	−15.2
Suspicion	1,125	480	−57.3	203	127	−37.4	922	353	−61.7
Curfew and loitering law violations	60,669	19,454	−67.9	60,669	19,454	−67.9	—	—	—

[a]Does not include suspicion.
[b]The 2006 rape figures are based on the legacy definition, and the 2015 rape figures are aggregate totals based on both the legacy and revised Uniform Crime Reporting definitions. For this reason, a percent change is not provided.
[c]Violent crimes are offenses of murder and nonnegligent manslaughter, rape, robbery, and aggravated assault. Property crimes are offenses of burglary, larceny-theft, motor vehicle theft, and arson.

SOURCE: "Table 32. Ten-Year Arrest Trends, Totals, 2006–2015," in *Crime in the United States, 2015*, Federal Bureau of Investigation, September 2016, https://ucr.fbi.gov/crime-in-the-u.s/2015/crime-in-the-u.s.-2015/tables/table-32 (accessed October 10, 2016).

TABLE 5.22
Arrests, by race, 2015

	Total arrests						Percent distribution[a]						Total arrests			Percent distribution[a]		
		Race						Race						Ethnicity			Ethnicity	
Offense charged	Total	White	Black or African American	American Indian or Alaska Native	Asian	Native Hawaiian or other Pacific Islander	Total	White	Black or African American	American Indian or Alaska Native	Asian	Native Hawaiian or other Pacific Islander	Total[b]	Hispanic or Latino	Not Hispanic or Latino	Total	Hispanic or Latino	Not Hispanic or Latino
Total	8,248,709	5,753,212	2,197,140	174,020	101,064	23,273	100.0	69.7	26.6	2.1	1.2	0.3	6,546,220	1,204,862	5,341,358	100.0	18.4	81.6
Murder and nonnegligent manslaughter	8,508	3,908	4,347	102	126	25	100.0	45.9	51.1	1.2	1.5	0.3	6,594	1,370	5,224	100.0	20.8	79.2
Rape[c]	17,370	11,809	4,907	301	271	82	100.0	68.0	28.2	1.7	1.6	0.5	13,117	3,516	9,601	100.0	26.8	73.2
Robbery	73,023	32,439	39,052	562	668	302	100.0	44.4	53.5	0.8	0.9	0.4	61,475	12,788	48,687	100.0	20.8	79.2
Aggravated assault	287,566	184,024	92,237	5,836	4,631	838	100.0	64.0	32.1	2.0	1.6	0.3	243,360	59,883	183,477	100.0	24.6	75.4
Burglary	165,948	112,992	48,907	1,686	1,963	400	100.0	68.1	29.5	1.0	1.2	0.2	140,692	28,267	112,425	100.0	20.1	79.9
Larceny-theft	893,707	621,585	244,722	14,311	11,078	2,011	100.0	69.6	27.4	1.6	1.2	0.2	701,523	104,678	596,845	100.0	14.9	85.1
Motor vehicle theft	59,533	40,000	17,499	917	763	354	100.0	67.2	29.4	1.5	1.3	0.6	47,712	13,137	34,575	100.0	27.5	72.5
Arson	6,762	4,952	1,519	177	88	26	100.0	73.2	22.5	2.6	1.3	0.4	5,354	898	4,456	100.0	16.8	83.2
Violent crime[d]	386,467	232,180	140,543	6,801	5,696	1,247	100.0	60.1	36.4	1.8	1.5	0.3	324,546	77,557	246,989	100.0	23.9	76.1
Property crime[d]	1,125,950	779,529	312,647	17,091	13,892	2,791	100.0	69.2	27.8	1.5	1.2	0.2	895,281	146,980	748,301	100.0	16.4	83.6
Other assaults	826,920	544,870	254,600	15,039	9,910	2,501	100.0	65.9	30.8	1.8	1.2	0.3	654,038	118,135	535,903	100.0	18.1	81.9
Forgery and counterfeiting	42,436	27,419	14,001	293	644	79	100.0	64.6	33.0	0.7	1.5	0.2	33,969	5,488	28,481	100.0	16.2	83.8
Fraud	101,556	67,594	31,495	1,157	1,171	139	100.0	66.6	31.0	1.1	1.2	0.1	81,207	8,785	72,422	100.0	10.8	89.2
Embezzlement	12,169	7,378	4,482	83	203	23	100.0	60.6	36.8	0.7	1.7	0.2	9,916	1,083	8,833	100.0	10.9	89.1
Stolen property; buying, receiving, possessing	68,057	44,561	21,783	751	832	130	100.0	65.5	32.0	1.1	1.2	0.2	53,831	10,638	43,193	100.0	19.8	80.2
Vandalism	146,090	101,481	39,778	2,816	1,667	348	100.0	69.5	27.2	1.9	1.1	0.2	120,293	21,509	98,784	100.0	17.9	82.1
Weapons; carrying, possessing, etc.	110,822	63,967	44,284	1,099	1,244	228	100.0	57.7	40.0	1.0	1.1	0.2	89,055	20,668	68,387	100.0	23.2	76.8
Prostitution and commercialized vice	31,362	17,084	12,513	111	1,599	55	100.0	54.5	39.9	0.4	5.1	0.2	28,234	5,275	22,959	100.0	18.7	81.3
Sex offenses (except rape and prostitution)	39,184	28,650	9,058	614	729	133	100.0	73.1	23.1	1.6	1.9	0.3	31,410	7,798	23,612	100.0	24.8	75.2
Drug abuse violations	1,136,950	803,809	307,140	11,717	12,436	1,848	100.0	70.7	27.0	1.0	1.1	0.2	919,876	186,841	733,035	100.0	20.3	79.7
Gambling	3,597	1,296	2,028	23	230	20	100.0	36.0	56.4	0.6	6.4	0.6	2,271	463	1,808	100.0	20.4	79.6
Offenses against the family and children	71,865	47,312	21,425	2,552	538	38	100.0	65.8	29.8	3.6	0.7	0.1	53,976	6,575	47,401	100.0	12.2	87.8
Driving under the influence	825,218	681,628	108,590	16,042	16,421	2,537	100.0	82.6	13.2	1.9	2.0	0.3	649,432	147,423	502,009	100.0	22.7	77.3
Liquor laws	201,550	160,772	27,742	9,788	2,749	499	100.0	79.8	13.8	4.9	1.4	0.2	149,364	22,338	127,026	100.0	15.0	85.0
Drunkenness	313,390	239,556	44,676	25,660	3,037	461	100.0	76.4	14.3	8.2	1.0	0.1	291,879	66,293	225,586	100.0	22.7	77.3
Disorderly conduct	295,835	189,321	91,716	11,912	2,311	575	100.0	64.0	31.0	4.0	0.8	0.2	206,368	25,996	180,372	100.0	12.6	87.4
Vagrancy	19,268	13,247	5,143	668	195	15	100.0	68.8	26.7	3.5	1.0	0.1	15,329	2,314	13,015	100.0	15.1	84.9
All other offenses (except traffic)	2,455,238	1,683,297	688,146	49,114	25,197	9,484	100.0	68.6	28.0	2.0	1.0	0.4	1,906,052	316,689	1,589,363	100.0	16.6	83.4
Suspicion	1,085	479	297	293	11	5	100.0	44.1	27.4	27.0	1.0	0.5	768	44	724	100.0	5.7	94.3
Curfew and loitering law violations	33,700	17,782	15,053	396	352	117	100.0	52.8	44.7	1.2	1.0	0.3	29,125	5,970	23,155	100.0	20.5	79.5

[a]Because of rounding, the percentages may not add to 100.0.
[b]The ethnicity totals are representative of those agencies that provided ethnicity breakdowns. Not all agencies provide ethnicity data; therefore, the race and ethnicity totals will not equal. The figures in this table are aggregate totals of the data submitted based on both the legacy and revised Uniform Crime Reporting definitions.
[c]The rape figures in this table are aggregate totals of the data submitted based on both the legacy and revised Uniform Crime Reporting definitions.
[d]Violent crimes are offenses of murder and nonnegligent manslaughter, rape, robbery, and aggravated assault. Property crimes are offenses of burglary, larceny-theft, motor vehicle theft, and arson.

SOURCE: "Table 43A. Arrests, by Race, 2015," in *Crime in the United States, 2015*, Federal Bureau of Investigation, September 2016, https://ucr.fbi.gov/crime-in-the-u.s/2015/crime-in-the-u.s.-2015/tables/table-43 (accessed October 10, 2016)

CHAPTER 6
GUN-RELATED INJURIES AND FATALITIES

The public health community, which is represented at the national level by the Centers for Disease Control and Prevention (CDC), believes that collecting comprehensive data on firearm injuries and deaths—such as who was shot, under what circumstances, and with what kind of weapon—is the first step in reducing these injuries and deaths. The next step, from a public health approach, would logically be a campaign similar to those that eradicated polio and reduced traffic fatalities.

However, the CDC's ability to treat firearm injuries as a public health problem is limited. Christine Jamieson explains in "Gun Violence Research: History of the Federal Funding Freeze" (February 2013, http://www.apa.org/science/about/psa/2013/02/gun-violence.aspx) that the CDC's ability to research firearm violence has been tightly restricted since 1996. During the mid-1990s, after having funded research demonstrating that people who lived in gun-owning households were at a greater risk of homicide than those who lived in nongun households, the CDC came under attack from the National Rifle Association of America (NRA) and other gun rights groups, who charged that the research was driven by an antigun agenda. As a consequence of an NRA campaign to prohibit federally funded research that might lead to increased gun control, Congress passed legislation in 1996 stating that "none of the funds made available for injury prevention and control at the Centers for Disease Control and Prevention may be used to advocate or promote gun control."

Two decades later, gun rights groups remained staunchly opposed to CDC research on gun violence, charging that the government agency was biased toward gun control and guilty of manipulating data to support its predetermined conclusion that even legally owned guns are a threat to public health. Although President Barack Obama (1961–) issued an executive order in January 2013, just weeks after the massacre at Sandy Hook Elementary School, directing the CDC to resume research on gun violence, Congress (many of whose members receive significant campaign contributions from the NRA) subsequently refused to allocate funding for this purpose.

Increasingly, health care professionals have become a leading voice in the call for renewed and robust government funding for gun violence research. In "Reducing Firearm-Related Harms: Time for Us to Study and Speak Out" (February 24, 2015, http://nhcva.org/files/2015/04/Taichman-2015.pdf), an editorial in *Annals of Internal Medicine*, Darren B. Taichman and Christine Laine explain why research is essential to improving public health outcomes: "When public health crises arise, our powerful health care complex responds by doing what our scientific training and duty to help others require. We formulate questions that need answers, collect and analyze data to answer them, test hypotheses to discover remedies, study how to implement them, and monitor progress. This is how polio was nearly eliminated, automobile-related injury and death rates were reduced, tobacco-related illness decreased, and an Ebola epidemic is being curtailed." They go on to argue the urgent need for a similar scientific response to what they see as a gun violence epidemic. Erin Schumaker reports in "Doctors Condemn the NRA-Fueled Ban on Gun Violence Research" (HuffingtonPost.com, June 14, 2016) that the American Medical Association, Doctors for America, the American College of Preventive Medicine, the American Academy of Pediatrics, and others insist that gun violence is a "public health crisis" that is "100 percent preventable."

Nevertheless, the CDC does supply much of the basic, authoritative data regarding victims of firearm violence. Its National Center for Injury Prevention and Control administers a system that tracks the numbers of firearm-related injuries, and these data are made available through the Web-Based Injury Statistics Query and Reporting System (WISQARS; http://www.cdc.gov/injury/wisqars/index.html). This interactive database provides reports of injury-related data of all types, including firearm injuries, both fatal and nonfatal.

The CDC also has a state-based system, the National Violent Death Reporting System (http://www.cdc.gov/ViolencePrevention/NVDRS/index.html), that collects information about violent deaths, including firearm deaths, from a variety of sources in some states. These sources include law enforcement, medical examiners and coroners, crime laboratories, and death certificates. These data help detail the circumstances that might have contributed to the firearm death. Furthermore, they help each participating state design and implement prevention and intervention efforts tailored to that state's needs.

NONFATAL GUNSHOT INJURIES

Table 6.1 shows WISQARS data for the numbers of nonfatal gunshot injuries and the rates per 100,000 population between 2005 and 2014. Both the number and rate of nonfatal gunshot injuries were higher in 2014 than in 2005, but there was considerable fluctuation in the interim, and no clear trend was discernible. In any event, the CDC indicates in "Ten Leading Causes of Death and Injury" (2016, http://www.cdc.gov/injury/wisqars/LeadingCauses.html) that gunshot injuries were not among the 10 leading causes of nonfatal injury for people of any age who were treated in emergency departments between 2005 and 2014. Table 6.2 shows the 10 leading causes of nonfatal injuries for all age groups in 2013. Although there was variation among different age groups regarding the most common causes of nonfatal injury, the most common injury causes for all age groups included falls, being struck by or against someone or something, and overexertion.

Table 6.3 shows nonfatal firearm injury numbers and rates among children between the ages of zero and 19 years in 2014. Both the numbers and the rates per 100,000 of nonfatal firearm injuries were low across all age groups, except for males aged 15 to 19 years. Of the 10.8 million males in this age group, 10,213 were injured by firearms, for a rate of 94.7 per 100,000. This rate was more than six times higher than that of females in the same age group, who were the next most likely cohort to be injured by a firearm, and it was more than eight times higher than the rate for the next youngest cohort of males (aged 10 to 14 years). Overall, children were less likely to be injured by a BB or pellet gun. (See Table 6.4.) The rate of injury from BB or pellet guns was slightly lower for males aged 15 to 19 years than for males aged 10 to 14 years, but the rate for the 15 to 19 cohort, at 24.74 per 100,000, was nearly four times lower than that cohort's firearm injury rate.

FIREARM FATALITIES

Figure 6.1 shows the trend in firearm homicide numbers and rates between 1993 and 2014. As is noted in Chapter 5, firearm homicides decreased dramatically, both in raw numbers and in rates per 100,000 people, after peaking during the early 1990s. According to Michael Planty and Jennifer L. Truman of the Bureau of Justice Statistics, in *Firearm Violence, 1993–2011* (May 2013, http://www.bjs.gov/content/pub/pdf/fv9311.pdf), the number of firearm homicides nationwide in 2011, 11,101, represented a 39% decline from the 1993 peak of 18,253 firearm homicides. The number of firearm homicides reached its lowest point in 1999, at 10,828, and then rose to 12,791 in 2006, before declining to the 2011 level.

Table 6.5 provides a demographic overview of firearm deaths not only from homicides but also from suicides and unintentional shootings. This overall firearm-related death rate is broken down by sex, race, Hispanic origin, and age, as well as by date for selected years between 1970 and 2014. The overall age-adjusted death rate from firearm-related deaths remained somewhat stable between 1970 and 1990 at about 14 deaths per 100,000 population. The rate then dropped to 13.4 deaths per 100,000 population in 1995 and continued to drop to 10.2 deaths per 100,000 population in 2000. Thereafter, it stabilized at about 10 deaths per 100,000 population.

Nevertheless, this overall rate disguises drastic fluctuations by demographic characteristics. Females have consistently had a much lower firearm-related death rate than males. Although the age-adjusted female firearm-related death rate fell roughly in tandem with the male rate, it was at all times approximately five to six times lower than the age-adjusted rate for males. (See Table 6.5.) The age-adjusted rate for females ranged from a high of 4.8 per 100,000 in 1970 to a low of 2.7 in 2010, whereas the age-adjusted rate for males ranged from a high of 26.1 per 100,000 in 1990 to a low of 17.9 in 2010.

There were also significant variations in the firearm-related death rate for males and females depending on age. For example, in 2014 the male rate was 1.3 at ages

TABLE 6.1

Nonfatal gunshot injuries and rates per 100,000 people, 2005–14

Year	Injuries	Population	Age-adjusted rate
2005	69,825	295,516,599	22.62
2006	71,417	298,379,912	23.61
2007	69,863	301,231,207	23.04
2008	78,622	304,093,966	25.77
2009	66,769	306,771,529	21.68
2010	73,505	308,745,538	23.97
2011	73,883	311,721,632	23.63
2012	81,396	314,112,078	25.85
2013	84,258	316,497,531	26.76
2014	81,034	318,857,056	25.49

SOURCE: Adapted from "Nonfatal Injury Reports, 2001–2014," in *Web-Based Injury Statistics Query and Reporting System (WISQARS)*, Centers for Disease Control and Prevention, 2016, http://webappa.cdc.gov/sasweb/ncipc/nfirates2001.html (accessed September 9, 2016)

TABLE 6.2

10 leading causes of nonfatal injuries treated in hospital emergency departments, by age group, 2013

Rank	<1	1–4	5–9	10–14	15–24	25–34	35–44	45–54	55–64	65+	Total
1	Unintentional fall 134,229	Unintentional fall 852,884	Unintentional fall 624,890	Unintentional struck by/against 561,690	Unintentional struck by/against 905,659	Unintentional fall 742,177	Unintentional fall 704,264	Unintentional fall 913,871	Unintentional fall 930,521	Unintentional fall 2,495,397	Unintentional fall 8,771,656
2	Unintentional struck by/against 28,786	Unintentional struck by/against 336,917	Unintentional struck by/against 403,522	Unintentional fall 558,177	Unintentional fall 814,829	Unintentional overexertion 638,745	Unintentional overexertion 530,422	Unintentional overexertion 461,114	Unintentional overexertion 266,126	Unintentional struck by/against 281,279	Unintentional struck by/against 4,214,125
3	Unintentional other bite/sting 12,186	Unintentional other bite/sting 158,587	Unintentional cut/pierce 112,633	Unintentional overexertion 294,669	Unintentional overexertion 672,946	Unintentional struck by/against 599,340	Unintentional struck by/against 444,089	Unintentional struck by/against 390,931	Unintentional struck by/against 261,840	Unintentional overexertion 212,293	Unintentional overexertion 3,256,567
4	Unintentional foreign body 10,650	Unintentional foreign body 139,597	Unintentional other bite/sting 107,975	Unintentional cut/pierce 114,285	Unintentional MV-occupant 627,565	Unintentional MV-occupant 526,303	Unintentional MV-occupant 374,231	Unintentional other specified 385,221	Unintentional MV-occupant 227,620	Unintentional MV-occupant 197,646	Unintentional MV-occupant 2,462,684
5	Unintentional other specified 10,511	Unintentional cut/pierce 83,575	Unintentional overexertion 93,612	Unintentional pedal cyclist 84,732	Unintentional cut/pierce 431,691	Unintentional cut/pierce 402,197	Unintentional other specified 300,154	Unintentional MV-occupant	Unintentional other specified 212,168	Unintentional cut/pierce 156,693	Unintentional cut/pierce 2,077,775
6	Unintentional fire/burn 9,816	Unintentional overexertion 81,588	Unintentional pedal cyclist 74,831	Unintentional unknown/unspecified 84,668	Other assault[a] struck by/against 381,522	Other assault[a] struck by/against 342,514	Unintentional cut/pierce 297,769	Unintentional cut/pierce 282,353	Unintentional cut/pierce 189,440	Unintentional poisoning 100,988	Unintentional other specified 1,767,630
7	Unintentional[b] inhalation/suffocation 8,294	Unintentional other specified 65,120	Unintentional foreign body 63,450	Unintentional MV-occupant 73,692	Unintentional other specified 321,914	Unintentional other specified 336,990	Other assault[a] struck by/against 207,287	Unintentional poisoning 237,328	Unintentional poisoning 153,767	Unintentional other bite/sting 90,850	Other assault[a] struck by/against 1,291,100
8	Unintentional cut/pierce 7,139	Unintentional fire/burn 52,884	Unintentional MV-occupant 58,114	Unintentional other bite/sting 64,848	Unintentional other bite/sting 177,665	Unintentional other bite/sting 180,922	Unintentional poisoning 175,870	Other assault[a] struck by/against 169,688	Unintentional other bite/sting 97,474	Unintentional other specified 86,729	Unintentional other bite/sting 1,174,267
9	Unintentional unknown/unspecified 5,735	Unintentional unknown/unspecified 41,297	Unintentional dog bite 43,499	Other assault[a] struck by/against 62,829	Unintentional unknown/unspecified 163,923	Unintentional poisoning 180,448	Unintentional other bite/sting 138,410	Unintentional other bite/sting 145,349	Other assault[a] struck by/against 73,674	Unintentional unknown/unspecified 74,864	Unintentional poisoning 1,055,960
10	Unintentional overexertion 4,985	Unintentional poisoning 32,443	Unintentional unknown/unspecified 35,303	Unintentional other transport 35,609	Unintentional poisoning 152,962	Unintentional unknown/unspecified 129,308	Unintentional unknown/unspecified 106,498	Unintentional unknown/unspecified 110,102	Unintentional unknown/unspecified 67,974	Unintentional other transport 68,022	Unintentional unknown/unspecified 819,878

[a]The "Other Assault" category includes all assaults that are not classified as sexual assault. It represents the majority of assaults.
[b]Injury estimate is unstable because of small sample size.

SOURCE: Adapted from "National Estimates of the 10 Leading Causes of Nonfatal Injuries Treated in Hospital Emergency Departments, United States—2013," in *Ten Leading Causes of Death and Injury*, Centers for Disease Control and Prevention, 2016, http://www.cdc.gov/injury/wisqars/pdf/leading_causes_of_nonfatal_injury_2013-a.pdf (accessed September 8, 2016)

TABLE 6.3

Nonfatal firearm injuries among children aged 0–19 years and rates per 100,000, 2014

Age group	Sex	Number of injuries	Population	Crude rate
0 to 4	Both sexes	129	19,876,883	0.65
	Males	105	10,155,728	1.04
	Females	24	9,721,155	0.25
5 to 9	Both sexes	334	20,519,566	1.63
	Males	66	10,478,429	0.63
	Females	268	10,041,137	2.67
10 to 14	Both sexes	1,436	20,671,506	6.94
	Males	1,179	10,551,219	11.17
	Females	257	10,120,287	2.54
15 to 19	Both sexes	11,677	21,067,647	55.43
	Males	10,213	10,784,023	94.7
	Females	1,464	10,283,624	14.24

SOURCE: Adapted from "Nonfatal Injury Reports, 2001–2014," in *Web-Based Injury Statistics Query and Reporting System (WISQARS)*, Centers for Disease Control and Prevention, 2016, http://webappa.cdc.gov/sasweb/ncipc/nfirates2001.html (accessed September 9, 2016)

TABLE 6.4

BB/pellet gun injuries among children aged 0–19 years and rates per 100,000, 2014

Age group	Sex	Number of injuries	Population	Crude rate
0 to 4	Both sexes	198	19,876,883	0.99
	Males	167	10,155,728	1.64
	Females	31	9,721,155	0.32
5 to 9	Both sexes	610	20,519,566	2.97
	Males	436	10,478,429	4.16
	Females	174	10,041,137	1.73
10 to 14	Both sexes	3,794	20,671,506	18.36
	Males	3,108	10,551,219	29.46
	Females	686	10,120,287	6.78
15 to 19	Both sexes	3,078	21,067,647	14.61
	Males	2,668	10,784,023	24.74
	Females	410	10,283,624	3.98

SOURCE: Adapted from "Nonfatal Injury Reports, 2001–2014," in *Web-Based Injury Statistics Query and Reporting System (WISQARS)*, Centers for Disease Control and Prevention, 2016, http://webappa.cdc.gov/sasweb/ncipc/nfirates2001.html (accessed September 9, 2016)

FIGURE 6.1

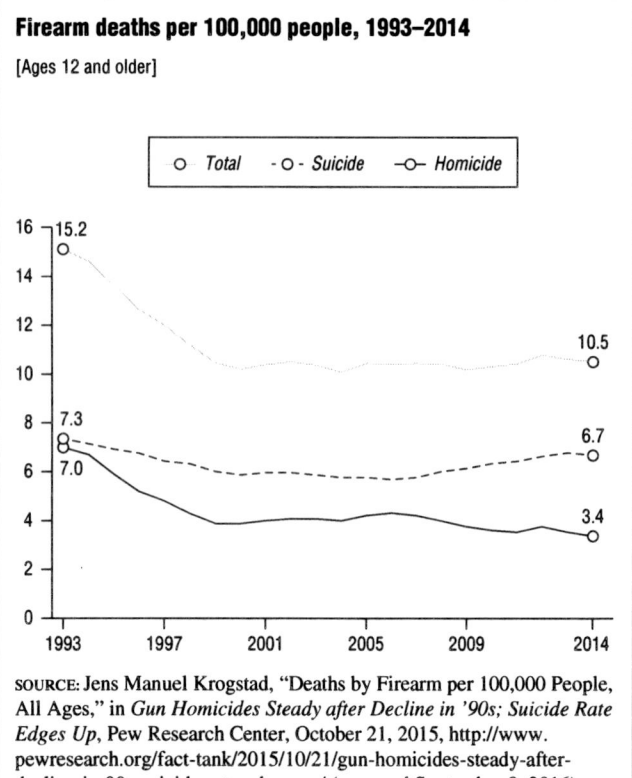

Firearm deaths per 100,000 people, 1993–2014

[Ages 12 and older]

Total, Suicide, Homicide

1993: 15.2 (Total), 7.3 (Suicide), 7.0 (Homicide)
2014: 10.5 (Total), 6.7 (Suicide), 3.4 (Homicide)

SOURCE: Jens Manuel Krogstad, "Deaths by Firearm per 100,000 People, All Ages," in *Gun Homicides Steady after Decline in '90s; Suicide Rate Edges Up*, Pew Research Center, October 21, 2015, http://www.pewresearch.org/fact-tank/2015/10/21/gun-homicides-steady-after-decline-in-90s-suicide-rate-edges-up/ (accessed September 8, 2016)

---

five to 14 years, 24.3 at ages 15 to 24 years, 25.2 at ages 25 to 34 years, and around 20 for all males between the ages of 35 and 74 years. (See Table 6.5.) However, males aged 75 years and older had a higher firearm-related death rate than males of any other age group.

This dramatic increase in the firearm-related death rate among elderly males is in large part a consequence of this group's high rate of suicide. As Table 6.6 shows, at all points between 1950 and 2014 the suicide rate for males aged 65 years and older jumped considerably. In 2014, as in prior years, males were increasingly likely to commit suicide once they reached age 75. Males aged 65 to 74 years had a lower suicide rate (26.6 per 100,000) than middle-aged males, but males aged 75 to 84 years had a higher suicide rate (34.8) than all other younger males, and males aged 85 years and older had the highest suicide rate (49.9) of any demographic subgroup.

As Table 6.5 shows, the firearm-related death rate for females has varied less over time, and in all years the patterns among females by age differed in important ways from the patterns among males. As with males, the firearm-related death rate for females typically increases with age at the transition from childhood to adolescence. Females aged 15 to 24 years had a firearm-related death rate of 3.1 per 100,000 in 2014, which was higher than 0.5 for females aged one to 14 years but much lower than the rate for adolescent males. As with males, the rate for females remained elevated through the middle years of adulthood, at 4 for females aged 25 to 44 years and 4.2 for females aged 45 to 64 years. However, in sharp contrast to the pattern among males the firearm-related death rate decreased for females at ages 75 years and beyond. Females aged 85 years and older had the lowest rate of any adult females, at 1.6.

Part of the reason for the disparity between the firearm-related death rate among elderly males and females is the difference in the two groups' suicide rates. Unlike their elderly male counterparts, women over the age of 75 years have one of the lowest suicide rates of all female age groups. The suicide rate for females aged 85 years and older, 3.4 per 100,000, was three times lower than the highest rate for females aged 45 to 54 years, 10.7 per 100,000. (See Table 6.6.)

TABLE 6.5

Death rates for firearm-related injuries, by sex, race, Hispanic origin, and age, selected years 1970–2014

[Data are based on death certificates]

Sex, race, Hispanic origin, and age	1970[a]	1980[a]	1990[a]	1995[a]	2000[b]	2010[b]	2013[b]	2014[b]
All persons				Deaths per 100,000 resident population				
All ages, age-adjusted[c]	14.3	14.8	14.6	13.4	10.2	10.1	10.4	10.3
All ages, crude	13.1	14.9	14.9	13.5	10.2	10.3	10.6	10.5
Under 1 year	†	†	†	†	†	†	†	†
1–14 years	1.6	1.4	1.5	1.6	0.7	0.6	0.7	0.8
1–4 years	1.0	0.7	0.6	0.6	0.3	0.4	0.4	0.4
5–14 years	1.7	1.6	1.9	1.9	0.9	0.7	0.8	0.9
15–24 years	15.5	20.6	25.8	26.7	16.8	14.2	14.1	14.0
15–19 years	11.4	14.7	23.3	24.1	12.9	10.6	9.7	9.9
20–24 years	20.3	26.4	28.1	29.2	20.9	17.9	18.1	17.7
25–44 years	20.9	22.5	19.3	16.9	13.1	13.3	13.9	13.4
25–34 years	22.2	24.3	21.8	19.6	14.5	15.0	15.3	14.7
35–44 years	19.6	20.0	16.3	14.3	11.9	11.7	12.3	12.1
45–64 years	17.6	15.2	13.6	11.7	10.0	11.6	11.9	11.8
45–54 years	18.1	16.4	13.9	12.0	10.5	12.0	12.3	12.2
55–64 years	17.0	13.9	13.3	11.3	9.4	11.1	11.5	11.4
65 years and over	13.8	13.5	16.0	14.1	12.2	11.7	12.5	12.7
65–74 years	14.5	13.8	14.4	12.8	10.6	10.7	11.3	11.5
75–84 years	13.4	13.4	19.4	16.3	13.9	12.7	14.1	13.9
85 years and over	10.2	11.6	14.7	14.4	14.2	13.2	13.9	15.0
Male								
All ages, age-adjusted[c]	24.8	25.9	26.1	23.8	18.1	17.9	18.3	18.0
All ages, crude	22.2	25.7	26.2	23.6	17.8	18.0	18.5	18.3
Under 1 year	†	†	†	†	†	†	†	†
1–14 years	2.3	2.0	2.2	2.3	1.1	1.0	1.0	1.1
1–4 years	1.2	0.9	0.7	0.8	0.4	0.6	0.6	0.5
5–14 years	2.7	2.5	2.9	2.9	1.4	1.1	1.2	1.3
15–24 years	26.4	34.8	44.7	46.5	29.4	25.0	24.4	24.3
15–19 years	19.2	24.5	40.1	41.6	22.4	18.4	17.0	17.2
20–24 years	35.1	45.2	49.1	51.5	37.0	31.8	31.2	30.8
25–44 years	34.1	38.1	32.6	28.4	22.0	22.9	23.7	22.8
25–34 years	36.5	41.4	37.0	33.2	24.9	26.4	26.3	25.2
35–44 years	31.6	33.2	27.4	23.6	19.4	19.3	20.8	20.1
45–64 years	31.0	25.9	23.4	20.0	17.1	19.9	20.1	19.8
45–54 years	30.7	27.3	23.2	20.1	17.6	20.3	20.4	19.9
55–64 years	31.3	24.5	23.7	19.8	16.3	19.3	19.8	19.8
65 years and over	29.7	29.7	35.3	30.7	26.4	24.1	25.3	25.7
65–74 years	29.5	27.8	28.2	25.1	20.3	20.0	20.9	21.5
75–84 years	31.0	33.0	46.9	37.8	32.2	27.5	29.8	29.2
85 years and over	26.2	34.9	49.3	47.1	44.7	37.4	38.3	40.7
Female								
All ages, age-adjusted[c]	4.8	4.7	4.2	3.8	2.8	2.7	3.0	3.0
All ages, crude	4.4	4.7	4.3	3.8	2.8	2.7	3.0	3.0
Under 1 year	†	†	†	†	†	†	†	†
1–14 years	0.8	0.7	0.8	0.8	0.3	0.3	0.4	0.5
1–4 years	0.9	0.5	0.5	0.5	†	0.3	0.3	0.4
5–14 years	0.8	0.7	1.0	0.9	0.4	0.3	0.4	0.5
15–24 years	4.8	6.1	6.0	5.9	3.5	2.9	3.2	3.1
15–19 years	3.5	4.6	5.7	5.6	2.9	2.3	2.0	2.3
20–24 years	6.4	7.7	6.3	6.1	4.2	3.5	4.3	3.9
25–44 years	8.3	7.4	6.1	5.5	4.2	3.8	4.0	4.0
25–34 years	8.4	7.5	6.7	5.8	4.0	3.5	4.1	3.9
35–44 years	8.2	7.2	5.4	5.2	4.4	4.1	4.0	4.1
45–64 years	5.4	5.4	4.5	3.9	3.4	3.7	4.1	4.2
45–54 years	6.4	6.2	4.9	4.2	3.6	3.8	4.5	4.7
55–64 years	4.2	4.6	4.0	3.5	3.0	3.4	3.8	3.6
65 years and over	2.4	2.5	3.1	2.8	2.2	2.2	2.4	2.5
65–74 years	2.8	3.1	3.6	3.0	2.5	2.6	2.8	2.7
75–84 years	1.7	1.7	2.9	2.8	2.0	2.1	2.3	2.4
85 years and over	†	1.3	1.3	1.8	1.7	1.5	1.5	1.6

As Table 6.5 shows, African American males at all points between 1970 and 2014 had the highest firearm-related death rate among all male demographic subgroups. The age-adjusted firearm-related death rate for African American males fell steadily between 1970 and 2014, from 70.8 per 100,000 to 31.5 per 100,000. Throughout this period, however, African American males typically died from firearm-related injuries at rates two to three times higher than Hispanic males, white males, and Native American or Alaskan Native males, and at rates five to seven times higher than Asian or Pacific Islander males, who had the lowest firearm-related death rate of any group.

TABLE 6.5
Death rates for firearm-related injuries, by sex, race, Hispanic origin, and age, selected years 1970–2014 [CONTINUED]

[Data are based on death certificates]

Sex, race, Hispanic origin, and age	1970[a]	1980[a]	1990[a]	1995[a]	2000[b]	2010[b]	2013[b]	2014[b]
White male[d]				Deaths per 100,000 resident population				
All ages, age-adjusted[c]	19.7	22.1	22.0	20.1	15.9	16.1	16.5	16.3
All ages, crude	17.6	21.8	21.8	19.9	15.6	16.5	17.1	17.0
1–14 years	1.8	1.9	1.9	1.9	1.0	0.8	1.0	1.0
15–24 years	16.9	28.4	29.5	30.8	19.6	16.2	16.0	16.1
25–44 years	24.2	29.5	25.7	23.2	18.0	18.6	18.9	18.4
25–34 years	24.3	31.1	27.8	25.2	18.1	19.1	19.1	18.7
35–44 years	24.1	27.1	23.3	21.2	17.9	18.0	18.8	18.2
45–64 years	27.4	23.3	22.8	19.5	17.4	21.3	21.6	21.4
65 years and over	29.9	30.1	36.8	32.2	28.2	26.5	28.0	28.3
Black or African American male[d]								
All ages, age-adjusted[c]	70.8	60.1	56.3	49.2	34.2	31.8	32.1	31.5
All ages, crude	60.8	57.7	61.9	52.9	36.1	33.4	33.5	32.9
1–14 years	5.3	3.0	4.4	4.4	1.8	1.9	1.4	1.9
15–24 years	97.3	77.9	138.0	138.7	89.3	73.2	69.9	68.9
25–44 years	126.2	114.1	90.3	70.2	54.1	57.3	60.2	57.4
25–34 years	145.6	128.4	108.6	92.3	74.8	78.2	76.6	71.3
35–44 years	104.2	92.3	66.1	46.3	34.3	35.2	41.5	41.0
45–64 years	71.1	55.6	34.5	28.3	18.4	16.5	16.7	16.9
65 years and over	30.6	29.7	23.9	21.8	13.8	9.4	8.8	10.9
American Indian or Alaska Native male[d]								
All ages, age-adjusted[c]	—	24.0	19.4	19.4	13.1	11.7	12.9	12.9
All ages, crude	—	27.5	20.5	20.9	13.2	12.5	12.5	12.4
15–24 years	—	55.3	49.1	40.9	26.9	26.0	18.4	17.7
25–44 years	—	43.9	25.4	31.2	16.6	16.9	18.8	19.3
45–64 years	—	f	f	14.2	12.2	11.1	13.5	12.6
65 years and over	—	f	f	f	f	f	f	f
Asian or Pacific Islander male[d]								
All ages, age-adjusted[c]	—	7.8	8.8	9.2	6.0	4.2	4.1	3.7
All ages, crude	—	8.2	9.4	10.0	6.2	4.4	4.3	3.8
15–24 years	—	10.8	21.0	24.3	9.3	6.8	5.9	6.3
25–44 years	—	12.8	10.9	10.6	8.1	6.0	6.4	4.8
45–64 years	—	10.4	8.1	8.2	7.4	4.4	4.6	4.1
65 years and over	—	f	f	f	f	3.9	f	3.4
Hispanic or Latino male[d, e]								
All ages, age-adjusted[c]	—	—	27.6	23.8	13.6	10.5	9.4	9.4
All ages, crude	—	—	29.9	26.2	14.2	10.5	9.4	9.4
1–14 years	—	—	2.6	2.8	1.0	0.6	0.5	0.5
15–24 years	—	—	55.5	61.7	30.8	20.9	17.4	17.2
25–44 years	—	—	42.7	31.4	17.3	14.4	13.8	14.1
25–34 years	—	—	47.3	36.4	20.3	18.0	16.2	15.7
35–44 years	—	—	35.4	24.2	13.2	10.2	11.1	12.3
45–64 years	—	—	21.4	17.2	12.0	9.1	8.2	7.9
65 years and over	—	—	19.1	16.5	12.2	9.9	8.1	7.8
White, not Hispanic or Latino male[e]								
All ages, age-adjusted[c]	—	—	20.6	18.6	15.5	16.6	17.5	17.3
All ages, crude	—	—	20.4	18.5	15.7	17.6	18.8	18.7
1–14 years	—	—	1.6	1.6	1.0	0.9	1.2	1.2
15–24 years	—	—	24.1	23.5	16.2	14.2	15.1	15.4
25–44 years	—	—	23.3	21.4	17.9	19.4	20.2	19.4
25–34 years	—	—	24.7	22.5	17.2	18.9	19.7	19.2
35–44 years	—	—	21.6	20.4	18.4	19.9	20.8	19.7
45–64 years	—	—	22.7	19.5	17.8	22.8	23.5	23.4
65 years and over	—	—	37.4	32.5	29.0	27.6	29.6	29.9
White female[d]								
All ages, age-adjusted[c]	4.0	4.2	3.8	3.5	2.7	2.7	3.0	3.1
All ages, crude	3.7	4.1	3.8	3.5	2.7	2.8	3.1	3.2
15–24 years	3.4	5.1	4.8	4.5	2.8	2.3	2.7	2.6
25–44 years	6.9	6.2	5.3	4.9	3.9	3.7	3.9	4.0
45–64 years	5.0	5.1	4.5	4.0	3.5	4.1	4.7	4.7
65 years and over	2.2	2.5	3.1	2.8	2.4	2.5	2.7	2.7

Although African American men aged 25 to 34 years had the highest firearm-related death rate of any demographic subgroup in 1970 (at 145.6 per 100,000), as the firearm homicide rate for all groups peaked during the mid-1990s, African American males aged 15 to 24 years had the highest rate of death from firearm-related injuries, at 138 in

TABLE 6.5
Death rates for firearm-related injuries, by sex, race, Hispanic origin, and age, selected years 1970–2014 [CONTINUED]

[Data are based on death certificates]

Sex, race, Hispanic origin, and age	1970[a]	1980[a]	1990[a]	1995[a]	2000[b]	2010[b]	2013[b]	2014[b]
Black or African American female[d]				Deaths per 100,000 resident population				
All ages, age-adjusted[c]	11.1	8.7	7.3	6.2	3.9	3.3	3.4	3.2
All ages, crude	10.0	8.8	7.8	6.5	4.0	3.3	3.4	3.3
15–24 years	15.2	12.3	13.3	13.2	7.6	6.4	6.1	6.2
25–44 years	19.4	16.1	12.4	9.8	6.5	5.6	6.0	5.3
45–64 years	10.2	8.2	4.8	4.1	3.1	2.2	2.5	2.1
65 years and over	4.3	3.1	3.1	2.6	1.3	[f]	0.9	1.4
American Indian or Alaska Native female[d]								
All ages, age-adjusted[c]	—	5.8	3.3	3.8	2.9	2.6	2.0	2.4
All ages, crude	—	5.8	3.4	4.1	2.9	2.4	2.0	2.4
15–24 years	—	[f]	[f]	[f]	[f]	[f]	[f]	[f]
25–44 years	—	10.2	[f]	7.0	5.5	3.7	[f]	5.6
45–64 years	—	[f]	[f]	[f]	[f]	[f]	[f]	[f]
65 years and over	—	[f]	[f]	[f]	[f]	[f]	[f]	[f]
Asian or Pacific Islander female[d]								
All ages, age-adjusted[c]	—	2.0	1.9	2.0	1.1	0.6	0.9	0.7
All ages, crude	—	2.1	2.1	2.1	1.2	0.6	0.9	0.8
15–24 years	—	[f]	[f]	3.9	[f]	[f]	[f]	[f]
25–44 years	—	3.2	2.7	2.7	1.5	1.1	1.5	0.8
45–64 years	—	[f]	[f]	[f]	[f]	[f]	0.8	1.3
65 years and over	—	[f]	[f]	[f]	[f]	[f]	[f]	[f]
Hispanic or Latina female[d, e]								
All ages, age-adjusted[c]	—	—	3.3	3.1	1.8	1.3	1.3	1.4
All ages, crude	—	—	3.6	3.3	1.8	1.3	1.3	1.4
15–24 years	—	—	6.9	6.1	2.9	2.1	2.3	2.6
25–44 years	—	—	5.1	4.7	2.5	1.8	2.0	1.9
45–64 years	—	—	2.4	2.4	2.2	1.5	1.2	1.2
65 years and over	—	—	[f]	[f]	[f]	[f]	[f]	[f]
White, not Hispanic or Latina female[e]								
All ages, age-adjusted[c]	—	—	3.7	3.4	2.8	3.0	3.3	3.4
All ages, crude	—	—	3.7	3.5	2.9	3.1	3.5	3.6
15–24 years	—	—	4.3	4.1	2.7	2.3	2.8	2.6
25–44 years	—	—	5.1	4.8	4.2	4.2	4.4	4.6
45–64 years	—	—	4.6	4.1	3.6	4.4	5.2	5.2
65 years and over	—	—	3.2	2.8	2.4	2.6	2.9	2.9

—Data not available.
[a]Underlying cause of death was coded according to the 8th Revision of the *International Classification of Diseases* (ICD) in 1970 and 9th Revision in 1980–1998.
[b]Starting with 1999 data, cause of death is coded according to ICD–10.
[c]Age-adjusted rates are calculated using the year 2000 standard population. Prior to 2001, age-adjusted rates were calculated using standard million proportions based on rounded population numbers. Starting with 2001 data, unrounded population numbers are used to calculate age-adjusted rates.
[d]The race groups, white, black, Asian or Pacific Islander, and American Indian or Alaska Native, include persons of Hispanic and non-Hispanic origin. Persons of Hispanic origin may be of any race. Death rates for Hispanic, American Indian or Alaska Native, and Asian or Pacific Islander persons should be interpreted with caution because of inconsistencies in reporting Hispanic origin or race on the death certificate (death rate numerators) compared with population figures (death rate denominators). The net effect of misclassification is an underestimation of deaths and death rates for races other than white and black.
[e]Prior to 1997, data from states that did not report Hispanic origin on the death certificate were excluded.
[f]Rates based on fewer than 20 deaths are considered unreliable and are not shown.
Notes: Starting with *Health, United States, 2003*, rates for 1991–1999 were revised using intercensal population estimates based on the 1990 and 2000 censuses. For 2000, population estimates are bridged-race April 1 census counts. Starting with *Health, United States, 2012*, rates for 2001–2009 were revised using intercensal population estimates based on the 2000 and 2010 censuses. For 2010, population estimates are bridged-race April 1 census counts. Rates for 2011 and beyond were computed using 2010-based postcensal estimates. Age groups were selected to minimize the presentation of unstable age-specific death rates based on small numbers of deaths and for consistency among comparison groups. Starting with 2003 data, some states allowed the reporting of more than one race on the death certificate. The multiple-race data for these states were bridged to the single-race categories of the 1977 Office of Management and Budget standards, for comparability with other states.

SOURCE: "Table 31. Death Rates for Firearm-Related Injuries, by Sex, Race, Hispanic Origin, and Age: United States, Selected Years 1970–2014," in *Health, United States, 2015: With Special Feature on Racial and Ethnic Health Disparities*, Centers for Disease Control and Prevention, National Center for Health Statistics, May 2016, http://www.cdc.gov/nchs/data/hus/hus15.pdf (accessed September 9, 2016)

1990 and 138.7 in 1995. In 2014 African American males between the ages of 15 and 24 years had a firearm-related death rate of 68.9, and African American males aged 25 to 34 years had a rate of 71.3. In other words, African American males aged 15 to 34 years were almost four times more likely to be killed by firearms than the average American male, whose age-adjusted rate in 2014 was 18.

Table 6.7 shows firearm-related deaths and rates attributable to homicide and suicide between 2005 and 2014, and Table 6.8 shows the firearm-related deaths and rates attributable to unintentional shootings between 2005 and 2014. As is evident, intentional firearm-related deaths are much more common than unintentional ones. In terms of the age-adjusted rates of death, both intentional and

TABLE 6.6

Death rates for suicide, by sex and age, selected years 1950–2014

[Data are based on death certificates]

Sex, race, Hispanic origin, and age	1950[a,b]	1960[a,b]	1970[b]	1980[b]	1990[b]	2000[c]	2010[c]	2013[c]	2014[c]
All persons				Deaths per 100,000 resident population					
All ages, age-adjusted[d]	13.2	12.5	13.1	12.2	12.5	10.4	12.1	12.6	13.0
All ages, crude	11.4	10.6	11.6	11.9	12.4	10.4	12.4	13.0	13.4
Under 1 year	—	—	—	—	—	—	—	—	—
1–4 years	—	—	—	—	—	—	—	—	—
5–14 years	0.2	0.3	0.3	0.4	0.8	0.7	0.7	1.0	1.0
15–24 years	4.5	5.2	8.8	12.3	13.2	10.2	10.5	11.1	11.5
15–19 years	2.7	3.6	5.9	8.5	11.1	8.0	7.5	8.3	8.7
20–24 years	6.2	7.1	12.2	16.1	15.1	12.5	13.6	13.7	14.2
25–44 years	11.6	12.2	15.4	15.6	15.2	13.4	15.0	15.5	15.8
25–34 years	9.1	10.0	14.1	16.0	15.2	12.0	14.0	14.8	15.1
35–44 years	14.3	14.2	16.9	15.4	15.3	14.5	16.0	16.2	16.6
45–64 years	23.5	22.0	20.6	15.9	15.3	13.5	18.6	19.0	19.5
45–54 years	20.9	20.7	20.0	15.9	14.8	14.4	19.6	19.7	20.2
55–64 years	26.8	23.7	21.4	15.9	16.0	12.1	17.5	18.1	18.8
65 years and over	30.0	24.5	20.8	17.6	20.5	15.2	14.9	16.1	16.6
65–74 years	29.6	23.0	20.8	16.9	17.9	12.5	13.7	15.0	15.6
75–84 years	31.1	27.9	21.2	19.1	24.9	17.6	15.7	17.1	17.5
85 years and over	28.8	26.0	19.0	19.2	22.2	19.6	17.6	18.6	19.3
Male									
All ages, age-adjusted[d]	21.2	20.0	19.8	19.9	21.5	17.7	19.8	20.3	20.7
All ages, crude	17.8	16.5	16.8	18.6	20.4	17.1	19.9	20.6	21.1
Under 1 year	—	—	—	—	—	—	—	—	—
1–4 years	—	—	—	—	—	—	—	—	—
5–14 years	0.3	0.4	0.5	0.6	1.1	1.2	0.9	1.2	1.3
15–24 years	6.5	8.2	13.5	20.2	22.0	17.1	16.9	17.3	18.2
15–19 years	3.5	5.6	8.8	13.8	18.1	13.0	11.7	12.4	13.0
20–24 years	9.3	11.5	19.3	26.8	25.7	21.4	22.2	21.9	22.9
25–44 years	17.2	17.9	20.9	24.0	24.4	21.3	23.6	24.1	24.3
25–34 years	13.4	14.7	19.8	25.0	24.8	19.6	22.5	23.4	23.8
35–44 years	21.3	21.0	22.1	22.5	23.9	22.8	24.6	24.8	25.0
45–64 years	37.1	34.4	30.0	23.7	24.3	21.3	29.2	29.0	29.7
45–54 years	32.0	31.6	27.9	22.9	23.2	22.4	30.4	29.6	30.0
55–64 years	43.6	38.1	32.7	24.5	25.7	19.4	27.7	28.3	29.4
65 years and over	52.8	44.0	38.4	35.0	41.6	31.1	29.0	30.9	31.4
65–74 years	50.5	39.6	36.0	30.4	32.2	22.7	23.9	26.0	26.6
75–84 years	58.3	52.5	42.8	42.3	56.1	38.6	32.3	34.7	34.8
85 years and over	58.3	57.4	42.4	50.6	65.9	57.5	47.3	48.5	49.9
Female									
All ages, age-adjusted[d]	5.6	5.6	7.4	5.7	4.8	4.0	5.0	5.5	5.8
All ages, crude	5.1	4.9	6.6	5.5	4.8	4.0	5.2	5.7	6.0
Under 1 year	—	—	—	—	—	—	—	—	—
1–4 years	—	—	—	—	—	—	—	—	—
5–14 years	0.1	0.1	0.2	0.2	0.4	0.3	0.4	0.7	0.7
15–24 years	2.6	2.2	4.2	4.3	3.9	3.0	3.9	4.5	4.6
15–19 years	1.8	1.6	2.9	3.0	3.7	2.7	3.1	3.9	4.2
20–24 years	3.3	2.9	5.7	5.5	4.1	3.2	4.7	5.2	5.0
25–44 years	6.2	6.6	10.2	7.7	6.2	5.4	6.4	6.8	7.2
25–34 years	4.9	5.5	8.6	7.1	5.6	4.3	5.3	6.1	6.3
35–44 years	7.5	7.7	11.9	8.5	6.8	6.4	7.5	7.6	8.2
45–64 years	9.9	10.2	12.0	8.9	7.1	6.2	8.6	9.4	9.8
45–54 years	9.9	10.2	12.6	9.4	6.9	6.7	9.0	10.0	10.7
55–64 years	9.9	10.2	11.4	8.4	7.3	5.4	8.0	8.7	8.9
65 years and over	9.4	8.4	8.1	6.1	6.4	4.0	4.2	4.6	5.0
65–74 years	10.1	8.4	9.0	6.5	6.7	4.0	4.8	5.4	5.9
75–84 years	8.1	8.9	7.0	5.5	6.3	4.0	3.7	3.9	4.3
85 years and over	8.2	6.0	5.9	5.5	5.4	4.2	3.3	3.3	3.4

unintentional firearm deaths saw slight overall declines between 2005 and 2014. The raw number of firearm-related homicides and suicides rose slightly during this period, but the growth of the U.S. population outpaced this increase, so the rate declined slightly. (See Table 6.7.)

Firearm Suicides versus Homicides

Jens Manuel Krogstad of the Pew Research Center notes in "Gun Homicides Steady after Decline in '90s; Suicide Rate Edges Up" (October 21, 2015, http://www.pewresearch.org/fact-tank/2015/10/21/gun-homicides-steady-after-decline-in-90s-suicide-rate-edges-up/) that gun suicides have consistently outnumbered gun homicides for the past two decades. However, although gun homicide rates dropped by more than half (from 7 to 3.4) between 1993 and 2014, gun suicide rates have hardly improved. Gun suicide rates fell from 7.3 per 100,000 in 1993 to below 6 in 2000, but they have since begun to rise again, reaching 6.7 in 2014.

TABLE 6.6

Death rates for suicide, by sex and age, selected years 1950–2014 [CONTINUED]

[Data are based on death certificates]

—Category not applicable.
[a]Includes deaths of persons who were not residents of the 50 states and the District of Columbia (D.C.).
[b]Underlying cause of death was coded according to the 6th Revision of the *International Classification of Diseases* (ICD) in 1950, 7th Revision in 1960, 8th Revision in 1970, and 9th Revision in 1980–1998.
[c]Starting with 1999 data, cause of death is coded according to ICD–10.
[d]Age-adjusted rates are calculated using the year 2000 standard population. Prior to 2001, age-adjusted rates were calculated using standard million proportions based on rounded population numbers. Starting with 2001 data, unrounded population numbers are used to calculate age-adjusted rates.
Notes: Starting with *Health, United States, 2003*, rates for 1991–1999 were revised using intercensal population estimates based on the 1990 and 2000 censuses. For 2000, population estimates are bridged-race April 1 census counts. Starting with *Health, United States, 2012*, rates for 2001–2009 were revised using intercensal population estimates based on the 2000 and 2010 censuses. For 2010, population estimates are bridged-race April 1 census counts. Rates for 2011 and beyond were computed using 2010-based postcensal estimates. Figures for 2001 include September 11-related deaths for which death certificates were filed as of October 24, 2002. Age groups were selected to minimize the presentation of unstable age-specific death rates based on small numbers of deaths and for consistency among comparison groups. Starting with 2003 data, some states allowed the reporting of more than one race on the death certificate. The multiple-race data for these states were bridged to the single-race categories of the 1977 Office of Management and Budget standards, for comparability with other states. Rates based on fewer than 20 deaths are considered unreliable and are not shown.

SOURCE: Adapted from "Table 30. Death Rates for Suicide, by Sex, Race, Hispanic Origin, and Age: United States, Selected Years 1950–2014," in *Health, United States, 2015: With Special Feature on Racial and Ethnic Health Disparities*, Centers for Disease Control and Prevention, National Center for Health Statistics, May 2016, http://www.cdc.gov/nchs/data/hus/hus15.pdf (accessed September 9, 2016)

TABLE 6.7

Fatal gunshot (homicide and suicide) deaths and rates per 100,000 people, 2005–14

Year	Deaths	Population	Age-adjusted rate
2005	30,694	295,516,599	10.27
2006	30,896	298,379,912	10.22
2007	31,224	301,231,207	10.24
2008	31,593	304,093,966	10.23
2009	31,347	306,771,529	10.05
2010	31,672	308,745,538	10.07
2011	32,351	311,721,632	10.16
2012	33,563	314,112,078	10.44
2013	33,636	316,497,531	10.37
2014	33,599	318,857,056	10.25

SOURCE: Adapted from "Fatal Injury Reports, National and Regional, 1999–2014," in *Web-Based Injury Statistics Query and Reporting System (WISQARS)*, Centers for Disease Control and Prevention, 2016, http://webappa.cdc.gov/sasweb/ncipc/mortrate10_us.html (accessed September 9, 2016)

TABLE 6.8

Unintentional firearm deaths and rates per 100,000 people, 2005–14

Year	Deaths	Population	Age-adjusted rate
2005	789	295,516,599	0.27
2006	642	298,379,912	0.21
2007	613	301,231,207	0.20
2008	592	304,093,966	0.19
2009	554	306,771,529	0.18
2010	606	308,745,538	0.20
2011	591	311,721,632	0.19
2012	548	314,112,078	0.17
2013	505	316,497,531	0.16
2014	586	318,857,056	0.18

SOURCE: Adapted from "Fatal Injury Reports, National and Regional, 1999–2014," in *Web-Based Injury Statistics Query and Reporting System (WISQARS)*, Centers for Disease Control and Prevention, 2016, http://webappa.cdc.gov/sasweb/ncipc/mortrate10_us.html (accessed September 9, 2016)

As Table 6.9 shows, firearm homicides and firearm suicides were among the 10 leading causes of injury deaths (deaths caused by some external force, as opposed to those caused by diseases or congenital ailments) for almost all age groups in the United States in 2014. Firearm homicides were among the top-10 causes of injury deaths for every age group except those under age one and those aged 65 years and older, and firearm suicides were among the top-10 causes of injury deaths for every age group except those under age one, between one and four years, and between five and nine years. In all, firearm suicides accounted for 21,334 deaths in 2014, and firearm homicides accounted for 10,945 deaths.

THE COST OF FIREARM INJURIES

Although U.S. government agencies (including the U.S. Environmental Protection Agency, the U.S. Department of Transportation, and the U.S. Department of Health and Human Services) have thoroughly studied the economic toll of many major societal problems (such as air pollution, motor vehicle crashes, and domestic violence), the CDC and other government agency data on the annual cost of firearm injuries remains incomplete. As with other aspects of gun violence, government research in this area has been obstructed and inhibited by the powerful influence of the gun lobby. Thus, most of the available data have been compiled by private research organizations.

In *The Hospital Costs of Firearm Assaults* (September 2013, http://www.urban.org/UploadedPDF/412894-The-Hospital-Costs-of-Firearm-Assaults.pdf), Embry M. Howell and Peter Abraham of the Urban Institute attempt to quantify how much money is spent on emergency department (ED) and inpatient hospital care resulting from both fatal and nonfatal firearm assaults. The researchers

TABLE 6.9

10 leading causes of injury deaths, by age group, 2014

Rank	<1	1–4	5–9	10–14	15–24	25–34	35–44	45–54	55–64	65+	Total
1	Unintentional suffocation 991	Unintentional drowning 388	Unintentional MV traffic 345	Unintentional MV traffic 384	Unintentional MV traffic 6,531	Unintentional poisoning 9,334	Unintentional poisoning 9,116	Unintentional poisoning 11,009	Unintentional poisoning 7,013	Unintentional fall 27,044	Unintentional poisoning 42,032
2	Homicide unspecified 119	Unintentional MV traffic 293	Unintentional drowning 125	Suicide suffocation 225	Homicide firearm 3,587	Unintentional MV traffic 5,856	Unintentional MV traffic 4,308	Unintentional MV traffic 5,024	Unintentional MV traffic 4,554	Unintentional MV traffic 6,373	Unintentional MV traffic 33,736
3	Homicide other spec., classifiable 83	Homicide unspecified 149	Unintentional fire/burn 68	Suicide firearm 174	Unintentional poisoning 3,492	Homicide firearm 3,260	Suicide firearm 2,830	Suicide firearm 3,953	Suicide firearm 3,910	Suicide firearm 5,367	Unintentional fall 31,959
4	Unintentional MV traffic 61	Unintentional suffocation 120	Homicide firearm 58	Homicide firearm 115	Suicide firearm 2,270	Suicide firearm 2,829	Suicide suffocation 2,057	Suicide suffocation 2,321	Unintentional fall 2,558	Unintentional unspecified 4,590	Suicide firearm 21,334
5	Undetermined suffocation 40	Unintentional fire/burn 117	Unintentional other land transport 36	Unintentional drowning 105	Unintentional suffocation 2,010	Suicide suffocation 2,402	Homicide firearm 1,835	Suicide poisoning 1,795	Suicide poisoning 1,529	Unintentional suffocation 3,692	Suicide suffocation 11,208
6	Unintentional drowning 29	Unintentional pedestrian, other 107	Unintentional suffocation 34	Unintentional fire/burn 49	Unintentional drowning 507	Suicide poisoning 800	Suicide poisoning 1,274	Unintentional fall 1,340	Suicide suffocation 1,509	Unintentional poisoning 1,993	Homicide firearm 10,945
7	Homicide suffocation 26	Homicide other spec., classifiable 73	Unintentional natural/environment 22	Unintentional other land transport 49	Suicide poisoning 363	Undetermined poisoning 575	Undetermined poisoning 637	Homicide firearm 1,132	Undetermined poisoning 698	Adverse effects 1,554	Suicide poisoning 6,808
8	Unintentional natural/environment 17	Homicide firearm 47	Unintentional pedestrian, other 18	Unintentional suffocation 33	Homicide cut/pierce 314	Homicide cut/pierce 430	Unintentional fall 504	Undetermined poisoning 820	Undetermined poisoning 539	Unintentional fire/burn 1,151	Unintentional suffocation 6,580
9	Undetermined unspecified 16	Unintentional struck by/against 38	Unintentional struck by/against 16	Unintentional poisoning 22	Undetermined poisoning 229	Unintentional drowning 399	Unintentional drowning 363	Unintentional suffocation 452	Homicide firearm 538	Suicide poisoning 1,028	Unintentional unspecified 5,848
10	Unintentional fire/burn 15	Unintentional natural/environment 35	Homicide unspecified (tied) 14	Homicide cut/pierce 19	Unintentional other land transport 177	Unintentional fall 285	Homicide cut/pierce 313	Unintentional drowning 442	Unintentional unspecified 530	Suicide suffocation 880	Unintentional drowning 3,406

MV = motor vehicle.

SOURCE: Adapted from "10 Leading Causes of Injury Deaths by Age Group Highlighting Violence-Related Injury Deaths, United States—2014," in *Ten Leading Causes of Death and Injury*, Centers for Disease Control and Prevention, 2016, http://www.cdc.gov/injury/images/lc-charts/leading_causes_of_injury_deaths_violence_2014_1040w760h.gif (accessed October 16, 2016)

note that in 2010 there were 36,341 ED admissions and 25,024 hospitalizations resulting from firearm assault incidents. An average ED visit for firearm assault injuries cost $1,126 in 2010, for a total ED cost of $40.9 million; and an average inpatient hospital visit for firearm assault injuries cost $23,497, for a total inpatient cost of $588 million. The total ED and hospital bills generated by the victims of gun violence in 2010 was thus $628.9 million.

Howell and Abraham note that young males living in the poorest areas of the country account for much of the hospital spending on firearm assault injuries. Males accounted for 91% of hospital costs related to firearm injuries in 2010, with males aged 15 to 24 years accounting for 41% and males aged 25 to 34 years accounting for 28%. People living in the poorest zip codes, those in the lowest quartile (the bottom 25%) by household income, accounted for 51% of hospital costs from firearm assault injuries, and those living in the next lowest quartile accounted for 22% of hospital costs. People receiving health care through government programs such as Medicaid (a state and federal health insurance program for low-income people) and Medicare (a federal health insurance program for people aged 65 years and older and people with disabilities) accounted for more than half (52%) of total hospital costs relating to firearm injuries, and people without any form of insurance coverage accounted for 28% of these costs. Thus, 80% of firearm violence costs in hospitals are passed on to the federal government and ultimately to taxpayers.

Mark Follman et al. explain in "What Does Gun Violence Really Cost?" (MotherJones.com, May–June 2015) that costs associated with gun violence are both direct (including emergency services, short- and long-term medical and mental health care, police investigations, legal costs, and prison costs) and indirect (including lost income and productivity, as well as impact on quality of life, which is assigned a value by juries that award damages to victims for their pain and suffering). Based on research by Ted Miller of the Pacific Institute for Research and Evaluation, an independent nonprofit that studies public health, education, and safety issues, *Mother Jones* estimates that gun violence costs American taxpayers $229 billion per year. This total reflects $8.6 billion in direct costs (including an average cost of $400,000 for each gun homicide) and $221 billion in indirect costs (including $49 billion annually for victims' lost wages and $169 billion for the impact on victims' quality of life). All told, the researchers estimate that the average total cost of a single gun death is about $6 million and the average cost of each gun injury requiring hospitalization is $583,000.

GUNS AND SELF-DEFENSE

The frequency with which guns are used in self-defense is one of many topics hotly disputed by both gun control and gun rights advocates. Most research suggests that the phenomenon of firearm use in self-defense is exceptionally rare, but gun rights groups typically question the neutrality of such studies and contend that guns are used in self-defense with relative frequency.

As is discussed in Chapter 1, gun rights advocates often point to the research of Gary Kleck (1951–), a criminologist at Florida State University, in suggesting that guns prevent crime more often than they are used to commit crime. Kleck has consistently asserted, over the course of a number of articles and books written with various coauthors, that guns are used in self-defense approximately 2 million to 2.5 million times annually. Kleck's estimate suggests that guns are used more often in self-defense than to commit crimes, and if this assertion is accurate, it would largely substantiate gun rights advocates' belief that firearm ownership makes the United States safer, rather than more dangerous. Few other gun researchers have produced estimates anywhere near as high as Kleck's, however, and Kleck's critics, such as David Hemenway of the Harvard Injury Control Research Center, note many apparent methodological flaws in his research. Hemenway also argues that a comparison of Kleck's estimate with Federal Bureau of Investigation crime statistics and with data about the number of people admitted to hospitals for gunshot wounds shows the 2 million to 2.5 million number to be implausibly high. For his own part, Kleck considers Hemenway's critique to be ideologically motivated and methodologically dubious.

The Violence Policy Center (VPC), a nonprofit group that advocates for stricter gun control laws, is responsible for one of the most current estimates, as of December 2016, of how often guns are used in self-defense. The VPC observes in *Firearm Justifiable Homicides and Non-fatal Self-Defense Gun Use* (June 2015, http://www.vpc.org/studies/justifiable15.pdf), which is based on analyses of Federal Bureau of Investigation and Bureau of Justice Statistics data for the five-year period between 2007 and 2011, that a total of 29.6 million people were the victims of attempted or completed criminal violence and that firearms were used defensively in only 0.8% of incidents. Thus, the VPC calculates a five-year total of 235,700 people, or an average of 47,140 people per year, who used firearms in self-defense.

The VPC indicates that there were 259 justifiable gun homicides in 2012, compared with 8,342 criminal gun homicides. (See Table 6.10.) Thus, the ratio of criminal to justifiable gun homicides was 32 to 1. The data also indicate that between 2008 and 2012 African Americans were much more likely to be killed in a justifiable firearm homicide than any other race. Over this five-year period African Americans accounted for 646 (59.2%) of the total 1,091 justifiable firearms homicide fatalities, whereas whites accounted for 422 (38.7%) of these fatalities. (See Table 6.11.) Notably, in a significant majority of justifiable gun homicides, the shooter and the victim belonged to the same race: white shooters were responsible for 377

TABLE 6.10

Number of justifiable, compared to criminal, gun homicides, 2008–12

	Number of homicides											
Circumstance	2008		2009		2010		2011		2012		Total	
Criminal homicide	9,039	97.8%	8,697	97.6%	8,275	97.3%	8,066	97.6%	8,342	97.0%	42,419	97.5%
Justifiable homicide	205	2.2%	213	2.4%	230	2.7%	201	2.4%	259	3.0%	1,108	2.5%
Ratio of criminal homicide to justifiable homicide	44-1		41-1		36-1		40-1		32-1		38-1	
Total	9,244		8,910		8,505		8,267		8,601		43,527	

SOURCE: "Table Two. Circumstances for Homicides by Firearm, 2008–2012," in *Firearm Justifiable Homicides and Non-Fatal Self-Defense Gun Use*, Violence Policy Center, June 2015, http://www.vpc.org/studies/justifiable15.pdf (accessed September 10, 2016)

TABLE 6.11

Race of shooter and person killed in justifiable gun homicides, 2008–12

Race of shooter	Race of person killed	Number of justifiable homicides											
		2008		2009		2010		2011		2012		Total	
White	White	71	62.3%	77	68.8%	82	67.8%	62	72.1%	85	64.9%	377	66.8%
	Black	38	33.3%	34	30.4%	37	30.6%	23	26.7%	40	30.5%	172	30.5%
	Asian	1	0.9%	0	0.0%	0	0.0%	0	0.0%	1	0.8%	2	0.4%
	American Indian	3	2.6%	1	0.9%	1	0.8%	1	1.2%	3	2.3%	9	1.6%
	Unknown	1	0.9%	0	0.0%	1	0.8%	0	0.0%	2	1.5%	4	0.7%
Black	White	5	6.8%	8	8.8%	5	4.9%	3	2.8%	10	8.6%	31	6.3%
	Black	69	93.2%	83	91.2%	97	95.1%	102	96.2%	107	91.5%	458	93.5%
	Asian	0	0.0%	0	0.0%	0	0.0%	1	0.9%	0	0.0%	1	0.2%
	American Indian	0	0.0%	0	0.0%	0	0.0%	0	0.0%	0	0.0%	0	0.0%
	Unknown	0	0.0%	0	0.0%	0	0.0%	0	0.0%	0	0.0%	0	0.0%
Asian	White	5	45.5%	0	0.0%	1	20.0%	2	50.0%	1	16.7%	9	28.1%
	Black	5	45.5%	4	66.7%	4	80.0%	1	25.0%	2	33.3%	16	50.0%
	Asian	1	9.1%	1	16.7%	0	0.0%	1	25.0%	3	50.0%	6	18.8%
	American Indian	0	0.0%	1	16.7%	0	0.0%	0	0.0%	0	0.0%	1	3.1%
	Unknown	0	0.0%	0	0.0%	0	0.0%	0	0.0%	0	0.0%	0	0.0%
American Indian	White	1	100.0%	1	100.0%	0	0.0%	2	100.0%	1	100.0%	5	100.0%
	Black	0	0.0%	0	0.0%	0	0.0%	0	0.0%	0	0.0%	0	0.0%
	Asian	0	0.0%	0	0.0%	0	0.0%	0	0.0%	0	0.0%	0	0.0%
	American Indian	0	0.0%	0	0.0%	0	0.0%	0	0.0%	0	0.0%	0	0.0%
	Unknown	0	0.0%	0	0.0%	0	0.0%	0	0.0%	0	0.0%	0	0.0%
Total		200		210		228		198		255		1,091	

SOURCE: "Table Nine. Race of Shooter and Person Killed in Justifiable Homicides by Firearm, 2008–2012," in *Firearm Justifiable Homicides and Non-Fatal Self-Defense Gun Use*, Violence Policy Center, June 2015, http://www.vpc.org/studies/justifiable15.pdf (accessed September 10, 2016)

(89.3%) of all justifiable homicides of white people during this period, and African American shooters were responsible for 458 (70.9%) of all justifiable homicides of African American people. As with criminal homicides, handguns were overwhelmingly the most prevalent firearm used in justifiable homicides. Of a total of 1,108 justifiable firearms homicides committed between 2008 and 2012, handguns were used in 838 (75.6%) incidents. (See Table 6.12).

In spite of the lack of objective, indisputable evidence of how often firearms are used in self-defense, scholars overwhelmingly support estimates closer to Hemenway's and the VPC's than to Kleck's. In a survey conducted in June 2014 (http://www.hsph.harvard.edu/wp-content/uploads/sites/1264/2014/05/Expert-Survey2-Results.pdf), the Harvard Injury Control Research Center found that 73% of experts believe with a high degree of certainty that guns are used in crime more often than they are used in self-defense, but only 32% believe there is strong evidence for such a position. This lack of overwhelming evidence is likely to remain a persistent feature of the debate over defensive gun use, given the difficulty of determining how many times guns are used in self-defense without being reported to law enforcement agencies.

Self-Defense Narratives

Regardless of the statistical balance between the use of guns for self-defense and the use of guns to commit homicide and suicide, the fact remains that many people do successfully use guns to defend themselves and their property. In the eyes of many gun rights advocates, these stories show that the right to self-defense is a vital one no matter the aggregate totals of defensive gun use. Accordingly, the NRA maintains a regularly updated feature on

TABLE 6.12

Firearms used in justifiable homicides, by type, 2008–12

Weapon	2008		2009		2010		2011		2012		Total	
Firearm, type not stated	19	9.3%	21	9.9%	28	12.2%	26	12.9%	30	11.6%	124	11.2%
Handgun	162	79.0%	163	76.5%	166	72.2%	152	75.6%	195	75.3%	838	75.6%
Rifle	11	5.4%	9	4.2%	8	3.5%	12	6.0%	19	7.3%	59	5.3%
Shotgun	13	6.3%	19	8.9%	28	12.2%	11	5.5%	15	5.8%	86	7.8%
Other gun	0	0.0%	1	0.5%	0	0.0%	0	0.0%	0	0.0%	1	0.1%
Total	205		213		230		201		259		1,108	

SOURCE: "Table Eleven. Type of Firearms Used in Justifiable Homicides, 2008–2012," in *Firearm Justifiable Homicides and Non-Fatal Self-Defense Gun Use*, Violence Policy Center, June 2015, http://www.vpc.org/studies/justifiable15.pdf (accessed September 10, 2016)

its website called "Armed Citizen" (http://www.nraila.org/gun-laws/armed-citizen.aspx) that summarizes local news accounts of law-abiding gun owners successfully defending themselves with guns. The organization posts two to four stories in a typical week, which demonstrates that such uses of firearms are indeed common by most definitions. For example, typical stories from December 2016 included an account of a home invader in Pennsylvania who was fatally shot by an 81-year-old man after physically assaulting him; an account of a Virginia homeowner who fatally shot an intruder with his own gun after the two got into a struggle; and a Chicago store employee with a valid concealed carry license who shot two men after they tried robbing the store.

EFFORTS TO PROMOTE GUN SAFETY
Safe Storage of Guns in the Home

If a gun is to be used for self-defense, does it make sense to keep it unloaded and locked up? This is a question asked by people who oppose safe-storage laws and laws that hold gun owners criminally liable for any injury caused by a child gaining unsupervised access to a gun. The Law Center to Prevent Gun Violence notes in "Child Access Prevention" (http://smartgunlaws.org/gun-laws/policy-areas/consumer-child-safety/child-access-prevention/) that as of 2016, 27 states and the District of Columbia had child access prevention (CAP) laws. (See Figure 6.2.) Gun control advocates have long pressed for such legislation at the federal level.

Lisa Hepburn et al. provide in "The Effect of Child Access Prevention Laws on Unintentional Child Firearm Fatalities, 1979–2000" (*Journal of Trauma Injury, Infection, and Critical Care*, vol. 61, no. 2, August 2006) some of the most current research available on the impact of CAP laws in various states. The researchers conclude that CAP laws may have influenced the continued reduction in unintentional firearm death rates that occurred between 1979 and 2000 nationally among children. They determine that "the decrease in rates of unintentional firearm deaths for children aged 0 to 14 in CAP law states exceeded the average for states without CAP laws in 9 of the 14 states for which data were available." According to Hepburn et al., statistical analyses of the data "showed a significant association between CAP laws and rates of unintentional firearm deaths for children 0 to 14 years old."

Some argue that if society chooses to hold people accountable for negligent actions or child endangerment, it should do so with all such actions and not single out firearms. In *Deaths: Leading Causes for 2014* (June 30, 2016; http://www.cdc.gov/nchs/data/nvsr/nvsr65/nvsr65_05.pdf), Melonie Heron of the CDC indicates that unintentional injuries (a category that includes drowning, falls, automobile accidents, fires, and many other events that occur both inside and outside of the home) were the number-one cause of death among young people in 2014, accounting for 31.5% of the deaths among children aged one to nine years and 39.7% of those aged 10 to 24 years. Accidental shootings are extremely rare, accounting for only 586 total U.S. deaths in 2014. (See Table 6.8.) Thus, according to the line of reasoning put forward by those who oppose CAP laws, gun control advocates unfairly focus on guns when a variety of parental behaviors are more deadly than negligence having to do with guns.

CAP laws are intended not only to protect children from unintentional injury from firearms but also from gaining access to firearms to commit suicide and homicide. Heron indicates that among youths aged 10 to 24 years, suicide (17.4%) was the second most common cause of death in 2014, followed by homicide (13.6%). Significantly, the CDC notes in "Suicide Prevention: Youth Suicide" (March 10, 2015, http://www.cdc.gov/violenceprevention/suicide/youth_suicide.html) that firearms are the leading method of suicide among young people (as among adults). However, the effectiveness of preventing youth suicide and homicide by restricting their access to guns has not been firmly established.

According to Jeffrey A. Bridge et al., in "Changes in Suicide Rates by Hanging and/or Suffocation and Firearms among Young Persons Aged 10–24 Years in the United States: 1992–2006" (*Journal of Adolescent*

FIGURE 6.2

Child access prevention laws, by state

- ☐ No child access prevention statute
- ▢ Charges only if owner intentionally, knowingly, or recklessly gives a gun to a child
- ▣ Charges if child accesses a carelessly stored gun
- ■ Charges if child may or is likely to access carelessly stored gun

CAP = Child Access Prevention.

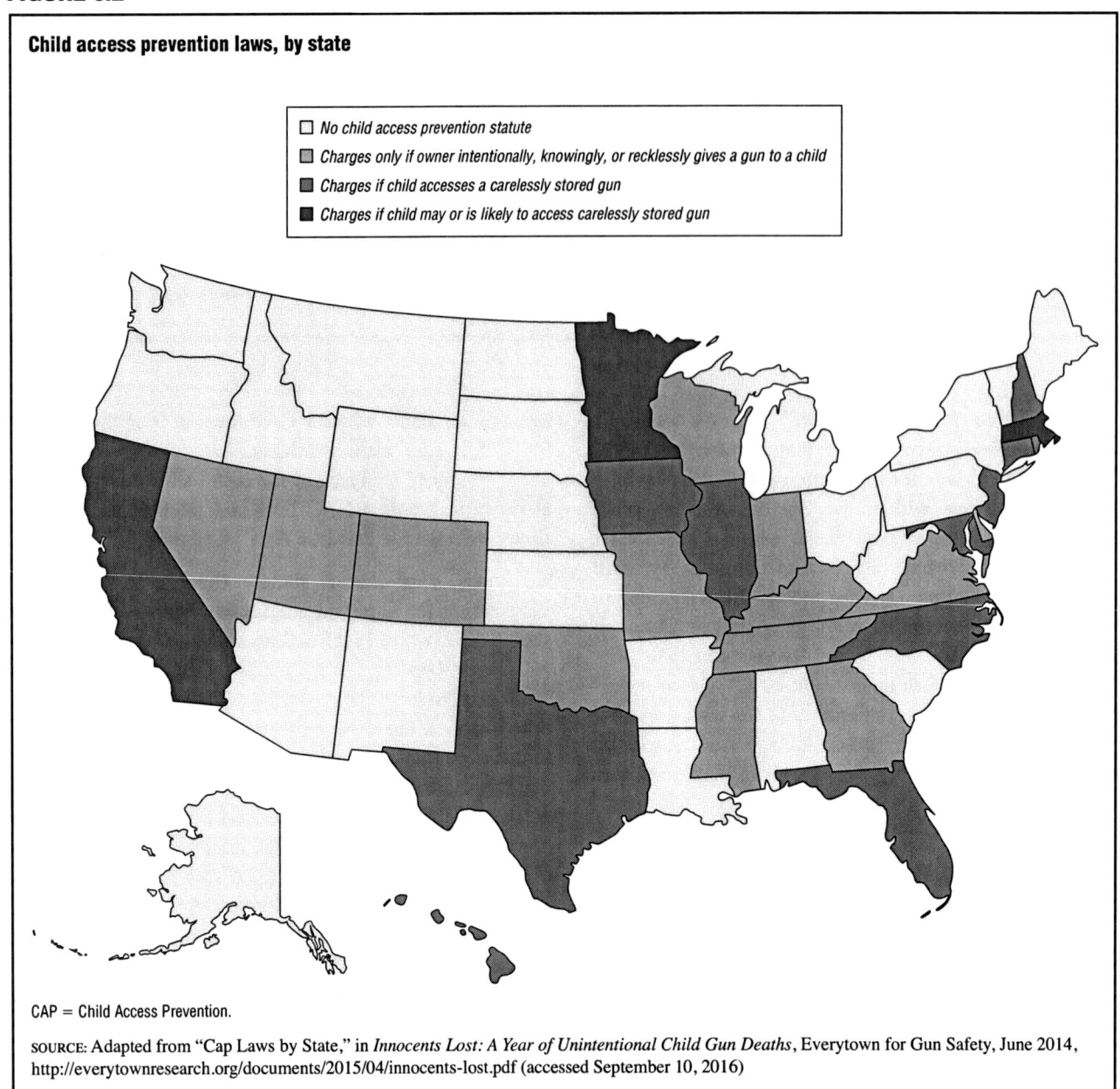

SOURCE: Adapted from "Cap Laws by State," in *Innocents Lost: A Year of Unintentional Child Gun Deaths*, Everytown for Gun Safety, June 2014, http://everytownresearch.org/documents/2015/04/innocents-lost.pdf (accessed September 10, 2016)

Health, vol. 46, no. 5, May 2010), the number of youth suicides committed with firearms decreased between 1992 and 2006; however, the researchers do not know if this decrease was due to CAP laws. They note that suicide by poisoning and other methods declined simultaneously in this age group, while suicide by hanging and suffocation increased concurrently. These data suggest a selection of suffocation and hanging rather than a thwarting of other methods. Bridge et al. indicate that hanging and suffocation were the leading methods of suicide among youth in many countries, such as Australia, England, and New Zealand, and recommend that future research investigate the reasons for this preference so that public health programs might be developed to curb this method of suicide.

Safety Programs to Protect Young Children

In 1988 the NRA created the Eddie Eagle GunSafe Program (https://eddieeagle.nra.org/) for gun safety. Eddie Eagle is a school-based program that teaches gun safety to young children (preschool to third grade). With the help of the cartoon character Eddie the Eagle, kids are taught that when they find a gun they should not touch it, but instead leave the area and tell an adult. The VPC contends in "Joe Camel with Feathers: How the NRA with Gun and Tobacco Industry Dollars Uses Its Eddie Eagle Program to Market Guns to Kids" (November 1997, http://www.vpc.org/studies/eddiecon.htm) that the program is a marketing tool for the NRA, allowing it to whitewash the dangers of guns and recruit a new generation of customers for the firearms industry. A segment on the ABC News program *20/20*, which aired in 1999, made a similar argument. Michael B. Himle et al. assess in "An Evaluation of Two Procedures for Training Skills to Prevent Gun Play in Children" (*Pediatrics*, vol. 113, no. 1, January 1, 2004) the effectiveness of the Eddie Eagle program and a separate program of behavioral skills training. They find that both programs increased children's awareness of the tenets of gun safety, that the behavioral training program alone (and not the Eddie Eagle program) was effective in teaching gun-safety skills as observed during supervised role play, and that neither program was effective in teaching children to apply gun safety skills outside of the controlled environment of the training sessions.

In "Peer Tutoring to Prevent Firearm Play: Acquisition, Generalization, and Long-Term Maintenance of Safety Skills" (*Journal of Applied Behavior Analysis*, vol. 41, no. 1, 2008), Candace M. Jostad et al. support the behavioral skills training approach and show that six- and seven-year-old children acquire and maintain firearm safety skills when behavioral skills training is used and is taught by other children trained in the techniques.

CHAPTER 7
GUNS AND YOUTH

Young people are often at the center of the gun control debate. Gun rights advocates maintain that responsible firearm use is the key to gun safety, but children, adolescents, and many young adults lack the capacity for sound judgment that is the necessary precondition to responsible firearm use. A patchwork of federal and state laws is designed to prevent the purchase and possession of firearms by minors (in some cases defined as those under the age of 21 years and in other cases as those under the age of 18 years) except under special circumstances, but young people who commit murder or suicide often do so with guns owned by their parents or family members, and guns used by underage criminals are often obtained illegally. The question of how to balance gun owners' rights with the public welfare thus inevitably arises when young people kill one another or themselves with firearms.

During the late 1980s and early 1990s a confluence of trends contributed to a dramatic increase in the prevalence of firearm use among youths. For the first time, the firearms industry began mass producing and popularizing semiautomatic pistols, spurring an unprecedented proliferation of handguns in the United States. Meanwhile, shifts in the labor market led to high rates of youth joblessness, particularly among urban minorities, while increases in youth gang activity paralleled the rise in illegal drug markets, especially for crack cocaine. During this period the rate of firearm-related deaths (including homicides, suicides, and deaths from unintended injuries) among youths aged 15 to 19 years more than doubled—from 12.5 per 100,000 in 1984 to 27.8 per 100,000 in 1994. (See Figure 7.1.) Together, these trends were among gun control advocates' most powerful reasons to push for stricter gun laws. During the late 1990s and into the next decade, however, the homicide and violent crime rates nationwide and in many previously violent urban areas declined dramatically. By 2004 the rate of firearm-related deaths among youths aged 15 to 19 years had fallen to 11.8 per 100,000, its lowest rate since the early 1970s. The reasons for the decline in youth firearm-related violence have not been conclusively established, but little evidence exists to suggest that it was a result of gun control efforts because the period was not marked by any substantial shift in gun laws.

Youth firearm violence remains a major concern in many urban areas, however, and mass shootings in schools and other public areas have by most accounts increased in regularity since the first decade of the 21st century. Furthermore, as is discussed in Chapter 6, although the firearm homicide rate has declined nationally, the firearm suicide rate remains high. The suicide rate for the U.S. population has neither increased nor decreased dramatically since the mid-20th century, but the suicide rate for young people, and for young males in particular, has increased significantly since that time.

DEADLY ASSAULTS

In *Homicide Trends in the United States, 1980–2008* (November 2011, http://bjs.ojp.usdoj.gov/content/pub/pdf/htus8008.pdf), the most comprehensive study on this topic as of December 2016, Alexia Cooper and Erica L. Smith of the Bureau of Justice Statistics attribute the increase in homicides during the late 1980s and early 1990s and the subsequent decline to gun violence among teenagers and young adults; homicide victimization rates for teens (aged 14 to 17 years) peaked in 1993 at 12 homicides per 100,000 and declined after that time to 5.1 homicides per 100,000 in 2008. Similarly, homicide offending reached a high of 30.7 teen offenders per 100,000 in that age group in 1993; since 2000 the rate of homicide offenders among teenagers has remained relatively stable at a level similar to that which occurred in 1985 (9.5 per 100,000).

Melissa Sickmund and Charles Puzzanchera of the National Center for Juvenile Justice indicate in *Juvenile Offenders and Victims: 2014 National Report* (December

FIGURE 7.1

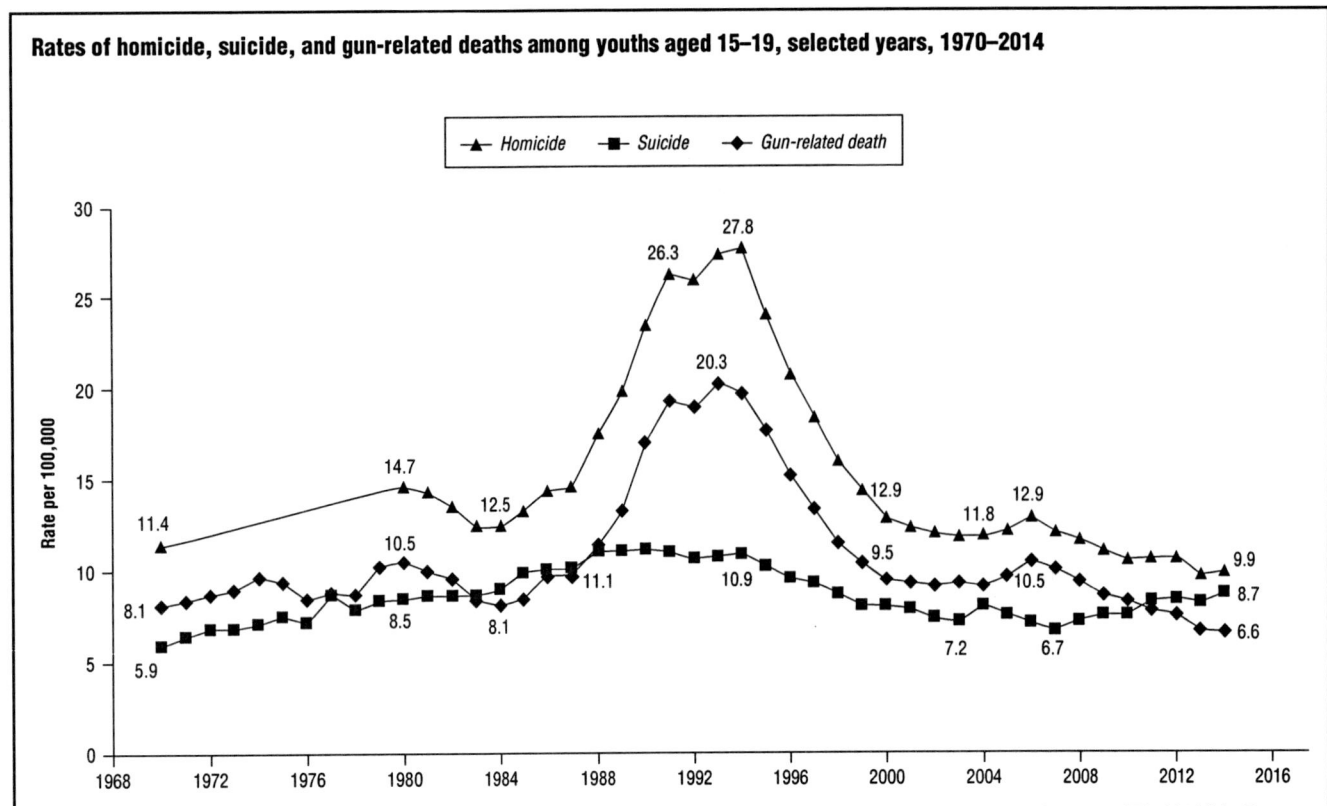

Rates of homicide, suicide, and gun-related deaths among youths aged 15–19, selected years, 1970–2014

SOURCE: "Figure 1. Rates (per 100,000) for Homicide, Suicide, and Firearm-Related Deaths of Youth, Ages 15–19: Selected Years, 1970–2014," in *Teen Homicide, Suicide, and Firearm Deaths*, Child Trends, December 2015, http://www.childtrends.org/wp-content/uploads/2014/10/70_Homicide_Suicide_Firearms.pdf (accessed September 12, 2016)

TABLE 7.1

Juveniles murdered between 1980 and 2010, by age and weapon type

Weapon	Age of victim 0–17	0–5	6–11	12–14	15–17	Victim ages 0–17 Male	Female
Total	100%	100%	100%	100%	100%	100%	100%
Firearm	50	10	42	66	78	60	30
Knife/blunt object	14	11	19	17	14	12	16
Personal*	19	48	11	5	2	16	28
Other/unknown	17	31	28	12	6	13	26

*Personal includes hands, fists, or feet.
Note: Detail may not total 100% because of rounding.

SOURCE: Melissa Sickmund and Charles Puzzanchera, editors, "Of the 58,900 Juveniles Murdered between 1980 and 2010, Half Were Murdered with a Firearm," in *Juvenile Offenders and Victims: 2014 National Report*, National Center for Juvenile Justice, December 2014, http://www.ojjdp.gov/ojstatbb/nr2014/downloads/NR2014.pdf (accessed September 12, 2016)

2014, http://www.ojjdp.gov/ojstatbb/nr2014/downloads/NR2014.pdf) that between 1980 and 2010 approximately 58,900 juveniles were murdered. Of these, half were murdered with a firearm. (See Table 7.1.) By comparison, 14% were killed with a knife or blunt object, 19% were killed by "personal" means (including hands, fists, and feet), and 17% were killed by other or unknown means. Among all juvenile homicide victims, the likelihood of being killed by a firearm increased with age. Those between the ages of 15 and 17 years were by far the most likely to be killed by a firearm. Indeed, 78% of all homicide victims in this age bracket were killed by firearms.

Looking at data from 1980 to 2010, Sickmund and Puzzanchera note significant disparities in firearm homicide victimization rates for males and females in adolescence and early adulthood. Whereas adolescent girls under the age of five years were just as likely to be killed with a firearm as boys of that age, boys became much more likely to be killed with a firearm when they entered adolescence. Among homicide victims between the ages of 14 and 17 years, 83% of boys were killed with a firearm, compared with 56% of girls. Not until late middle age did these gender disparities begin to narrow in a significant way.

Table 7.2 presents the number and rate of firearm deaths for those aged zero to 19 years in 2014. The highest victimization rate occurred among male teenagers aged 15 to 19 years, at 21.48 firearm deaths per 100,000 population. Females in this same age group registered the highest rate among girls, at 2.39 per 100,000. With the exception of boys aged 10 to 14 years, whose firearm death rate was slightly elevated, at 1.97 per 100,000, children under the age of 14 years were among the least likely of all population groups to be killed by firearms.

Figure 7.2 shows homicide trends between 1980 and 2008 by victim/offender relationship and weapon use. The rise in gun-related homicides between 1984 and 1994 is

TABLE 7.2

Firearm deaths of children aged 0–19 years and rates per 100,000, 2014

Age group	Sex	Deaths	Population	Rate
0 to 4	Males	49	10,175,595	0.48
	Females	20	9,741,805	0.21
	Total	**69**	**19,917,400**	**0.35**
5 to 9	Males	38	9,915,167	0.38
	Females	24	9,473,900	0.25
	Total	**62**	**19,389,067**	**0.32**
10 to 14	Males	214	10,864,197	1.97
	Females	59	10,348,382	0.57
	Total	**273**	**21,212,579**	**1.29**
15 to 19	Males	2,374	11,051,425	21.48
	Females	249	10,434,789	2.39
	Total	**2,623**	**21,486,214**	**12.21**

SOURCE: Adapted from "Fatal Injury Reports, National and Regional, 1999–2014," in *Web-Based Injury Statistics Query and Reporting System (WISQARS)*, Centers for Disease Control and Prevention, 2016, http://webappa.cdc.gov/sasweb/ncipc/mortrate10_us.html (accessed September 9, 2016)

FIGURE 7.2

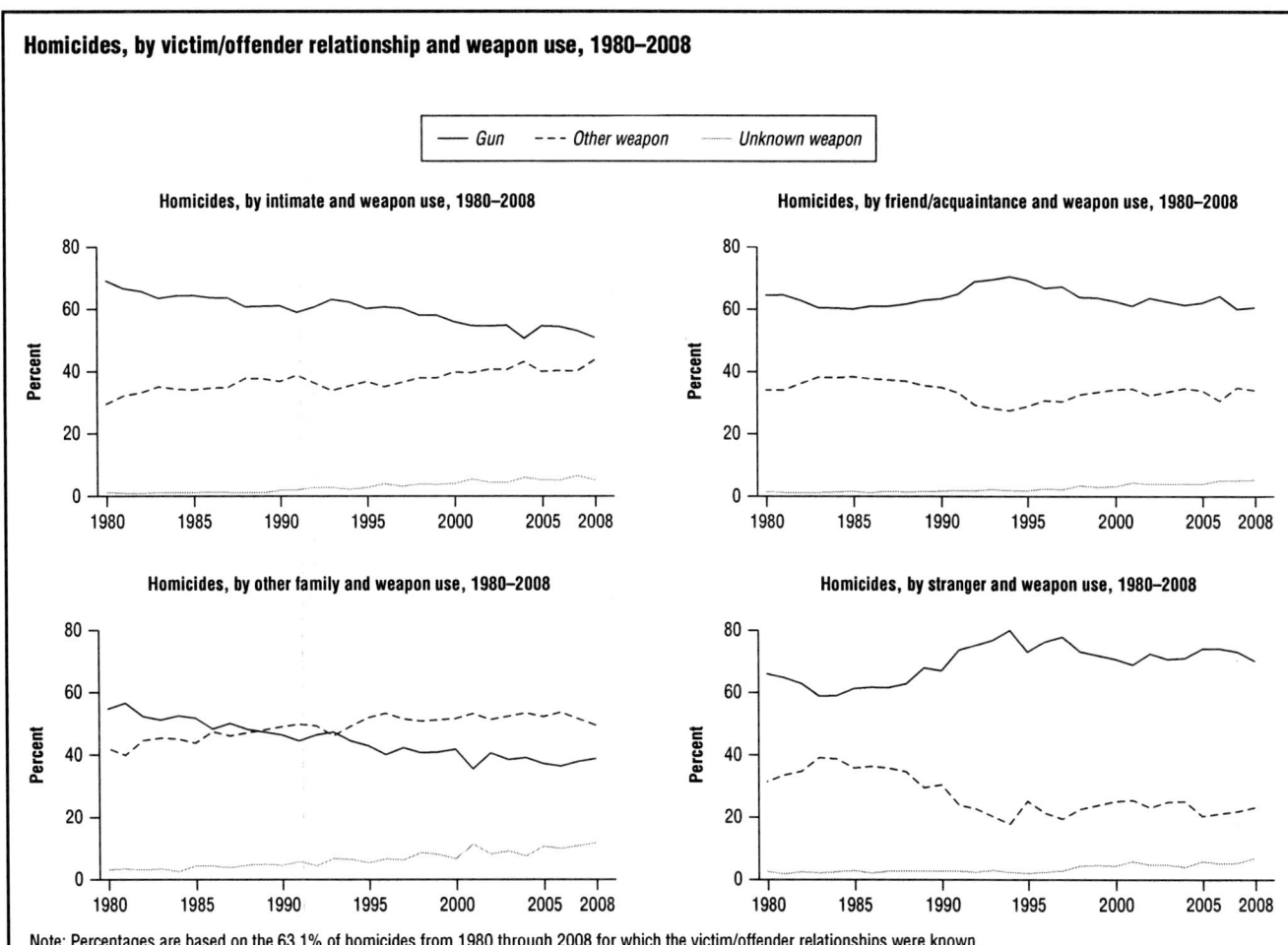

Homicides, by victim/offender relationship and weapon use, 1980–2008

Note: Percentages are based on the 63.1% of homicides from 1980 through 2008 for which the victim/offender relationships were known.

SOURCE: Adapted from Alexia Cooper and Erica L. Smith, "Figure 25a. Homicides, by Intimate and Weapon Use, 1980–2008," "Figure 25b. Homicides, by Other Family and Weapon Use, 1980–2008," "Figure 25c. Homicides, by Friend/Acquaintance and Weapon Use, 1980–2008," and "Figure 25d. Homicides, by Stranger and Weapon Use, 1980–2008," in *Homicide Trends in the United States, 1980–2008*, U.S. Department of Justice, Bureau of Justice Statistics, November 2011, http://www.bjs.gov/content/pub/pdf/htus8008.pdf (accessed September 12, 2016)

evident in homicides by friends and acquaintances and is even more pronounced in homicides by strangers. Gun violence among family members and those in intimate relationships showed a gradual decline during the 1980s and 1990s. Among family members, other weapons (a category that includes knives, sharp instruments, and blunt objects as well as personal weapons such as hands and feet) overtook guns as the most likely type of weapon to be used in a homicide.

Youthful Offenders

Table 7.3 illustrates demographic characteristics of juvenile homicide offenders in 1994 (the peak year for number of juvenile homicide offenders) and 2010, along with data concerning weapons use and relationship of offender to victim. Although the proportion of juvenile homicide offenders aged 15 years and younger declined from 1994 to 2010, the proportion of juvenile homicide offenders aged 16 and 17 years rose, with 17-year-old offenders seeing the greatest increase, from 41% in 1994 to 46% in 2010. The most significant change reflected in Table 7.3 is the decline in firearm-related homicides from 1994 to 2010. Specifically, the presence of firearms in homicide incidents declined by 11 percentage points, from 81% to 70%.

Figure 7.3 reflects the pronounced fluctuations between 1980 and 2010 of homicides in which a juvenile offender killed an acquaintance or a stranger. Although the number of homicides in which a juvenile offender killed a family member remained fairly stable between 1980 and 1994, instances of a juvenile offender killing an acquaintance and a stranger both more than doubled during the same period. After the 1994 peak, the decline in these nonfamily homicides was similarly dramatic: in 2010 juvenile killing of acquaintances reached its lowest level of the 30-year period, while juvenile killing of strangers reached its lowest level since 2003. Figure 7.4 shows the number of juveniles who committed homicide with and without guns between 1980 and 2014. The number of juveniles who committed homicide with a gun rose precipitously after 1984, peaked in 1994, and then returned close to 1984 levels by 2014.

Why did juvenile gun-related homicides rise between 1984 and 1994? The Virginia Youth Violence Project at the University of Virginia's Curry School of Education suggests in "Juvenile Homicide" (2010) that violent juvenile crime during that time was linked to many factors, including the introduction of crack cocaine, the availability of cheap handguns, inadequate after-school supervision, and the prevalence of violence in the media. Reasons for the subsequent fall have been much debated, and experts disagree about the relative importance of different factors. Some of the most widely acknowledged factors explaining the trend toward improved safety in urban areas include increases in the number of police as well as the introduction of new police techniques; the gentrification of urban areas, whereby wealthier residents moved into previously impoverished neighborhoods and businesses opened to serve their needs, leading to increasing real estate prices and the dispersal of criminal activity; and the receding crack epidemic.

JUVENILE ARRESTS. Table 7.4 shows the estimated number of juvenile arrests in 2014 and the percentage change in juvenile arrests over various time spans. Juvenile crime fell 50% overall between 2005 and 2014, and violent crime perpetrated by juveniles fell 44%. Decreases during this period were significant in most types of crimes in which firearms were typically used, including murder and nonnegligent manslaughter (−40%) and aggravated assault (−51%). Meanwhile, weapons offenses fell 54% over the course of those 10 years.

RACE AND GENDER. Young males are far more likely than their female counterparts to be homicide offenders. Sickmund and Puzzanchera of the National Center for Juvenile Justice indicate that in 2010 males accounted for the overwhelming majority (91%) of all juvenile homicide offenders. (See Table 7.3.) Among all young males, African American males are disproportionately likely to be homicide offenders. In 2010, among the total population of juveniles aged 10 and 17 years, 76% were white and 17% were African American. Whites, however, accounted for 35% of juvenile homicide offenders (less

TABLE 7.3

Demographics of juvenile homicide offenders, selected years 1994 and 2010

Characteristic	1994	2010
Age	100%	100%
Younger than 15	12	9
Age 15	18	15
Age 16	29	30
Age 17	41	46
Gender	100%	100%
Male	94	91
Female	6	9
Race	100%	100%
White	36	35
Black	61	63
Other race	3	3
Weapon presence	100%	100%
Firearm	81	70
No firearm	19	30
Relationship to victim	100%	100%
Family	7	11
Acquaintance	55	48
Stranger	37	42

Note: 1994 was the peak year for number of juvenile homicide offenders. Detail may not total 100% because of rounding.

SOURCE: Melissa Sickmund and Charles Puzzanchera, editors, "Profile of Juvenile Homicide Offenders Known to Law Enforcement," in *Juvenile Offenders and Victims: 2014 National Report*, National Center for Juvenile Justice, December 2014, http://www.ojjdp.gov/ojstatbb/nr2014/downloads/NR2014.pdf (accessed September 12, 2016)

FIGURE 7.3

Juvenile homicide offenders, by relationship to victim, 1980–2010

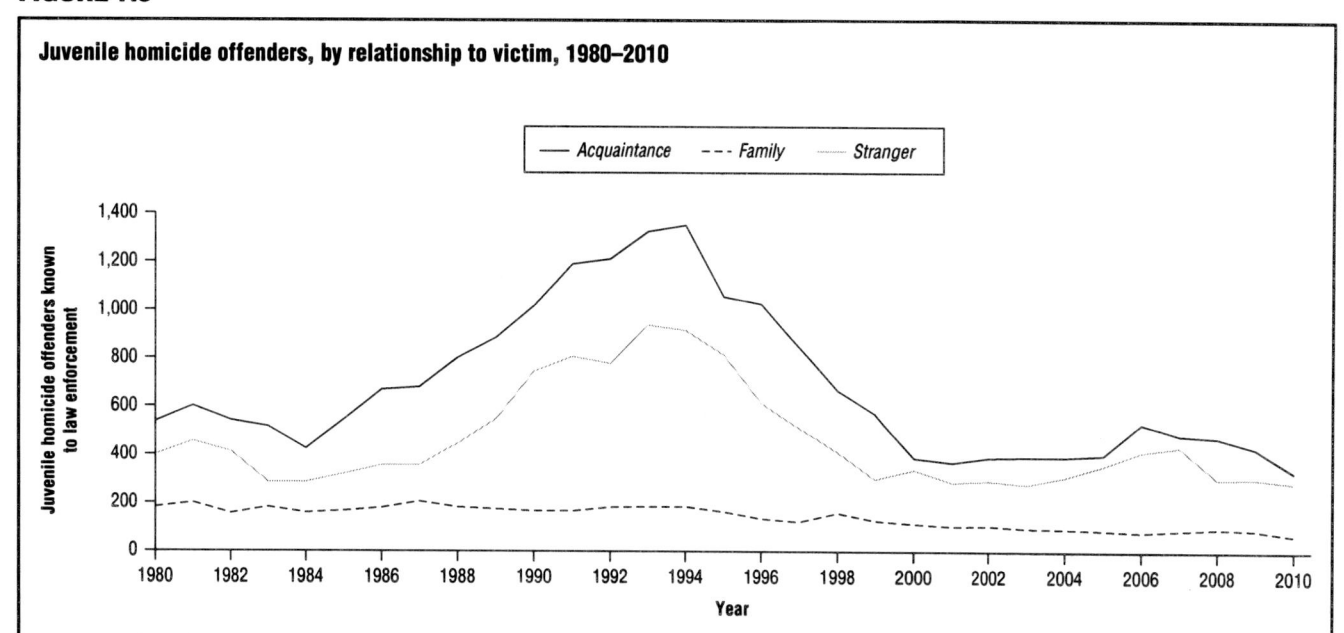

SOURCE: Melissa Sickmund and Charles Puzzanchera, editors, "The Number of Juvenile Offenders Who Killed Acquaintances and Strangers Varied Considerably between 1980 and 2010," in *Juvenile Offenders and Victims: 2014 National Report*, National Center for Juvenile Justice, December 2014, http://www.ojjdp.gov/o-jstatbb/nr2014/downloads/NR2014.pdf (accessed September 12, 2016)

FIGURE 7.4

Homicides committed by juveniles, by firearm involvement, 1980–2014

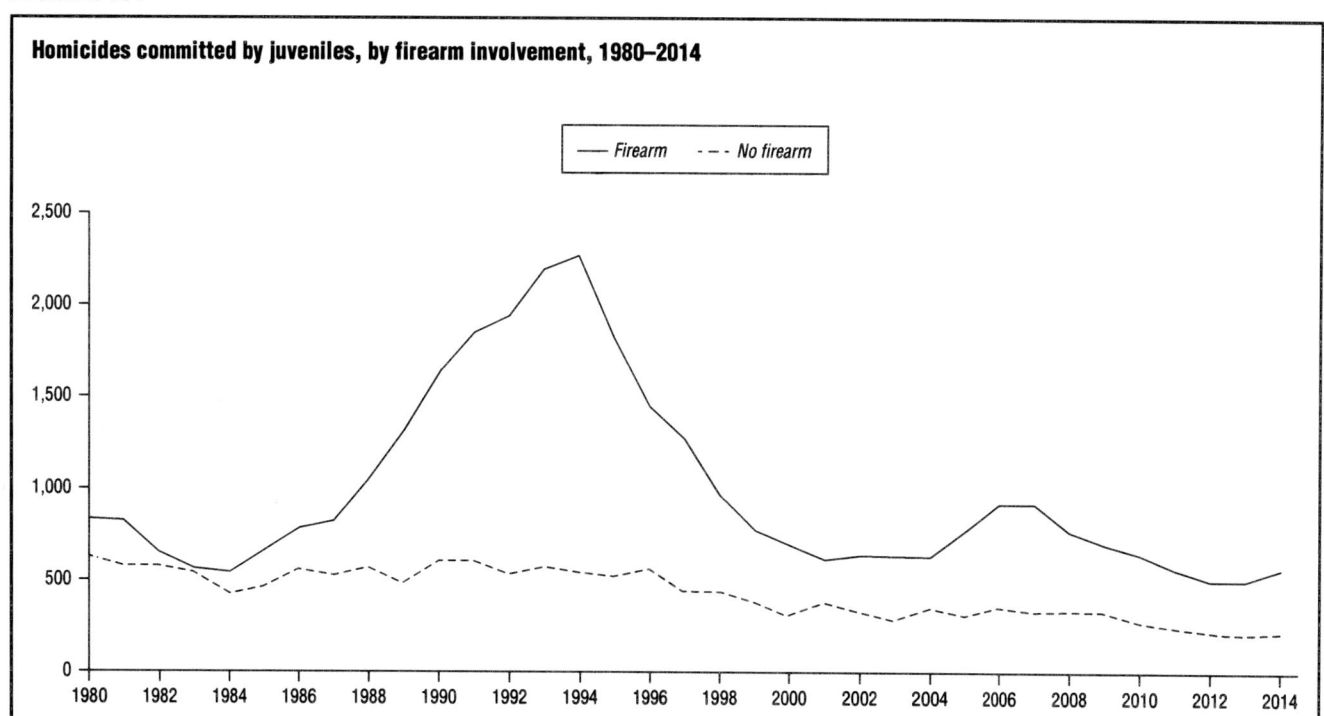

Notes: Between 1984 and 1994, the number of firearm related homicides committed by known juvenile offenders quadrupled. The sharp decline in homicides by known juvenile offenders between 1994 and 2001 was attributable entirely to a decline in homicides by firearm. Between 2002 and 2014, the number of nonfirearm-related homicides committed by known juvenile offenders was relatively stable. Firearm-related murders by juveniles increased between 2001 and 2007, then declined 39% through 2014. The number of firearm-related homicides involving known juvenile offenders in 2013 was at its lowest level since at least 1980, then increased 13% by 2014. In the mid 1990s, about 80% of known juvenile homicide offenders were involved in firearm-related homicides; this percentage fell to 62% in 2001 then rose to 72% by 2014.

SOURCE: "Known Juvenile Homicide Offenders by Firearm Involvement, 1980–2014," in *OJJDP Statistical Briefing Book*, U.S. Department of Justice, Office of Juvenile Justice and Delinquency Prevention, May 25, 2016, http://www.ojjdp.gov/ojstatbb/offenders/qa03103.asp?qaDate=2014 (accessed September 10, 2016)

TABLE 7.4

Juvenile arrests, 2014, and percentage change, 2005–14

Most serious offense	Number of juvenile arrests	Percent change 2005–14	Percent change 2010–14	Percent change 2013–14
Total	1,024,000	−50%	−38%	−6%
Violent crime index	53,500	−44%	−29%	1%
Murder and nonnegligent manslaughter	800	−40%	−25%	1%
Rape[a]	3,300			
Robbery	19,400	−33%	−29%	2%
Aggravated assault	30,100	−51%	−33%	−2%
Property crime index	234,200	−44%	−36%	−5%
Burglary	40,300	−48%	−38%	−6%
Larceny-theft	178,000	−39%	−37%	−5%
Motor vehicle theft	12,700	−66%	−19%	9%
Arson	3,200	−61%	−31%	−14%
Nonindex				
Other assaults	139,100	−44%	−34%	−5%
Forgery and counterfeiting	1,200	−72%	−31%	10%
Fraud	4,300	−45%	−25%	−4%
Embezzlement	500	−59%	5%	18%
Stolen property (buying, receiving, possessing)	10,400	−53%	−29%	0%
Vandalism	45,200	−57%	−41%	−3%
Weapons (carrying, possessing, etc.)	20,700	−54%	−34%	1%
Prostitution and commercialized vice	700	−54%	−29%	−2%
Sex offenses (except forcible rape and prostitution)	9,400	−44%	−27%	−9%
Drug abuse violations	112,600	−42%	−34%	−4%
Gambling	600	−72%	−57%	−20%
Offenses against the family and children	3,400	−39%	−9%	20%
Driving under the influence	7,000	−61%	−42%	−8%
Liquor law violations	53,300	−57%	−44%	−12%
Drunkenness	6,500	−59%	−49%	−12%
Disorderly conduct	80,800	−61%	−48%	−15%
Vagrancy	900	−75%	−56%	8%
All other offenses (except traffic)	186,000	−49%	−37%	−5%
Curfew and loitering	53,700	−62%	−43%	−5%
Runaways[b]	n/a	n/a	n/a	n/a

[a]The new definition of rape went into effect in 2013. The revised definition expands rape to include both male and female victims, and reflects the various forms of sexual penetration understood to be rape, especially nonconsenting acts of sodomy, and sexual assaults with objects. The percent change in the number of rape arrests is not displayed since the new definition is not compatible with the prior definition.
[b]As of 2010, the Federal Bureau of Investigation (FBI) no longer reports arrests for running away.
Note: Arrest estimates for 2014 developed by the National Center for Juvenile Justice based on data published in the FBI's 2014 Crime in the United States report. These are preliminary estimates that will be updated upon release of final estimates on the Bureau of Justice Statistics' Arrest Data Analysis Tool. OJJDP = Office of Juvenile Justice and Delinquency Prevention.

SOURCE: "Estimated Number of Juvenile Arrests, 2014," in *OJJDP Statistical Briefing Book*, U.S. Department of Justice, Law Enforcement and Juvenile Crime, December 13, 2015, http://www.ojjdp.gov/ojstatbb/crime/qa05101.asp?qaDate=2014 (accessed September 10, 2016)

---

than their share of the total population) in 2010, whereas African Americans accounted for 63% (significantly more than their share of the total population).

As reflected in Figure 7.5, the number of known white juvenile homicide offenders doubled between 1984 and 1994, while the number of African American offenders quadrupled. Sickmund and Puzzanchera note that the number of known white juvenile homicide offenders declined 68% between 1994 and 2010, while the number of known African American juvenile homicide offenders fell 67% during the same period.

Youthful Victims

Among youths under the age of 18 years, the risk of becoming a victim of violent crime varies considerably depending on age, gender, and race. According to Sickmund and Puzzanchera, 1,450 juveniles were murdered in 2010, accounting for 10% of all U.S. homicides that year. Among these victims, 42% were under six years of age, 6% were aged six to 11 years, 7% were aged 12 to 14 years, and 45% were aged 15 to 17 years. As Table 7.5 shows, young females (31%) were much more likely than young males (17%) to be killed by a parent or stepparent, whereas young males (33%) were somewhat more likely than young females (28%) to be killed by an acquaintance.

Sickmund and Puzzanchera report there were approximately 58,900 juvenile homicide victims between 1980 and 2010, half of whom were killed with a firearm. As Figure 7.6 illustrates, the number of juveniles killed with firearms rose dramatically from 1984, when there were fewer than 600 juvenile gun homicides, to 1993, when the number had nearly tripled to about 1,700. Approximately 60% of juvenile homicide victims were killed with a

FIGURE 7.5

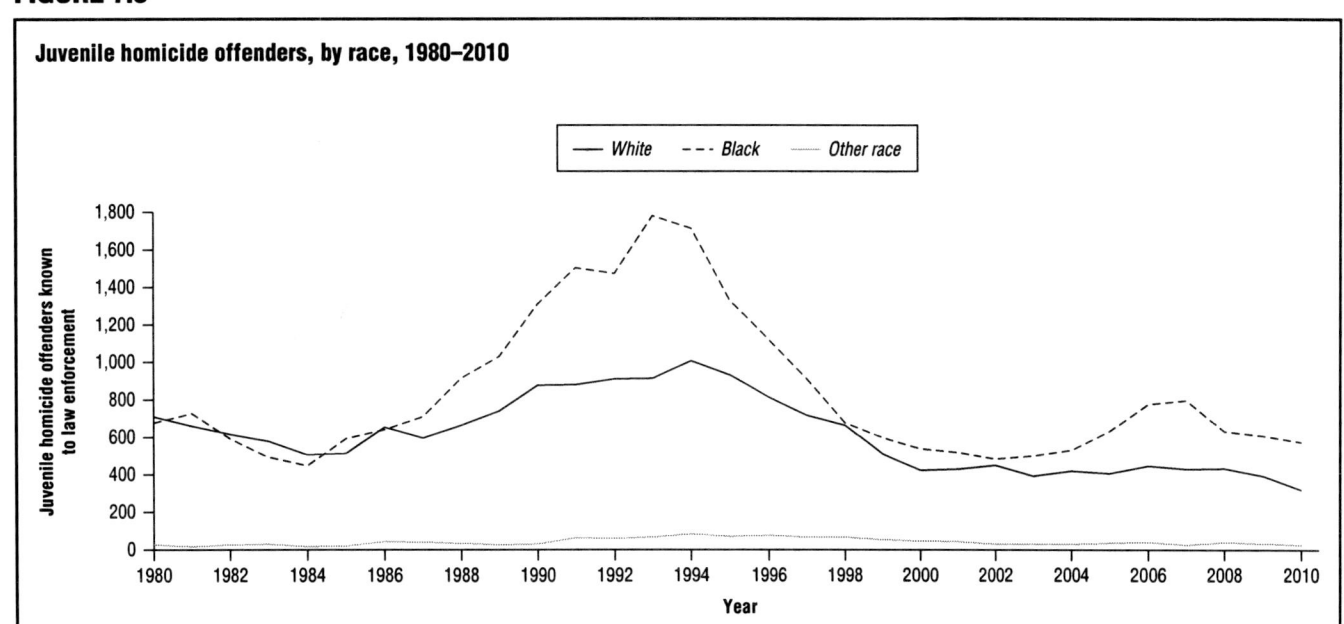

Juvenile homicide offenders, by race, 1980–2010

SOURCE: Melissa Sickmund and Charles Puzzanchera, editors, "The Number of Juvenile Homicide Offenders in 2010 Was about One-Third the Number in 1994 for Both White Youth and Black Youth," in *Juvenile Offenders and Victims: 2014 National Report*, National Center for Juvenile Justice, December 2014, http://www.ojjdp.gov/ojstatbb/nr2014/downloads/NR2014.pdf (accessed September 12, 2016)

TABLE 7.5

Juveniles murdered between 1980 and 2010, by age and relationship to offender

Offender relationship to victim	Age of victim 0–17	0–5	6–11	12–14	15–17	Victim ages 0–17 Male	Female
Offender known	67%	82%	60%	62%	58%	65%	71%
Total	100	100	100	100	100	100	100
Parent/stepparent	21	51	24	6	2	17	31
Other family member	4	5	8	6	3	4	6
Acquaintance	31	23	18	37	38	33	28
Stranger	10	2	10	13	16	12	7
Offender unknown	33	18	40	38	33	35	29

Note: Detail may not total 100% because of rounding.

SOURCE: Melissa Sickmund and Charles Puzzanchera, editors, "Of the 58,900 Juveniles Murdered between 1980 and 2010, Most Victims under Age 6 Were Killed by a Parent, While Parents Were Rarely Involved in the Killing of Juveniles Ages 15–17," in *Juvenile Offenders and Victims: 2014 National Report*, National Center for Juvenile Justice, December 2014, http://www.ojjdp.gov/ojstatbb/nr2014/downloads/NR2014.pdf (accessed September 12, 2016)

firearm each year between 1992 and 1995. After falling considerably from the mid-1990s into the next decade, the number of juvenile gun homicides increased again from about 700 in 2003 to more than 900 in 2006. Between 2006 and 2010 the number again declined, although it remained higher than the 1984 level.

According to Sickmund and Puzzanchera, juveniles aged 15 to 17 years were more likely than any other age group to be killed with a firearm. Indeed, between 1980 and 2010, four out of five homicide victims in this age bracket were killed with a firearm. (See Figure 7.7.) Although the percentage of female firearm homicides increased from about 40% at the age of nine to more than 55% at the age of 17, the increase was much more dramatic for young males. Whereas firearms caused 45% of the homicides among nine-year-old boys between 1980 and 2010, they were responsible for nearly 85% of homicides among 17-year-old boys. For purposes of comparison, note that less than half of homicide victims of both sexes aged 63 years and older or aged 11 years and younger were killed with guns during the same period.

Young children are less frequently the victims of gun violence than older children. Table 7.6 shows that both the number and the rate of homicide firearm deaths for children aged 12 years and younger remained steady between 2005 and 2014. The number and rate of firearm

FIGURE 7.6

Juvenile homicide victims killed with and without firearms, 1980–2010

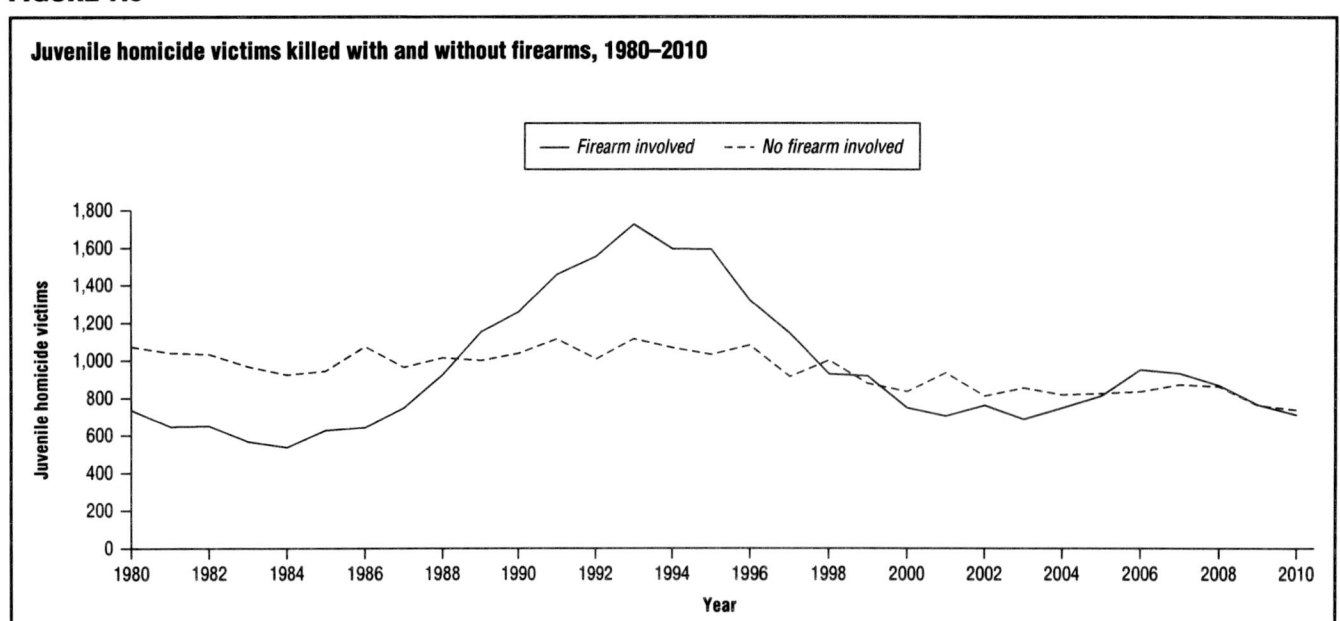

SOURCE: Melissa Sickmund and Charles Puzzanchera, editors, "The Growth in the Number of Juveniles Murdered Using a Firearm That Began in 2003 Was Reversed between 2006 and 2010 as the Number Fell 25% over the Past 4 Years," in *Juvenile Offenders and Victims: 2014 National Report*, National Center for Juvenile Justice, December 2014, http://www.ojjdp.gov/ojstatbb/nr2014/downloads/NR2014.pdf (accessed September 12, 2016)

FIGURE 7.7

Percentage of homicides involving guns, by age of victim, 1980–2010

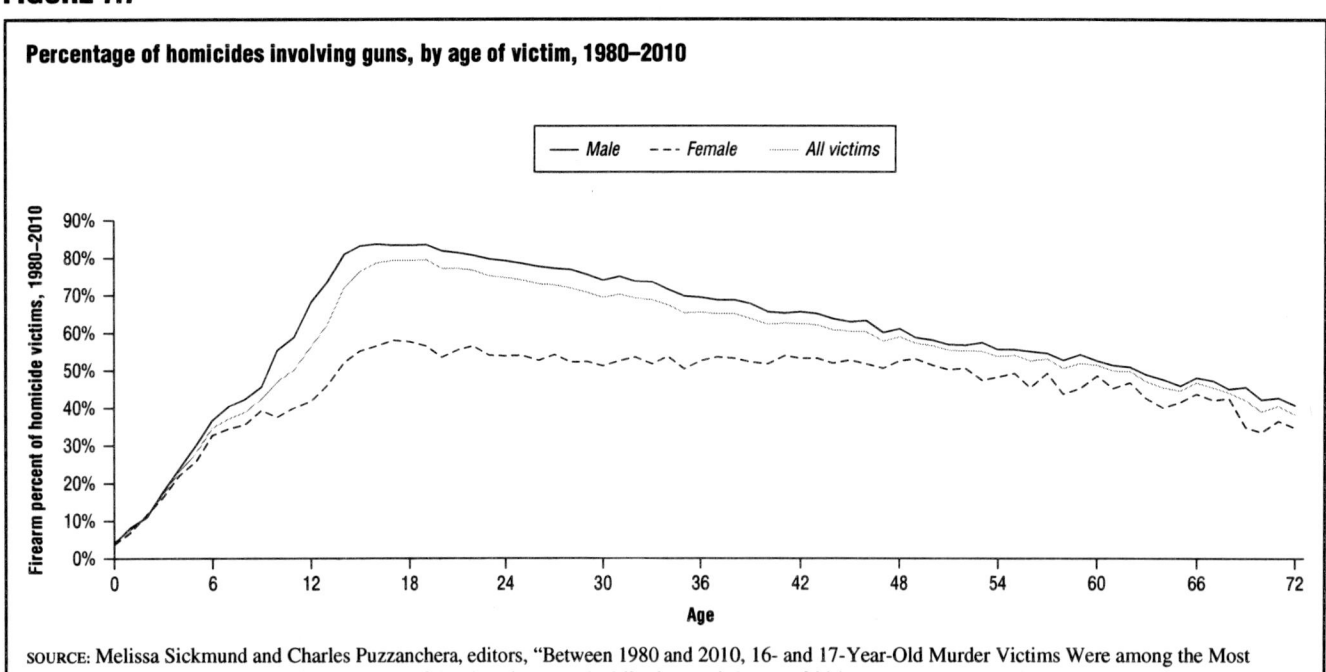

SOURCE: Melissa Sickmund and Charles Puzzanchera, editors, "Between 1980 and 2010, 16- and 17-Year-Old Murder Victims Were among the Most Likely to Be Killed with Firearms, Regardless of Gender," in *Juvenile Offenders and Victims: 2014 National Report*, National Center for Juvenile Justice, December 2014, http://www.ojjdp.gov/ojstatbb/nr2014/downloads/NR2014.pdf (accessed September 12, 2016)

homicide deaths in this age group varied from a low of 126 deaths and a rate of 0.24 per 100,000 in 2005 to a high of 161 deaths in 2008, with a rate of 0.31. There were fluctuations within this range but no general observable tendency in either direction over the course of the decade.

RACE AND GENDER. Besides being disproportionately likely to be homicide offenders, young males, and young African American males in particular, are disproportionately likely to be homicide victims. Sickmund and Puzzanchera indicate that in 2010 male juveniles accounted for roughly twice the number of homicide victims as their female counterparts. Meanwhile, African American youth accounted for 47% of juvenile homicide victims between 1980 and 2010, even though they represented only about 16% of the juvenile population. As Figure 7.8 illustrates, the racial disparity between juvenile homicide victims peaked in 1993, when, given their much smaller share of the juvenile population as a whole, African Americans were about six times more likely to be homicide victims than their white counterparts. In 2010 the disparity had dropped to a ratio of just under five to one.

Smith and Cooper observe in *Homicide in the U.S. Known to Law Enforcement, 2011* (December 2013, http://www.bjs.gov/content/pub/pdf/hus11.pdf) that more than 75% of young white male homicide victims were killed with firearms between 2008 and 2011, slightly down from the more than 80% killed with firearms between 1992 and 1995, the period when the homicide rate peaked. Meanwhile, more than 90% of young African American male homicide victims were killed with firearms between 2008 and 2011, a proportion that was unchanged from the period of peak homicide rates.

According to Smith and Cooper, between 2002 and 2011 young white females were less likely to be homicide victims than young white males, and young African American females were less likely to be homicide victims than young African American males. For young white females, the homicide rate peaked in infancy: girls younger than one had a homicide rate of 4.5 per 100,000. African American girls under age one also had a notably high homicide rate of 10.3 per 100,000, but the peak rate for young African American females came at age 22, at 11.8 per 100,000. Thus, although young males in general were far more likely to be homicide victims than young females, African American females aged 15 to 24 years were approximately as likely as white males of the same age to be homicide victims.

Additionally, young African American female homicide victims were more likely than young white female victims to be killed with guns. Among young white females, the percentage of homicides involving guns peaked at 60% at around age 15 between 2008 and

TABLE 7.6

Homicide firearm deaths of children aged 12 and under and rate per 100,000, 2005–14

Year	Deaths	Population	Rate
2005	126	51,856,979	0.24
2006	160	51,943,526	0.31
2007	156	52,214,785	0.30
2008	161	52,509,274	0.31
2009	157	52,737,288	0.30
2010	142	52,943,218	0.27
2011	145	52,920,122	0.27
2012	160	52,851,442	0.30
2013	140	52,731,709	0.27
2014	151	52,666,129	0.29
Total	1,498	525,374,472	0.29

SOURCE: Adapted from "Fatal Injury Reports, National and Regional, 1999–2014," in *Web-Based Injury Statistics Query and Reporting System (WISQARS)*, Centers for Disease Control and Prevention, 2016, http://webappa.cdc.gov/sasweb/ncipc/mortrate10_us.html (accessed September 13, 2016)

FIGURE 7.8

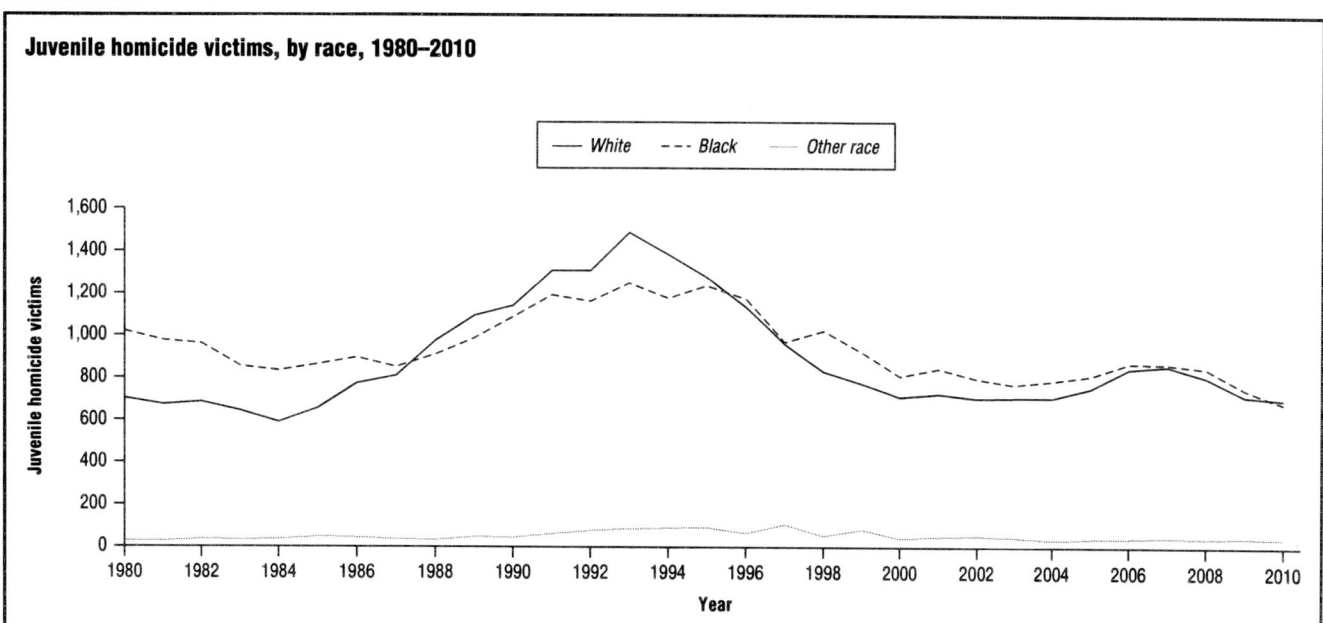

Juvenile homicide victims, by race, 1980–2010

SOURCE: Melissa Sickmund and Charles Puzzanchera, editors, "Between 1984 and 1993, While the Number of Homicides of White Juveniles Increased 50%, Homicides of Black Juveniles Increased 150%," in *Juvenile Offenders and Victims: 2014 National Report*, National Center for Juvenile Justice, December 2014, http://www.ojjdp.gov/ojstatbb/nr2014/downloads/NR2014.pdf (accessed September 12, 2016)

2011, little different from observable trends during the peak homicide rate between 1992 and 1995. Among young African American females between 2008 and 2011, the percentage of homicides involving guns peaked at the same age but at a much higher level of around 75%. Furthermore, among African American female homicide victims in their late teens and mid-20s, the percentage who were killed with guns had risen significantly from the peak homicide period between 1992 and 1995.

YOUNG, ARMED, AND DANGEROUS
Youth Gangs

Much of the violent activity among young people can be attributed to youth gangs, which tend to be concentrated in poor, urban neighborhoods. The Office of Juvenile Justice and Delinquency Prevention's National Youth Gang Center surveyed law enforcement agencies annually between 1996 and 2012 to determine the extent of gang problems nationwide and to investigate demographic and other gang-related data. The National Youth Gang Surveys (NYGS; http://www.nationalgangcenter.gov/Survey-Analysis) reveal that the prevalence of gang problems nationwide declined between 1996 and 2001 but rose again through 2012. The survey found that there were more than 30,000 gangs in the United States in 2012, the highest annual estimate since 1996 (and up 15% since 2006). The number of individual gang members was also on the rise: in 2012 there were an estimated 850,000 gang members nationally, an 8.6% increase over 2011.

Over time, the results of the NYGS have consistently indicated that gang activity is concentrated in metropolitan areas (areas with an urban core and suburbs whose populations exceed 100,000), and especially in the larger cities themselves (cities of more than 100,000 people). In 2012 law enforcement agencies in 3,100 U.S. jurisdictions, or 30% of all responding jurisdictions, reported gang problems in 2012. The percentage of metropolitan jurisdictions reporting gang activity was considerably higher, at 86%. Overall, 67.4% of all gangs and 81.7% of all gang members were located in metropolitan areas, and these urban/suburban areas accounted for 84.5% of all gang-related homicides. There were 2,363 gang homicides in 2012, up significantly from 2011, when there were 1,824 gang homicides. More than two-thirds (67.2%) of these homicides occurred in cities with populations of more than 100,000, and another 17.3% occurred in suburban areas. Smaller cities (those with populations between 50,000 and 100,000) and rural areas together accounted for 15.6% of all gang homicides in 2012.

Sickmund and Puzzanchera report on the Centers for Disease Control and Prevention (CDC), which examined homicide data that were collected between 2003 to 2008 from five cities with high homicide levels: Oklahoma City, Oklahoma; Los Angeles, Long Beach, and Oakland, California; and Newark, New Jersey. After analyzing gang and nongang homicides in these five cities, the CDC found that homicide victims aged 15 to 19 years were three times more likely to be killed in gang-related incidents than in nongang incidents. Furthermore, 90% of gang-related homicides involved firearms, compared with between 57% and 86% of nongang homicides. In all five cities, gang homicide offenders were predominantly males between the ages of 22 and 25 years.

Criminally Active Nongang Groups

Gangs are not always the source of gun violence, however. Anthony A. Braga of the Malcolm Wiener Center for Social Policy at Harvard University reports in *Gun Violence among Serious Young Offenders* (March 29, 2004, http://www.cops.usdoj.gov/pdf/pop/e01042199.pdf) that in some cities criminally active nongang groups are major gun offenders.

Braga explains that gun violence and murders in gangs are usually related to rivalries among gangs; offenders often become victims and vice versa. Gun violence and murders in criminally active groups that are not gangs are usually related to "business interests," such as drug dealing. Murders tied to these groups usually occur in or near "street" drug markets, and many of the victims are part of the drug organization or criminal network.

In "Gun Carrying and Drug Selling among Young Incarcerated Men and Women" (*Journal of Urban Health*, vol. 83, no. 2, March 2006), Deborah Kacanek and David Hemenway study the correlation between illegal drug dealing and guns. Based on interviews with 204 state prison inmates between the ages of 18 and 25 years, the researchers find that 45% of the incarcerated men and 16% of the women reported carrying a weapon in the 12 months prior to their arrest. Kacanek and Hemenway conclude that survey participants who sold drugs were more likely to have carried a gun: 65% of men and 22% of women who sold crack cocaine reported carrying a gun, as did 49% of men and 27% of women who sold drugs but not crack, compared with just 16% of men and 3% of women who did not sell drugs at all.

How Do Young Offenders Acquire Guns?

GUN TRACE DATA. As is discussed in Chapter 2, the Bureau of Alcohol, Tobacco, Firearms, and Explosives (ATF) has been prevented by law from publicly releasing certain categories of gun trace statistics since 2003. The ATF currently releases state-by-state data on the total number of guns recovered by law enforcement agencies, but it is prevented from releasing crucial data regarding how those guns were obtained. This has hampered research into the understanding of how young people obtain guns, and the most informative national studies currently available date from before that time. The ATF's Youth Crime Gun Interdiction Initiative, which analyzed guns recovered from crimes and traced them to their original sources, was the foundation for much of the valuable gun trace analysis that existed in the public domain prior to 2003.

According to the ATF, in "Firearms Trace Data—2015" (July 12, 2016, https://www.atf.gov/about/firearms-trace-data-2015), 190,538 firearms were recovered and traced back to a purchaser in the United States in 2015. Table 7.7 shows where these firearms were recovered, and

TABLE 7.7

Guns recovered after use in crimes, by state and age of possessor, 2015

State/territory	17 & Under	18 to 21	22 to 24	25 to 30	31 to 40	41 to 50	Over 50	Average age
Alabama	86	381	388	696	710	344	483	35
Alaska	6	54	57	157	136	89	52	34
Arizona	113	620	578	1,023	1,286	688	986	36
Arkansas	28	115	79	157	236	182	62	34
California	659	2,479	2,351	4,400	5,588	3,573	5,927	39
Colorado	85	218	263	467	616	314	951	39
Connecticut	12	62	81	106	146	116	99	37
Delaware	44	130	119	162	224	66	320	42
District of Columbia	36	138	116	148	111	63	44	30
Florida	511	1,514	1,424	2,180	2,271	1,410	2,670	37
Georgia	348	1,009	1,048	1,721	1,616	964	701	33
Guam and Northern Mariana Islands	1	1	0	2	7	3	12	46
Hawaii	0	0	7	8	16	8	14	40
Idaho	11	39	61	142	187	134	240	42
Illinois	529	1,105	1,017	1,314	1,269	781	1,088	34
Indiana	228	725	631	951	877	488	418	32
Iowa	38	82	106	176	197	244	196	39
Kansas	31	96	121	169	461	130	92	34
Kentucky	91	388	288	444	521	374	338	34
Louisiana	293	843	715	960	1,227	542	907	35
Maine	2	17	17	46	71	38	67	40
Maryland	155	536	518	991	1,288	1,061	1,254	38
Massachusetts	51	181	173	239	244	113	158	33
Michigan	168	579	572	869	911	733	512	34
Minnesota	69	261	212	397	456	219	189	33
Mississippi	59	305	281	466	459	230	188	33
Missouri	146	550	599	772	964	438	477	33
Montana	2	11	25	82	119	29	454	55
Nebraska	54	146	166	219	237	111	252	36
Nevada	24	124	157	295	436	249	244	38
New Hampshire	2	17	9	30	40	28	28	37
New Jersey	111	341	268	420	465	256	443	36
New Mexico	30	180	193	410	417	159	255	35
New York	287	720	658	1,144	1,323	646	735	34
North Carolina	243	1,074	1,096	1,500	1,478	905	1,201	35
North Dakota	1	26	21	88	117	42	23	34
Ohio	370	1,050	918	1,397	1,517	935	1,011	34
Oklahoma	50	132	170	236	301	134	166	35
Oregon	45	186	202	514	689	434	687	40
Pennsylvania	176	586	558	795	831	644	951	37
Puerto Rico	0	26	49	68	82	38	26	33
Rhode Island	6	47	44	85	75	36	26	32
South Carolina	114	342	336	556	488	331	2,194	41
South Dakota	4	11	19	18	54	66	15	39
Tennessee	232	721	626	924	957	706	507	33
Texas	290	1,229	1,321	2,484	2,998	1,465	1,432	35
US Virgin Islands	0	17	22	27	29	13	4	31
Utah	22	72	79	150	232	134	161	37
Vermont	0	2	8	19	15	12	114	49
Virginia	196	749	945	1,141	1,264	697	951	35
Washington	78	161	251	471	615	429	588	39
West Virginia	8	74	52	144	239	199	409	43
Wisconsin	100	455	577	705	568	254	250	31
Wyoming	4	6	23	15	40	17	25	38

Notes: The national average possessor age is 36 years old. Includes firearms recovered and traced between 1/1/2015–12/31/2015, or, if the recovery date was blank, the trace entry date was between 1/1/2015–12/31/2015. Duplicate traces, firearms not recovered, gun buyback and firearms turned in are not included in the statistics. Includes traces that provide a possessor and the possessor's date of birth. Possessor's age is calculated by subtracting the possessor's date of birth from the recovery date. Statistics are based on a query of the Firearms Tracing System (FTS) on April 11, 2016. All traces may not have been submitted or completed at the time of this analysis.

SOURCE: "Firearms Recovered and Traced in the United States and Territories, Age of Possessor, January 1, 2015–December 31, 2015" in *Firearms Trace Data—2015*, U.S. Department of Justice, Bureau of Alcohol, Tobacco, Firearms, and Explosives, July 12, 2016, https://www.atf.gov/docs/finalpossessorage-cy2015xlsx/download (September 14, 2016)

the age of the person in possession of the weapon at the time of its recovery. Of the total number of recovered firearms, 6,249 (3.3%) were held by juveniles aged 17 years and younger and 20,933 (11%) were held by youths aged 18 to 21 years. The states with the highest number of guns recovered from youths aged 21 years and younger were California, where 3,138 guns were recovered from this age bracket, followed by Florida (2,025), Illinois (1,634), Texas (1,519), and Ohio (1,420).

Garen J. Wintemute et al. analyze in "The Life Cycle of Crime Guns: A Description Based on Guns Recovered from Young People in California" (*Annals of Emergency Medicine*, vol. 43, no. 6, June 2004) data from ATF firearms tracing records to follow the life cycle of 2,121 crime guns recovered in California in 1999. The researchers make several interesting conclusions:

- Guns recovered from individuals younger than 18 years old were most often purchased by people aged 45 years and older.
- Small-caliber handguns made up 41% of handguns recovered from this group.
- For 17.3% of crime guns recovered from teenagers, the median time from sale to recovery was less than three years, which indicates deliberate gun trafficking (the illegal selling of guns and ammunition).
- A minority of retailers and straw purchasers (people buying guns for someone else) are disproportionately linked to the sale or transfer of crime guns.

OTHER STUDIES. Other studies have used research resources other than gun trace data to determine how young people who use guns in crimes obtain their weapons. Daniel W. Webster et al. of Johns Hopkins University interviewed 45 youths incarcerated in a juvenile justice facility to determine how they obtained guns. The researchers reported their findings in "How Delinquent Youths Acquire Guns: Initial versus Most Recent Gun Acquisitions" (*Journal of Urban Health*, vol. 79, no. 1, March 2002), which is still a frequently cited study. Of the 45 youths, 30 had acquired at least one gun and 22 had acquired multiple guns. Approximately 50% of their first guns were given to them by friends or family, or they found discarded guns. Those who acquired more than one gun usually got them from acquaintances or drug addicts. If they bought new guns, the youths generally purchased them from gun traffickers (people who are in the business of selling guns illegally). Webster et al. conclude that a way to reduce the number of guns in the hands of young offenders is to stop high-volume gun traffickers and recover discarded guns from areas in which illicit drug sales take place.

In "Source of Firearms Used by Students in School-Associated Violent Deaths—United States, 1992–1999" (*Morbidity and Mortality Weekly Report*, vol. 52, no. 9, March 7, 2003), another frequently cited study, the CDC investigates how students obtained firearms used in serious school-associated crimes such as homicide and suicide. Most students obtained guns from home (37.5%), with the next likely source being a friend or relative (23.4%). Only 7% of guns used in school-related crimes were purchased and 5.5% were stolen.

Students and Guns

Since 1991 the CDC has regularly surveyed young people as part of its Youth Risk Behavior Surveillance System, which collects information about dangerous behaviors and risk factors among school-aged young people. Figure 7.9 shows the percentage of students by gender in grades nine to 12 who reported carrying a weapon anywhere and on school property between 1993 and 2013. The prevalence of weapon-carrying declined considerably both away from school (anywhere) and on school property over this period. Among males, the proportion who reported carrying a weapon anywhere fell from 34.3% in 1993 to 28.1% in 2013 and the proportion who reported carrying a weapon at school fell from 17.9% to 7.6%. The proportion of females who carried weapons either anywhere or at school was much lower throughout these years and showed only a slight downward trend.

Table 7.8 breaks down the prevalence of weapon-carrying among high school students by sex, race, ethnicity, grade, and location. Native American or Alaskan Native students were the most likely demographic subgroup to carry a weapon anywhere, both in 1993, when 34.2% reported carrying a weapon, and in 2011, when 27.6% reported carrying a weapon. In 2013, however, with a steep decline in this demographic subgroup's likelihood to carry a weapon anywhere (to 17.8%), whites became the most likely to carry a weapon (20.8%). Among all demographic subgroups, weapon-carrying among African American students also saw a significant decline over this period, from 28.5% to 12.5%.

Weapon-carrying on school property varied less by race and ethnicity than did weapon-carrying anywhere. In 1993 Native American or Alaskan Native students (17.6%) had the highest prevalence of weapon-carrying on school property, followed by African American students (15%), Hispanic students (13.3%), and white students (10.9%). (See Table 7.8.) In 2013 Native American or Alaskan Native students (7%) still had a slightly higher prevalence rate of weapon-carrying on school property than other groups, followed by students of two or more races (6.3%), white students (5.7%), and Hispanic students (4.7%). The prevalence rates of weapon-carrying among African American, Asian American, and Pacific Islander students were between 3.8% and 4%. Figure 7.10 presents a visual interpretation of weapon-carrying among students by race, ethnicity, and location in 2013.

FIGURE 7.9

Percentage of students in grades 9–12 who reported carrying a weapon at least 1 day in the past 30 days, by location and sex, 1993–2013

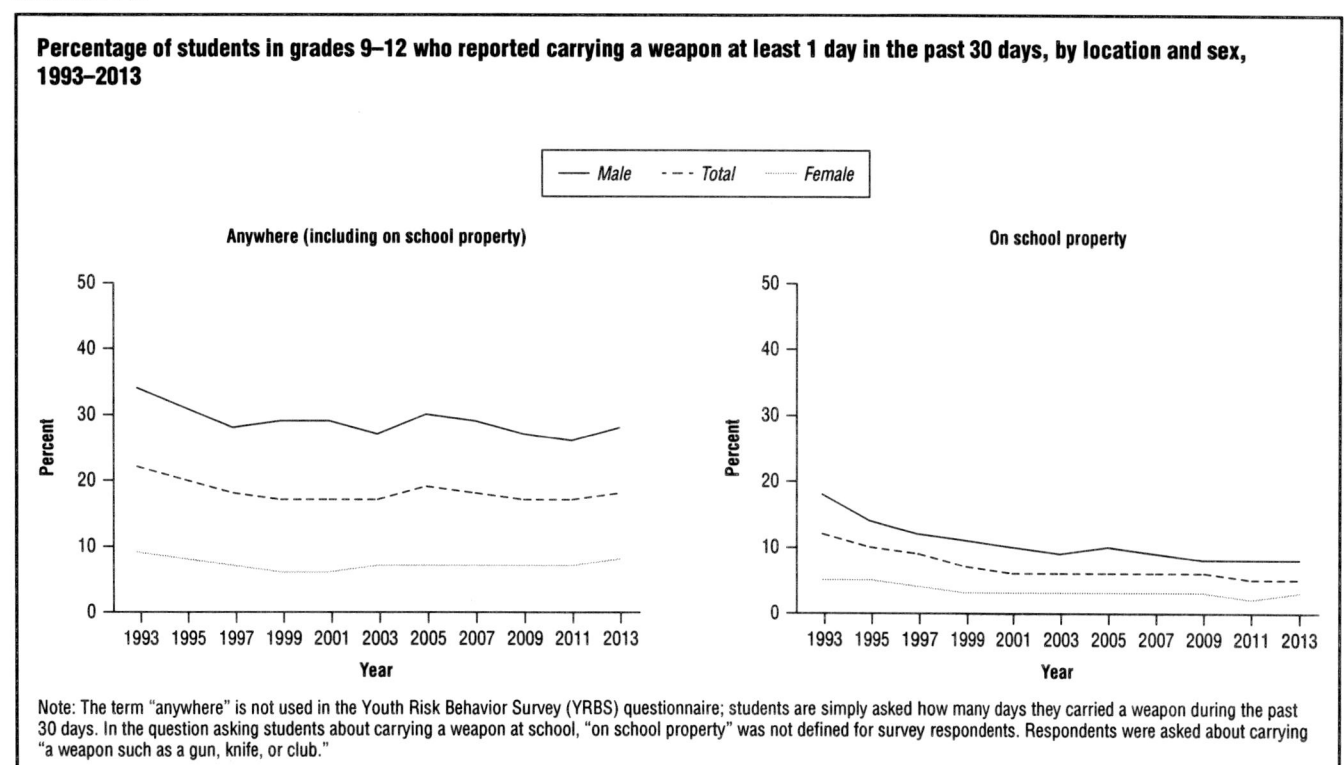

Note: The term "anywhere" is not used in the Youth Risk Behavior Survey (YRBS) questionnaire; students are simply asked how many days they carried a weapon during the past 30 days. In the question asking students about carrying a weapon at school, "on school property" was not defined for survey respondents. Respondents were asked about carrying "a weapon such as a gun, knife, or club."

SOURCE: Anlan Zhang, Lauren Musu-Gillette, and Barbara A. Oudekerk, "Figure 14.1. Percentage of Students in Grades 9–12 Who Reported Carrying a Weapon at Least 1 Day during the Previous 30 Days, by Location and Sex: Selected Years, 1993–2013," in *Indicators of School Crime and Safety: 2015*, National Center for Education Statistics and Bureau of Justice Statistics, May 2016, http://www.bjs.gov/content/pub/pdf/iscs15.pdf (accessed September 14, 2016)

Table 7.9 provides information on weapon-carrying among high school students in 2015, distinguishing between the carrying of any weapon (such as a gun, knife, or club) and the carrying of a gun specifically. Among white male students, 28% reported carrying a weapon in the 30 days preceding the survey, and 9.6% reported carrying a gun. Among African American male students, 17.6% reported carrying a weapon and 9.6% reported carrying a gun; and among Hispanic male students, 20.2% of reported carrying a weapon and 6.5% reported carrying a gun. As with other statistics concerning weapons and violence, females were far less likely than males to carry either any weapon or a gun.

There was also considerable variation in weapon- and gun-carrying by state and urban location in 2015. As Table 7.10 shows, students in rural and southern states were considerably more likely to carry either weapons or guns than were students in coastal states or states with large urban populations. This pattern of variation applied to both males and females, so that states with high percentages of weapon- and gun-carrying males also tended to have high percentages of weapon- and gun-carrying females, but male rates were far higher in all states. The states in which male students were the most likely to carry weapons of any kind were Idaho, where 40.9% of high school males reported carrying weapons, Wyoming (40.8%), West Virginia (38.3%), Montana (37.8%), Alabama (34.6%), Kentucky (34%), and Missouri (33.8%). Students in large urban school districts were in general among the least likely to carry weapons or guns in 2015. The highest rates of weapon-carrying among male students in urban areas were found in Baltimore, Maryland (28.1%), Duval County, Florida (26.2%), the District of Columbia (23.6%), and Cleveland, Ohio (23.4%).

Student Reports of Threats or Injuries

Table 7.11 shows trend data on weapons threats and injuries on school property between 1993 and 2013. The percentage of ninth-to-12th-grade students who were threatened or injured with a weapon, such as a gun, knife, or club, on school property ranged from 7.3% in 1993 to 9.2% in 2003. Threat reports gradually declined in the decade that followed, reaching a low of 6.9% in 2013. Males were more likely than females to be threatened or injured with a weapon. For example, 7.7% of male students were threatened or injured with a weapon in 2013, compared with 6.1% of female students. In all the survey years, students in lower grades were more likely to be threatened than students in higher grades. In 2013, 18.5% of Native American or Alaskan Native students, 8.7% of Pacific Islander students, 8.5% of Hispanic students, 8.4% of African American students, 7.7% of mixed-race

TABLE 7.8

Percentage of students in grades 9–12 who reported carrying a weapon at least 1 day in the past 30 days, by location and selected student characteristics, selected years 1993–2013

Location and student characteristic	1993	1995	1997	1999	2001	2003	2005	2007	2009	2011	2013
Anywhere (including on school property)[a]											
Total	22.1	20.0	18.3	17.3	17.4	17.1	18.5	18.0	17.5	16.6	17.9
Sex											
Male	34.3	31.1	27.7	28.6	29.3	26.9	29.8	28.5	27.1	25.9	28.1
Female	9.2	8.3	7.0	6.0	6.2	6.7	7.1	7.5	7.1	6.8	7.9
Race/ethnicity[b]											
White	20.6	18.9	17.0	16.4	17.9	16.7	18.7	18.2	18.6	17.0	20.8
Black	28.5	21.8	21.7	17.2	15.2	17.3	16.4	17.2	14.4	14.2	12.5
Hispanic	24.4	24.7	23.3	18.7	16.5	16.5	19.0	18.5	17.2	16.2	15.5
Asian[c]	—	—	—	13.0	10.6	11.6	7.0	7.8	8.4	9.1	8.7
Pacific Islander[c]	—	—	—	25.3	17.4	16.3[f]	20.0[f]	25.5	20.3	20.7	12.6[f]
American Indian/Alaska Native	34.2	32.0	26.2	21.8	31.2	29.3	25.6	20.6	20.7	27.6	17.8
Two or more races[c]	—	—	—	22.2	25.2	29.8	26.7	19.0	17.9	23.7	18.8
Grade											
9th	25.5	22.6	22.6	17.6	19.8	18.0	19.9	20.1	18.0	17.3	17.5
10th	21.4	21.1	17.4	18.7	16.7	15.9	19.4	18.8	18.4	16.6	17.8
11th	21.5	20.3	18.2	16.1	16.8	18.2	17.1	16.7	16.2	16.2	17.9
12th	19.9	16.1	15.4	15.9	15.1	15.5	16.9	15.5	16.6	15.8	18.3
Urbanicity[d]											
Urban	—	—	18.7	15.8	15.3	17.0	—	—	—	—	—
Suburban	—	—	16.8	17.0	17.4	16.5	—	—	—	—	—
Rural	—	—	22.3	22.3	23.0	18.9	—	—	—	—	—
On school property[e]											
Total	11.8	9.8	8.5	6.9	6.4	6.1	6.5	5.9	5.6	5.4	5.2
Sex											
Male	17.9	14.3	12.5	11.0	10.2	8.9	10.2	9.0	8.0	8.2	7.6
Female	5.1	4.9	3.7	2.8	2.9	3.1	2.6	2.7	2.9	2.3	3.0
Race/ethnicity[b]											
White	10.9	9.0	7.8	6.4	6.1	5.5	6.1	5.3	5.6	5.1	5.7
Black	15.0	10.3	9.2	5.0	6.3	6.9	5.1	6.0	5.3	4.6	3.9
Hispanic	13.3	14.1	10.4	7.9	6.4	6.0	8.2	7.3	5.8	5.8	4.7
Asian[c]	—	—	—	6.5	7.2	6.6[f]	2.8[f]	4.1	3.6	4.3[f]	3.8
Pacific Islander[c]	—	—	—	9.3	10.0[f]	4.9[f]	15.4[f]	9.5[f]	9.8	10.9[f]	4.0[f]
American Indian/Alaska Native	17.6[f]	13.0[f]	15.9	11.6[f]	16.4	12.9	7.2	7.7	4.2[f]	7.5	7.0[f]
Two or more races[c]	—	—	—	11.4	13.2	13.3[f]	11.9	5.0	5.8	7.5	6.3
Grade											
9th	12.6	10.7	10.2	7.2	6.7	5.3	6.4	6.0	4.9	4.8	4.8
10th	11.5	10.4	7.7	6.6	6.7	6.0	6.9	5.8	6.1	6.1	4.8
11th	11.9	10.2	9.4	7.0	6.1	6.6	5.9	5.5	5.2	4.7	5.9
12th	10.8	7.6	7.0	6.2	6.1	6.4	6.7	6.0	6.0	5.6	5.3
Urbanicity[d]											
Urban	—	—	7.0	7.2	6.0	5.6	—	—	—	—	—
Suburban	—	—	8.7	6.2	6.3	6.4	—	—	—	—	—
Rural	—	—	11.2	9.6	8.3	6.3	—	—	—	—	—

—Not available.

[a]The term "anywhere" is not used in the Youth Risk Behavior Survey (YRBS) questionnaire; students were simply asked how many days they carried a weapon during the past 30 days.
[b]Race categories exclude persons of Hispanic ethnicity.
[c]Before 1999, Asian students and Pacific Islander students were not categorized separately, and students could not be classified as two or more races. Because the response categories changed in 1999, caution should be used in comparing data on race from 1993, 1995, and 1997 with data from later years.
[d]Refers to the Student Metropolitan Statistical Area (MSA) status of the respondent's household as defined in 2000 by the U.S. Census Bureau. Categories include "central city of an MSA (Urban)," "in MSA but not in central city (Suburban)," and "not MSA (Rural)."
[e]In the question asking students about carrying a weapon at school, "on school property" was not defined for survey respondents.
[f]Interpret data with caution. The coefficient of variation (CV) for this estimate is between 30 and 50 percent.
Note: Respondents were asked about carrying "a weapon such as a gun, knife, or club."

SOURCE: Adapted from Anlan Zhang, Lauren Musu-Gillette, and Barbara A. Oudekerk, "Table 14.1. Percentage of Students in Grades 9–12 Who Reported Carrying a Weapon at Least 1 Day during the Previous 30 Days, by Location and Selected Student Characteristics: Selected Years, 1993 through 2013," in *Indicators of School Crime and Safety: 2015*, National Center for Education Statistics and Bureau of Justice Statistics, May 2016, http://www.bjs.gov/content/pub/pdf/iscs15.pdf (accessed September 14, 2016)

students, 5.8% of white students, and 5.3% of Asian American students reported being threatened or injured with a weapon on school property.

The percentage of students who were threatened or injured with a weapon at school varied significantly by location. (See Table 7.12.) In 2013, 10.9% of Arkansas

FIGURE 7.10

Percentage of students in grades 9–12 who reported carrying a weapon at least 1 day in the past 30 days, by race/ethnicity and location, 2013

Race/ethnicity	Anywhere (including on school property)	On school property
Total	17.9	5.2
White	20.8	5.7
Black	12.5	3.9
Hispanic	15.5	4.7
Asian	8.7	3.8
Pacific Islander	12.6!	4.0*
American Indian/Alaska Native	17.8	7.0*
Two or more races	18.8	6.3

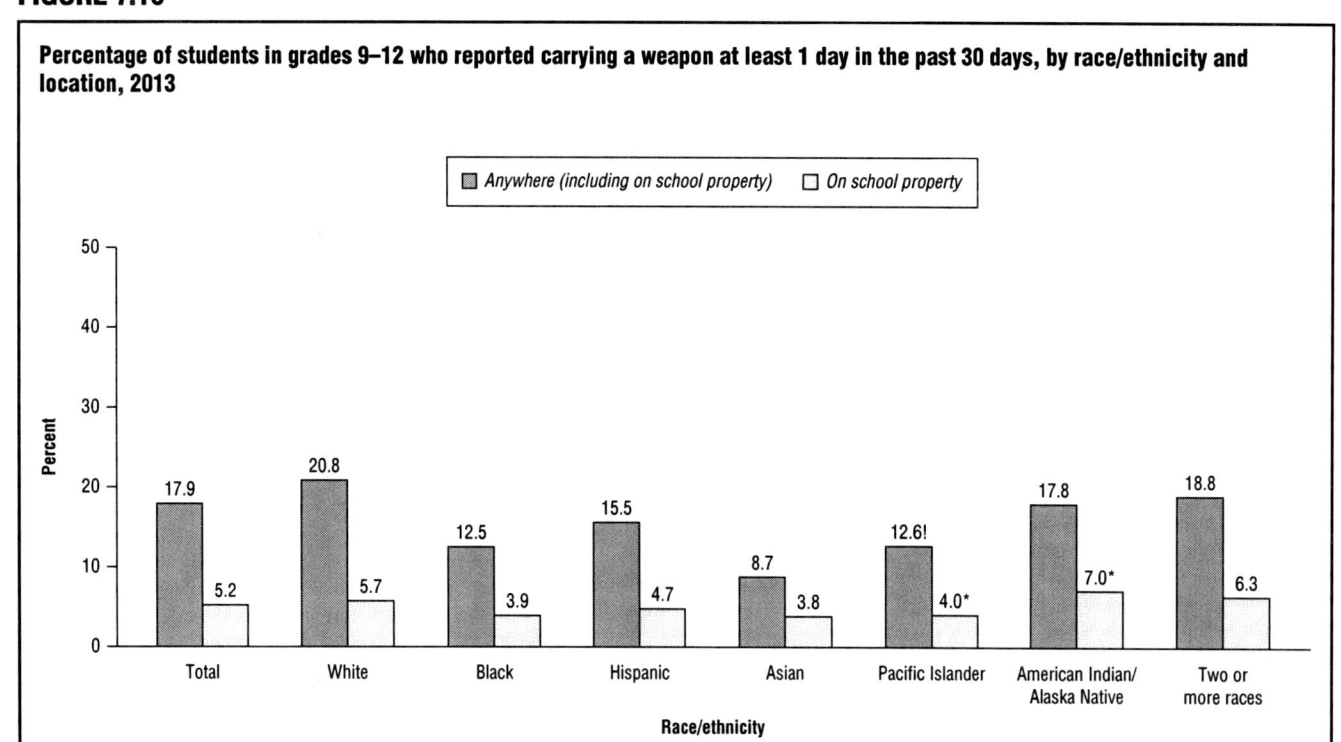

*Interpret data with caution. The coefficient of variation (CV) for this estimate is between 30 and 50 percent.
Note: Race categories exclude persons of Hispanic ethnicity. The term "anywhere" is not used in the Youth Risk Behavior Survey (YRBS) questionnaire; students were simply asked how many days they carried a weapon during the past 30 days. In the question asking students about carrying a weapon at school, "on school property" was not defined for survey respondents. Respondents were asked about carrying "a weapon such as a gun, knife, or club."

SOURCE: Anlan Zhang, Lauren Musu-Gillette, and Barbara A. Oudekerk, "Figure 14.2. Percentage of Students in Grades 9–12 Who Reported Carrying a Weapon at Least 1 Day during the Previous 30 Days, by Race/Ethnicity and Location: 2013," in *Indicators of School Crime and Safety: 2015*, National Center for Education Statistics and Bureau of Justice Statistics, May 2016, http://www.bjs.gov/content/pub/pdf/iscs15.pdf (accessed September 14, 2016)

TABLE 7.9

Percentage of students in grades 9–12 who carried a weapon and who carried a gun, by sex, race/ethnicity, and grade, 2015

	Carried a weapon			Carried a gun		
	Female	Male	Total	Female	Male	Total
Category	%	%	%	%	%	%
Race/ethnicity						
White*	8.1	28.0	18.1	1.4	9.6	5.5
Black*	6.2	17.6	12.4	1.7	9.6	6.0
Hispanic	7.1	20.2	13.7	1.9	6.5	4.3
Grade						
9	6.6	24.6	16.1	1.2	7.0	4.4
10	7.2	25.5	16.3	1.6	8.8	5.2
11	8.0	23.0	16.0	1.4	9.0	5.5
12	8.0	23.4	15.8	1.7	9.7	5.7
Total	7.5	24.3	16.2	1.6	8.7	5.3

*Non-Hispanic.
Notes: Weapons such as, a gun, knife, or club. Carried a weapon within at least 1 day during the 30 days before the survey.

SOURCE: "Table 9. Percentage of High School Students Who Carried a Weapon and Who Carried a Gun, by Sex, Race/Ethnicity, and Grade—United States, Youth Risk Behavior Survey, 2014," in "Youth Risk Behavior Surveillance—United States, 2015," *MMWR*, vol. 65, no. 6, June 10, 2016, http://www.cdc.gov/mmwr/volumes/65/ss/pdfs/ss6506.pdf (accessed September 14, 2016)

students and 10.5% of Louisiana students reported a threat or injury on school property, more than double the percentage of students who reported a threat or injury in Wisconsin (4.3%) and Massachusetts (4.4%). Alabama students (9.9%), Maryland students (9.4%), Tennessee students (9.3%), and Arizona students (9.1%) were among the other states with the highest prevalence of student-reported threats or injuries. Percentages for

TABLE 7.10

Percentage of students in grades 9–12 who carried a weapon and who carried a gun, by sex, state, and urban area, 2015

	Carried a weapon			Carried a gun		
	Female	Male	Total	Female	Male	Total
Site	%	%	%	%	%	%
State surveys						
Alabama	10.2	34.6	22.5	3.4	15.9	10.0
Alaska	—*	—	—	—	—	—
Arizona	10.2	25.5	18.0	2.8	6.9	4.9
Arkansas	9.3	32.1	21.0	3.6	15.7	9.8
California	4.6	13.2	8.9	0.7	4.8	2.8
Connecticut	—	—	—	—	—	—
Delaware	6.0	19.6	13.0	1.1	8.2	4.7
Florida	8.1	22.3	15.4	—	—	—
Hawaii	6.1	15.1	10.7	—	—	—
Idaho	14.7	40.9	28.2	—	—	—
Illinois	8.2	22.5	15.4	2.1	9.0	5.6
Indiana	8.4	30.2	19.6	1.6	10.6	6.2
Kentucky	11.6	34.0	23.1	—	—	—
Maine	—	—	—	—	—	—
Maryland	8.3	20.9	14.9	—	—	—
Massachusetts	5.0	19.9	12.6	0.9	4.5	2.7
Michigan	8.1	24.8	16.6	2.2	6.9	4.6
Mississippi	10.0	32.4	21.0	2.7	14.3	8.5
Missouri	10.8	33.8	22.1	—	—	—
Montana	14.2	37.8	26.4	4.4	17.2	11.1
Nebraska	—	—	—	—	—	—
Nevada	9.5	26.7	18.3	3.1	8.0	5.7
New Hampshire	—	—	—	—	—	—
New Mexico	12.8	32.1	22.5	3.3	11.9	7.7
New York	6.6	19.0	13.0	1.8	6.3	4.1
North Carolina	9.7	28.5	19.3	—	—	—
North Dakota	—	—	—	—	—	—
Oklahoma	9.0	30.5	19.5	3.3	10.5	6.8
Pennsylvania	8.1	26.4	17.4	2.7	12.7	7.9
Rhode Island	—	—	—	—	—	—
South Carolina	9.6	31.3	20.5	3.1	13.5	8.4
South Dakota	—	—	—	—	—	—
Tennessee	—	—	—	—	—	—
Vermont	—	—	—	—	—	—
Virginia	6.7	22.5	15.0	—	—	—
West Virginia	13.4	38.3	26.1	2.4	12.7	7.6
Wyoming	18.1	40.8	29.6	6.9	15.9	11.5
Median			28.5		10.6	6.8
Range						
Large urban school district surveys						
Baltimore, MD	14.9	28.1	21.9	2.1	7.0	5.4
Boston, MA	5.4	17.7	11.7	—	—	—
Broward County, FL	7.6	16.7	12.4	1.1	4.9	3.1
Cleveland, OH	13.8	23.4	19.2	—	—	—
DeKalb County, GA	6.0	15.3	10.7	1.7	6.9	4.3
Detroit, MI	10.2	19.5	14.4	2.4	8.7	5.4
District of Columbia	12.5	23.6	18.1	—	—	—
Duval County, FL	12.2	26.2	19.3	—	—	—
Ft. Worth, TX	6.6	18.4	12.5	1.6	7.5	4.5
Houston, TX	7.7	18.1	13.2	3.3	7.3	5.5
Los Angeles, CA	4.0	11.8	7.8	0.9	3.5	2.2
Miami-Dade County, FL	4.7	13.6	9.1	1.9	6.5	4.2
New York City, NY	4.6	10.4	7.7	1.0	3.3	2.3
Oakland, CA	9.6	18.3	14.4	2.2	9.2	5.9
Orange County, FL	5.1	18.3	11.7	1.6	7.8	4.7
Palm Beach County, FL	7.9	20.3	14.5	1.1	7.4	4.7
Philadelphia, PA	8.5	17.0	12.7	1.8	8.0	5.0
San Diego, CA	4.7	16.1	10.5	0.4	4.9	2.7
San Francisco, CA	6.1	12.2	9.2	1.2	3.7	2.6
Median	7.6	18.1	12.5	1.6	7.0	4.5
Range	(4.0–14.9)	(10.4–28.1)	(7.7–21.9)	(0.4–3.3)	(3.3–9.2)	(2.2–5.9)

*Not available.
Notes: Weapons such as, a gun, knife, or club. Carried a weapon on at least 1 day during the 30 days before the survey.

SOURCE: "Table 10. Percentage of High School Students Who Carried a Weapon and Who Carried a Gun, by Sex—Selected U.S. Sites, Youth Risk Behavior Survey, 2015," in "Youth Risk Behavior Surveillance—United States, 2015," *MMWR*, vol. 65, no. 6, June 10, 2016, http://www.cdc.gov/mmwr/volumes/65/ss/pdfs/ss6506.pdf (accessed September 14, 2016)

TABLE 7.11

Percentage of students in grades 9–12 who reported being threatened or injured with a weapon on school property during the last 12 months, by selected student characteristics and number of times threatened or injured, selected years 1993–2013

Number of times and year	Total	Sex Male	Sex Female	White	Black	Hispanic	Asian[b]	Pacific islander[b]	American Indian/ Alaska Native[b]	Two or more races[b]	9th grade	10th grade	11th grade	12th grade
At least once														
1993	7.3	9.2	5.4	6.3	11.2	8.6	[c]	[c]	11.7	[c]	9.4	7.3	7.3	5.5
1995	8.4	10.9	5.8	7.0	11.0	12.4	[c]	[c]	11.4[d]	[c]	9.6	9.6	7.7	6.7
1997	7.4	10.2	4.0	6.2	9.9	9.0	[c]	[c]	12.5[d]	[c]	10.1	7.9	5.9	5.8
1999	7.7	9.5	5.8	6.6	7.6	9.8	7.7	15.6	13.2[d]	9.3	10.5	8.2	6.1	5.1
2001	8.9	11.5	6.5	8.5	9.3	8.9	11.3	24.8	15.2[d]	10.3	12.7	9.1	6.9	5.3
2003	9.2	11.6	6.5	7.8	10.9	9.4	11.5	16.3	22.1	18.7	12.1	9.2	7.3	6.3
2005	7.9	9.7	6.1	7.2	8.1	9.8	4.6	14.5	9.8	10.7	10.5	8.8	5.5	5.8
2007	7.8	10.2	5.4	6.9	9.7	8.7	7.6[d]	8.1	5.9	13.3	9.2	8.4	6.8	6.3
2009	7.7	9.6	5.5	6.4	9.4	9.1	5.5	12.5	16.5	9.2	8.7	8.4	7.9	5.2
2011	7.4	9.5	5.2	6.1	8.9	9.2	7.0	11.3	8.2	9.9	8.3	7.7	7.3	5.9
2013	6.9	7.7	6.1	5.8	8.4	8.5	5.3	8.7	18.5	7.7	8.5	7.0	6.8	4.9
Number of times, 2013														
0 times	93.1	92.3	3.9	94.2	91.6	91.5	94.7	91.3	81.5	92.3	91.5	93.0	93.2	95.1
1 time	3.0	3.0	3.0	2.7	3.8	3.3	1.7[d]	2.0	9.6	3.1[d]	3.5	3.0	3.4	2.0
2 or 3 times	1.7	1.7	1.6	1.6	1.8	1.6	[e]	[e]	[e]	2.2[d]	2.5	1.7	1.3	1.1
4 to 11 times	1.3	1.7	0.9	0.8	1.7	2.3	[e]	[e]	4.9	1.5[d]	1.5	1.4	1.3	1.0
12 or more times	0.9	1.3	0.6	0.7	1.2	1.2	[e]	[e]	[e]	[e]	1.0	0.9	0.9	0.9

[a]Race categories exclude persons of Hispanic ethnicity.
[b]Before 1999, Asian students and Pacific Islander students were not categorized separately, and students could not be classified as two or more races. Because the response categories changed in 1999, caution should be used in comparing data on race from 1993, 1995, and 1997 with data from later years.
[c]Not available.
[d]Interpret data with caution. The coefficient of variation (CV) for this estimate is between 30 and 50 percent.
[e]Reporting standards not met. Either there are too few cases for a reliable estimate or the coefficient of variation (CV) is 50 percent or greater.
Note: Survey respondents were asked about being threatened or injured "with a weapon such as a gun, knife, or club on school property." "On school property" was not defined for respondents. Detail may not sum to totals because of rounding.

SOURCE: Adapted from Anlan Zhang, Lauren Musu-Gillette, and Barbara A. Oudekerk, "Table 4.1. Percentage of Students in Grades 9–12 Who Reported Being Threatened or Injured with a Weapon on School Property during the Previous 12 Months, by Selected Student Characteristics and Number of Times Threatened or Injured: Selected Years, 1993 through 2013," in *Indicators of School Crime and Safety: 2015*, National Center for Education Statistics and Bureau of Justice Statistics, May 2016, http://www.bjs.gov/content/pub/pdf/iscs15.pdf (accessed September 14, 2016)

---

individual states varied from year to year, however, so there were no clear geographical judgments to be drawn based on these data.

Barring Guns from Schools

The Gun-Free Schools Act of 1994 required states to pass laws forcing schools that receive funding under the Elementary and Secondary Education Act of 1965 to expel for at least one year any student who brings a firearm to school. The U.S. Department of Education provides detail on the implementation of the act in periodic reports, the most recent of which, as of December 2016, was *Report on the Implementation of the Gun-Free Schools Act of 1994 in the States and Outlying Areas: School Years 2005–06 and 2006–07* (September 2010, http://www2.ed.gov/about/reports/annual/gfsa/gfsarp100610.pdf). The Department of Education indicates that 2,695 students were expelled from school during the 2006–07 academic year for carrying a firearm to school, a rate of 5.5 per 100,000 students. More than half (53%) of these students were expelled for carrying handguns, 10% were expelled for carrying rifles or shotguns, and the remaining 37% were expelled for carrying other types of firearms or destructive devices, such as starter pistols, bombs, and grenades.

The Department of Education also notes that the number of students expelled for carrying a firearm to school dropped by 22%, from 3,477 during the 1998–99 academic year to 2,695 during the 2006–07 academic year. Some states experienced reductions in the number of students expelled for carrying a firearm to school between the 2005–06 and the 2006–07 academic years. The states experiencing the greatest percentage of decreases in expulsions during this period were Delaware, Iowa, Kansas, New Hampshire, New Jersey, Rhode Island, and Wisconsin. The states that experienced the greatest percentage of increases in expulsions during this period were Idaho, Indiana, Louisiana, and Minnesota.

SCHOOL SHOOTINGS

Mass shootings at U.S. schools became increasingly common during the 1990s, shocking Americans across the political spectrum and leading to an evolving, ongoing debate about how best to prevent such atrocities in the future. The Columbine High School shooting in Littleton, Colorado, in 1999 marked a turning point in the public consciousness of such crimes and seemed destined to lead to a more concerted effort to prevent further mass

TABLE 7.12

Percentage of students in grades 9–12 who reported being threatened or injured with a weapon on school property during the last 12 months, by state, selected years 2003–13

State	2003	2005	2007	2009	2011	2013
United States[a]	9.2	7.9	7.8	7.7	7.4	6.9
Alabama	7.2	10.6	—	10.4	7.6	9.9
Alaska	8.1	—	7.7	7.3	5.6	—
Arizona	9.7	10.7	11.2	9.3	10.4	9.1
Arkansas	—	9.6	9.1	11.9	6.3	10.9
California	—	—	—	—	—	—
Colorado	—	7.6	—	8.0	6.7	—
Connecticut	—	9.1	7.7	7.0	6.8	7.1
Delaware	7.7	6.2	5.6	7.8	6.4	5.6
District of Columbia	12.7	12.1	11.3	—	8.7	—
Florida	8.4	7.9	8.6	8.2	7.2	7.1
Georgia	8.2	8.3	8.1	8.2	11.7	7.2
Hawaii	—	6.8	6.4	7.7	6.3	—
Idaho	9.4	8.3	10.2	7.9	7.3	5.8
Illinois	—	—	7.8	8.8	7.6	8.5
Indiana	6.7	8.8	9.6	6.5	6.8	—
Iowa	—	7.8	7.1	—	6.3	—
Kansas	—	7.4	8.6	6.2	5.6	5.3
Kentucky	5.2	8.0	8.3	7.9	7.4	5.4
Louisiana	—	—	—	9.5	8.7	10.5
Maine	8.5	7.1	6.8	7.7	6.8	5.3
Maryland	—	11.7	9.6	9.1	8.4	9.4
Massachusetts	6.3	5.4	5.3	7.0	6.8	4.4
Michigan	9.7	8.6	8.1	9.4	6.8	6.7
Minnesota	—	—	—	—	—	—
Mississippi	6.6	—	8.3	8.0	7.5	8.8
Missouri	7.5	9.1	9.3	7.8	—	—
Montana	7.1	8.0	7.0	7.4	7.5	6.3
Nebraska	8.8	9.7	—	—	6.4	6.4
Nevada	6.0	8.1	7.8	10.7	—	6.4
New Hampshire	7.5	8.6	7.3	—	—	—
New Jersey	—	8.0	—	6.6	5.7	6.2
New Mexico	—	10.4	10.1	—	—	—
New York	7.2	7.2	7.3	7.5	7.3	7.3
North Carolina	7.2	7.9	6.6	6.8	9.1	6.9
North Dakota	5.9	6.6	5.2	—	—	—
Ohio[b]	7.7	8.2	8.3	—	—	—
Oklahoma	7.4	6.0	7.0	5.8	5.7	4.6
Oregon	—	—	—	—	—	—
Pennsylvania	—	—	—	5.6	—	—
Rhode Island	8.2	8.7	8.3	6.5	—	6.4
South Carolina	—	10.1	9.8	8.8	9.2	6.5
South Dakota[b]	6.5	8.1	5.9	6.8	6.1	5.0
Tennessee	8.4	7.4	7.3	7.0	5.8	9.3
Texas	—	9.3	8.7	7.2	6.8	7.1
Utah	7.3	9.8	11.4	7.7	7.0	5.5
Vermont	7.3	6.3	6.2	6.0	5.5	6.4
Virginia	—	—	—	—	7.0	6.1
Washington	—	—	—	—	—	—
West Virginia	8.5	8.0	9.7	9.2	6.6	5.6
Wisconsin	5.5	7.6	5.6	6.7	5.1	4.3
Wyoming	9.7	7.8	8.3	9.4	7.3	6.8

— = Not available.

[a]Data for the U.S. total include both public and private schools and were collected through a national survey representing the entire country.
[b]Data include both public and private schools.

Note: Survey respondents were asked about being threatened or injured "with a weapon such as a gun, knife, or club on school property." "On school property" was not defined for respondents. State-level data include public schools only, with the exception of data for Ohio and South Dakota. Data for the U.S. total, Ohio, and South Dakota include both public and private schools. For specific states, a given year's data may be unavailable (1) because the state did not participate in the survey that year; (2) because the state omitted this particular survey item from the state-level questionnaire; or (3) because the state had an overall response rate of less than 60 percent (the overall response rate is the school response rate multiplied by the student response rate).

SOURCE: Adapted from Anlan Zhang, Lauren Musu-Gillette, and Barbara A. Oudekerk, "Table 4.2. Percentage of Public School Students in Grades 9–12 Who Reported Being Threatened or Injured with a Weapon on School Property at Least One Time during the Previous 12 Months, by State: Selected Years, 2003 through 2013," in *Indicators of School Crime and Safety: 2015*, National Center for Education Statistics and Bureau of Justice Statistics, May 2016, http://www.bjs.gov/content/pub/pdf/iscs15.pdf (accessed September 14, 2016)

---

shootings. Nevertheless, the rate of school shootings increased during the early 21st century. In recent years the country has witnessed both the deadliest school shooting in history, the Virginia Polytechnic Institute and State University killings in 2007, and, arguably the most horrifying, the mass murder of first graders in Newtown, Connecticut, in 2012.

The following sections describe some of the better-known shooting incidents at U.S. high schools and universities in which multiple young people were shot. Although the stories and data related to school shootings in the United States are central to the current gun control debate, it should be noted that homicides at schools are exceedingly rare. Anlan Zhang, Lauren Musu-Gillette, and Barbara A. Oudekerk note in *Indicators of School Crime and Safety: 2015* (May 2016, http://www.bjs.gov/content/pub/pdf/iscs15.pdf) that in each year since the early 1990s, "youth homicides occurring at school remained at less than 3 percent of the total number of youth homicides." Thus, although high-profile mass murders in schools give many people the impression that American children are extremely vulnerable to violence, in fact the nation's schools are among the safest places children can be.

SECONDARY SCHOOL SHOOTINGS

Thurston High School, Springfield, Oregon

MAY 21, 1998. Fifteen-year-old Kip Kinkel (1982–) entered the crowded cafeteria at Thurston High School in Springfield, Oregon, and opened fire with a Ruger .22-caliber semiautomatic rifle. (See Table 5.8 in Chapter 5.) Two students were killed and 25 were injured. Kinkel's parents were later found shot to death at their home. Kinkel's father had bought his son the rifle a year earlier under the condition that he would use it only with adult supervision.

On September 24, 1999, as part of a plea agreement, Kinkel pleaded guilty to four counts of murder and 26 counts of attempted murder. On November 2, 1999, after a six-day sentencing hearing that included victims' statements and the testimony of psychiatrists and psychologists, Kinkel, by then aged 17, was sentenced to 111 years in prison without the possibility of parole.

Prompted by Kinkel's behavioral problems, Faith Kinkel had taken her son to see a psychologist in January 1997, just over a year before the shooting. In this meeting the psychologist concluded that Kinkel was depressed, had difficulty managing his anger, and had shown a pattern of acting out his anger. The Public Broadcasting Service's *Frontline* offers a profile of Kinkel, a timeline of the shooting, a transcript of his confession, and an explanation of his sentencing in the program *The Killer at Thurston High* (January 18, 2000, http://www.pbs.org/wgbh/pages/frontline/shows/kinkel/).

Columbine High School, Littleton, Colorado

APRIL 20, 1999. Eighteen-year-old Eric Harris (1981–1999) and 17-year-old Dylan Klebold (1982–1999) spent more than a year planning their attack on Columbine High School in Littleton, Colorado. (See Table 5.8 in Chapter 5.) When explosives they had planted in the cafeteria failed to detonate, the boys—armed with a 9mm semiautomatic rifle, a pistol, and two 12-gauge shotguns—entered the school and commenced a shooting spree that left 13 people, including one teacher, dead and 24 injured. The carnage finally ended when the attackers killed themselves with their own guns. Harris and Klebold left behind a trove of documentation, including their diaries and video recordings, testifying to their profound social alienation and their grandiose fantasies of becoming world famous for the revenge they would wreak against their school and society at large.

Within days, authorities had learned that three of the guns used in the massacre were purchased the year before by a straw purchaser—Klebold's girlfriend shortly after her 18th birthday. On May 3, felony charges were filed against 22-year-old Mark E. Manes for admittedly selling to Harris the TEC-DC9 semiautomatic handgun that he used in the shooting. On August 18, Manes pleaded guilty to the charge. The facts of this case as outlined came from *The Columbine High School Shootings: Jefferson County Sheriff Department's Investigation Report* (May 2000).

Red Lake High School, Red Lake, Minnesota

MARCH 21, 2005. The shooting that occurred at Red Lake High School was the nation's worst since the 1999 Columbine massacre. Red Lake High School is located on a Native American reservation in northern Minnesota. Jeffrey Weise (1988–2005), a 17-year-old junior, killed nine people and wounded seven in his shooting spree, and then shot and killed himself. Weise began his rampage by killing his grandfather—a tribal police officer—and his grandfather's female friend at their home, using a .22-caliber pistol of unknown origin. Weise then drove his grandfather's police cruiser to Red Lake High School. At the school, Weise used his grandfather's police-issued handguns and shotgun to kill a security guard, a teacher, and five students.

Weise came from a troubled background. He had lost his father to suicide in 1997. His mother was seriously brain-damaged in a car accident in 1999 and lived in a nursing home. Weise was thought to have posted messages on a neo-Nazi website. He called himself an "Angel of Death" and a "NativeNazi." He was often ridiculed and bullied by other students for his odd behavior.

Seven days after the shooting Louis Jourdain, the son of the tribal chairman, was arrested and charged with conspiracy. It was believed that he helped plot Weise's actions. Jourdain was tried as a juvenile, and in January 2006 he received a sentence, which, because of his juvenile status, was not made public. In July 2006 families of those injured and killed settled a lawsuit with the school district for $1 million.

Chardon High School, Chardon, Ohio

FEBRUARY 27, 2012. At about 7:30 a.m. on February 27, 2012, Thomas "T. J." Lane (1994–) opened fire with a Ruger .22-caliber semiautomatic pistol in the Chardon High School cafeteria as he and other students waited for bus transportation to other educational sites. Six students were hit by the gunfire; one died at the scene, and two other students died the following day from their wounds. As the incident happened, Lane was confronted by an unarmed football coach at the school and chased from the building. He was apprehended outside by law enforcement and detained in a juvenile facility as prosecutors sought to try him as an adult. At a hearing in June 2012 Lane registered a plea of not guilty by reason of insanity. He reportedly admitted to the shooting but told authorities that he did not know why he did it. The gun used in the case had been legally purchased by a relative and was allegedly stolen by Lane the night before the shooting. Lane later changed his plea, pleading guilty to the murders of his three classmates as well as to other charges, and in 2013 he was sentenced to life imprisonment without the possibility for parole.

Marysville-Pilchuck High School, Marysville, Washington

OCTOBER 24, 2014. Fifteen-year-old freshman Jaylen Fryberg (1999–2014) was a proud member of the Tulalip Tribes, a skilled hunter, and a popular football player. He had recently been crowned Freshman Homecoming Prince. In a planned attack, he asked several of his closest friends (two of them cousins) to meet him for lunch in the school cafeteria. When they were all assembled at the table, eating and chatting, he stood up, took a .40-caliber Beretta semiautomatic pistol out of his backpack, and began shooting his friends in the head, one by one. Fryberg killed four and wounded one before he shot and killed himself. The gun Fryberg used in the attack was legally registered to his father, who kept it unlocked, along with extra ammunition, in the center console of his pickup truck.

Unlike the majority of school shootings, in which the attacker (or attackers) shot their victims at random, Fryberg deliberately selected his close friends as victims. Analysis of the boy's text messages appeared to indicate that he was suicidal but did not want to die alone. Max Kutner notes in "What Led Jaylen Fryberg to Commit the Deadliest High School Shooting in a Decade?" (Newsweek.com, September 16, 2015) that Native American teens are exposed to more violence—including domestic and gang violence, sexual assault, and bullying—than any other racial or ethnic group of adolescents. Often afflicted by post-traumatic stress disorder, they also have the highest suicide rates of any demographic group in the United States.

POSTSECONDARY SCHOOL SHOOTINGS
Virginia Polytechnic Institute and State University, Blacksburg, Virginia

APRIL 16, 2007. The Virginia Polytechnic Institute and State University (Virginia Tech) massacre was, at the time it occurred, the deadliest shooting rampage by a single gunman in U.S. history. (See Table 5.8 in Chapter 5.) On the morning of April 16, 2007, Seung-Hui Cho (1984–2007) entered a residence hall on campus, where he shot and killed a female student and a male resident assistant. Hours later, after returning to his dorm room to change out of his bloodied clothes and to delete various files from his computer, Cho entered a classroom building on campus and began shooting students and professors. When he finished, Cho had killed 32 students and faculty members and had injured 15 others. Cho committed suicide by shooting himself in the head.

A Virginia judge had declared Cho mentally ill and ordered him to receive outpatient psychiatric treatment, but this information was never entered into the National Instant Criminal Background Check System database. (See Chapter 3 for a detailed discussion about this database.) Thus, he was able to legally purchase the two semiautomatic pistols—a Walther P32 and a Glock 19—that he used in his attack. The U.S. Departments of Health and Human Services, Education, and Justice provide a thorough analysis of the incident in *Report to the President on Issues Raised by the Virginia Tech Tragedy* (June 13, 2007, https://schoolshooters.info/sites/default/files/Report_to_President.pdf), a report commissioned by President George W. Bush (1946–).

Northern Illinois University, DeKalb, Illinois

FEBRUARY 14, 2008. Steven Phillip Kazmierczak (1980–2008), a once-successful student at Northern Illinois University (NIU), opened fire on an auditorium lecture hall, killing five students and wounding 21 others before killing himself. (See Table 5.8 in Chapter 5.) The gunman's weapons included a Remington 12-gauge shotgun and three semiautomatic pistols—a Glock 19, a Sig Sauer 9mm, and a HiPoint 380 caliber. Kazmierczak had a history of mental health problems, including multiple suicide attempts in adolescence. Officials determined no clear motive for the killings other than the shooter's mental illness. The NIU's Department of Safety details in *Report of the February 14, 2008 Shootings at Northern Illinois University* (March 2010, http://www.niu.edu/feb14report/Feb14report.pdf) the shooting incident.

Oikos University, Oakland, California

APRIL 2, 2012. One L. Goh (1968–) opened fire at about 10:30 a.m. on April 2, 2012, with a .45-caliber semiautomatic handgun in a nursing classroom at Oikos University in Oakland, California, killing seven people and wounding three others. The Korean-born Goh was identified as a former student at the private vocational school, which was founded in 2004 and is affiliated with the Praise God Korean Church of Oakland. Officials at the school were unable to identify a motive in the shooting. Goh was in financial difficulty and had dropped out of the school, which is located in a tight-knit Korean American community, but he had not been expelled. On the morning of the attack he attempted to meet with a school administrator, but when she was not in her office, he forced another worker at the school to accompany him into a classroom where he began shooting at the students. Goh fled the scene in a stolen vehicle and threw his gun, which he had purchased legally in California, into a nearby estuary before turning himself in to authorities that afternoon. In late 2013 a court-appointed psychiatrist found Goh mentally incompetent to stand trial, and after a hearing to confirm this assessment, he was sent to the Napa State Hospital in Northern California, where he would be treated until considered fit to stand trial.

Umpqua Community College, Roseburg, Oregon

OCTOBER 1, 2015. Twenty-six-year-old Chris Harper-Mercer went to the small-town community college (where he was a student) wearing body armor and carrying six guns, as well as a considerable amount of ammunition. Entering an introductory composition class in which he was enrolled, he shot the teacher at point-blank range and then began shooting the students. In the space of 10 minutes, the shooter killed nine people and injured nine others. When police arrived at the scene, he engaged them in a gunfight before fatally shooting himself.

Much of what was learned about Harper-Mercer after the incident was derived from writing contained on a flash drive, which he had entrusted to a student in the class with instructions to deliver it to the police. Other insights about his character and mental state were inferred from his online activities in chat rooms and on social media. An avid gun collector and a student of past mass shootings, Harper-Mercer professed white supremacist beliefs and a strong dislike of organized religion. He was socially isolated and sexually frustrated. All of the gunman's firearms were legally obtained by himself or his family members through a federally licensed firearms dealer.

WHEN ADULTS SHOOT SCHOOLCHILDREN

Although most school shootings are carried out by students, school property occasionally becomes the site of gun violence perpetrated by adults against children.

West Nickel Mines Amish School, Nickel Mines, Pennsylvania

OCTOBER 2, 2006. A 32-year-old milk truck driver entered a one-room Amish schoolhouse in Lancaster County, Pennsylvania. Armed with three guns, a stun gun, two knives, 600 rounds of ammunition, and a number of instruments believed to be intended for torture and sexual assault, the gunman ordered all male students and adult females to leave the building. He then lined up the remaining 10 children—all girls, aged six to 13 years—against the blackboard and bound their feet. After escaping, a teacher at the school ran to a farmhouse that had a telephone and called police. When the police arrived, the gunman shot all 10 girls and then himself. Five of the girls died.

Later identified as Charles C. Roberts IV (1973–2006), the gunman lived with his wife and children in Nickel Mines. Family and friends, including many of his Amish neighbors, were shocked by Roberts's actions, maintaining that he had appeared to be a devoted husband and father. Before taking his own life, Roberts called his wife and explained that he was plagued by guilt for having molested two young relatives when he was 12 years old and that he had recently experienced recurring dreams of molesting more girls. Police, however, were unable to confirm Roberts's story.

Deer Creek Middle School, Littleton, Colorado

FEBRUARY 23, 2010. As buses were leaving Deer Creek Middle School at the end of the school day, several students heard two shots ring out. Within seconds, 32-year-old Bruco Strongeagle Eastwood (1977?–) was tackled by a seventh-grade math teacher as Eastwood tried to reload his high-powered rifle. Lying wounded were eighth graders Matt Thieu and Reagan Webber. Eastwood, a former student of Deer Creek Middle School, had previously visited the school and was seen inside the school shortly before the shooting. A student standing nearby the victims said that Eastwood approached and asked the students if they attended the school. When Thieu and Webber responded affirmatively, Eastwood shot them. Both students survived the shooting. Although police revealed no motive for the shooting, they did note that Eastwood was apparently depressed about never graduating from high school and being unable to obtain a general equivalency diploma.

Sandy Hook Elementary School, Newtown, Connecticut

DECEMBER 14, 2012. Twenty-year-old Adam Lanza (1992?–2012) shot his mother to death at their home in Newtown and then proceeded to Sandy Hook Elementary School, where he shot his way inside. Armed with a variety of weapons, including two handguns and an assault rifle, he killed 20 first-grade students and six adult staff members before killing himself. (See Table 5.9 in Chapter 5.) In the aftermath of the atrocity, details about Lanza's mental health problems emerged, as did details of his thorough knowledge of handguns and assault weapons. His mother was a firearms collector and an avid target shooter, who passed her enthusiasm onto her son and took no measures to restrict his access to her weapons and ammunition.

The incident drew international media attention and prompted intense debate over the adequacy of the nation's gun control regulations. On January 16, 2013 (http://www.whitehouse.gov/the-press-office/2013/01/16/remarks-president-and-vice-president-gun-violence), President Barack Obama (1961–) presented a plan to reduce mass gun violence that included 23 executive actions and proposals for four major pieces of legislation. Vowing to spend political capital in what would undoubtedly be difficult negotiations with the opponents of policy reform, Obama said, "I will put everything I've got into this." Political opposition to increased gun control quickly scuttled reformers' hopes for change in the months after Newtown, and many gun rights advocates, in assessing the legacy of Newtown, emphasized the poor national mental health infrastructure as a more salient factor in the shooting than the ease of access to guns.

Post-Newtown School Safety Efforts

In contrast to those who saw the Sandy Hook massacre as a clear indication of the need for stricter gun control laws, Wayne LaPierre, the chief executive officer of the National Rifle Association of America (NRA), responded to the tragedy by calling for Congress to appropriate the necessary funding to immediately deploy armed officers in all 99,000 public schools in the United States. Although LaPierre's proposal was widely criticized, even by some NRA members, as being heavy-handed and misguided, the tragedy did bring about substantial investment in school security standards nationwide.

In "Spending on School Security Rises" (WSJ.com, May 21, 2015), Caroline Porter reports on Department of Education data from a survey of approximately 1,400 public schools around the country, which reflected a clear increase in the number of schools that enhanced their security measures during the 2013–14 school year, compared with 2009–10, when the survey was last conducted. The Department of Education found that in 2013–14, 75% of schools surveyed were using one or more security cameras, up from 61% in 2009–10. Similarly, many more schools reported having electronic notification systems in place to alert parents about a school emergency—82%, compared with 63% four years earlier. Citing figures from the research company HIS Technology, Porter notes that the total market in schools and universities for video-surveillance equipment, mass notification systems, and access-control equipment totaled approximately $768 million in 2014. This spending was projected to reach $833 million in 2015 and $907 million in 2016. There was also increased demand for school resource officers, as police officers assigned to public schools are called.

Although the most immediate goal of improving school safety is to prevent the bloodshed and loss of life associated with shooting incidents, there is also evidence to suggest that school violence has negative consequences on educational outcomes. Louis-Philippe Beland and Dongwoo Kim examine in "The Effect of High School Shootings on Schools and Student Performance" (*Educational Evaluation and Policy Analysis*, vol. 38, no. 1, March 2016) test performance, rates of enrollment, attendance, graduation, and other variables at California schools where a fatal shooting had occurred and compare these measures to schools in the same district where no shooting had occurred. The researchers find that in the aftermath of a deadly shooting, behavioral variables (such as enrollment, attendance, and graduation) were largely unaffected, but standardized test scores appeared to suffer, even as long as three years after the incident. According to Beland and Kim's data, in schools that experienced a shooting homicide, the proficiency rate (a measure of test performance) in math was reduced by 4.9 percentage points, while the English proficiency rate was reduced by 3.9 percentage points, relative to the comparison schools.

CHILDREN INJURED AND KILLED BY GUNFIRE

The Children's Defense Fund (CDF), a charitable organization that focuses on the needs of poor and minority children and those with disabilities, regularly reports on the number of children who lose their lives to gun violence. In "2014 Child Gun Deaths" (December 2015, http://www.childrensdefense.org/library/data/2015-protect childrennotgunsfactsheet.pdf), the CDF indicates that in 2014, 2,525 American children and teens died from guns. This represents one child or teen death every three hours and 28 minutes. The CDF further notes that "as shocked as the nation was by the 2012 Newtown massacre, more children and teens died from guns every three days in 2014 than died then" and that the total number of children and teens who died in 2014 "would fill 126 classrooms of 20 children."

FIREARMS AND YOUTH SUICIDE

Although homicide rates have declined dramatically since the 1990s, suicide rates have remained flat since that time. In 1950 the age-adjusted rate of death by suicide for all ages was 13.2 per 100,000. (See Table 6.6 in Chapter 6.) After dropping to 10.4 per 100,000 in 2000, it rebounded to 13 per 100,000 in 2014. In contrast, the suicide rate for young people rose precipitously over this same period. The rate of death by suicide nearly tripled for youth aged 15 to 24 years between 1950 (4.5 per 100,000) and 1990 (13.2 per 100,000) and then declined slightly to 11.5 per 100,000 in 2014. Between 1950 and 2014, then, the youth suicide rate increased by 156%. As Table 7.13 shows, suicide was the second-leading cause of death among children aged 10 to 14 years, young people aged 15 to 19 years, and young adults aged 20 to 24 years. Among young people aged 10 to 24 years, there were 5,504 deaths by suicide in 2014. As Figure 7.11 illustrates, suicide accounted for 17.4% of all deaths for this age group in 2014.

According to the CDC, in "Suicide Prevention: Youth Suicide" (March 10, 2015, http://www.cdc.gov/violenceprevention/suicide/youth_suicide.html), firearms account for 45% of suicides among young people aged 10 to 24 years, while suffocation accounts for 40% and poisoning for 8%. Firearms are different from other suicide methods in one key aspect: attempts are much more likely to be fatal. Other means of attempting suicide are less reliable, take time, and can be reversed, often through intervention from loved ones or through emergency medical care. In "The Epidemiology of Case Fatality Rates for Suicide in the Northeast" (*Annals of Emergency Medicine*, vol. 43, no. 6, June 2004), Matthew Miller, Deborah Azrael, and David Hemenway of the Harvard Injury Control Research Center find that the fatality rate of suicide attempts with a firearm exceeds 90%, compared with a fatality rate of less than 5% for other common suicide methods, such as drug overdoses and cutting/piercing.

The lethality of firearms is of particular concern in adolescence and young adulthood, a time of life when many people begin experiencing pronounced emotional and social difficulties without having the life experience or maturity to put them into perspective. The CDC indicates that approximately 157,000 youth are hospitalized for suicide attempts annually. This number, when compared with the 4,600 young people who die from suicide each year, suggests that adolescents and young adults are particularly likely to entertain the possibility of suicide without necessarily being resolute in their determination to die. Nevertheless, young people who attempt suicide with firearms are almost certain to die even if their suicidal impulses are temporary rather than deeply rooted.

TABLE 7.13

Deaths and death rates for the 10 leading causes of death among young people, 2014

[Rates are per 100,000 population in specified group. Data for races other than white and black should be interpreted with caution because of misreporting of race on death certificates.]

Rank[a]	Cause of death (based on ICD–10), race, sex, and age	Number[b]	Percent of total deaths	Death rate[b]
	All races, both sexes, 1–4 years			
—	All causes	3,830	100.0	24.0
1	Accidents (unintentional injuries) (V01–X59, Y85–Y86)	1,216	31.7	7.6
2	Congenital malformations, deformations and chromosomal abnormalities (Q00–Q99)	399	10.4	2.5
3	Assault (homicide) (ᶜU01–ᶜU02, X85–Y09, Y87.1)	364	9.5	2.3
4	Malignant neoplasms (C00–C97)	321	8.4	2.0
5	Diseases of heart (I00–I09, I11, I13, I20–I51)	149	3.9	0.9
6	Influenza and pneumonia (J09–J18)	109	2.8	0.7
7	Septicemia (A40–A41)	53	1.4	0.3
8	Chronic lower respiratory diseases (J40–J47)	53	1.4	0.3
9	In situ neoplasms, benign neoplasms and neoplasms of uncertain or unknown behavior (D00–D48)	38	1.0	0.2
10	Certain conditions originating in the perinatal period (P00–P96)	38	1.0	0.2
—	All other causes (residual)	1,090	28.5	6.8
	All races, both sexes, 5–9 years			
—	All causes	2,357	100.0	11.5
1	Accidents (unintentional injuries) (V01–X59, Y85–Y86)	730	31.0	3.6
2	Malignant neoplasms (C00–C97)	436	18.5	2.1
3	Congenital malformations, deformations and chromosomal abnormalities (Q00–Q99)	192	8.1	0.9
4	Assault (homicide) (ᶜU01–ᶜU02, X85–Y09, Y87.1)	123	5.2	0.6
5	Diseases of heart (I00–I09, I11, I13, I20–I51)	69	2.9	0.3
6	Chronic lower respiratory diseases (J40–J47)	68	2.9	0.3
7	Influenza and pneumonia (J09–J18)	57	2.4	0.3
8	Cerebrovascular diseases (I60–I69)	45	1.9	0.2
9	In situ neoplasms, benign neoplasms and neoplasms of uncertain or unknown behavior (D00–D48)	36	1.5	0.2
10	Septicemia (A40–A41)	33	1.4	0.2
—	All other causes (residual)	568	24.1	2.8
	All races, both sexes, 10–14 years			
—	All causes	2,893	100.0	14.0
1	Accidents (unintentional injuries) (V01–X59, Y85–Y86)	750	25.9	3.6
2	Intentional self-harm (suicide) (ᶜU03, X60–X84, Y87.0)	425	14.7	2.1
3	Malignant neoplasms (C00–C97)	416	14.4	2.0
4	Congenital malformations, deformations and chromosomal abnormalities (Q00–Q99)	156	5.4	0.8
5	Assault (homicide) (ᶜU01–ᶜU02, X85–Y09, Y87.1)	156	5.4	0.8
6	Diseases of heart (I00–I09, I11, I13, I20–I51)	122	4.2	0.6
7	Chronic lower respiratory diseases (J40–J47)	71	2.5	0.3
8	Cerebrovascular diseases (I60–I69)	43	1.5	0.2
9	Influenza and pneumonia (J09–J18)	41	1.4	0.2
10	In situ neoplasms, benign neoplasms and neoplasms of uncertain or unknown behavior (D00–D48)	38	1.3	0.2
—	All other causes (residual)	675	23.3	3.3
	All races, both sexes, 15–19 years			
—	All causes	9,586	100.0	45.5
1	Accidents (unintentional injuries) (V01–X59, Y85–Y86)	3,736	39.0	17.7
2	Intentional self-harm (suicide) (ᶜU03, X60–X84, Y87.0)	1,834	19.1	8.7
3	Assault (homicide) (ᶜU01–ᶜU02, X85–Y09, Y87.1)	1,397	14.6	6.6
4	Malignant neoplasms (C00–C97)	612	6.4	2.9
5	Diseases of heart (I00–I09, I11, I13, I20–I51)	299	3.1	1.4
6	Congenital malformations, deformations and chromosomal abnormalities (Q00–Q99)	179	1.9	0.8
7	Influenza and pneumonia (J09–J18)	64	0.7	0.3
8	Cerebrovascular diseases (I60–I69)	58	0.6	0.3
9	Chronic lower respiratory diseases (J40–J47)	55	0.6	0.3
10	Septicemia (A40–A41)	39	0.4	0.2
—	All other causes (residual)	1,313	13.7	6.2
	All races, both sexes, 20–24 years			
—	All causes	19,205	100.0	83.8
1	Accidents (unintentional injuries) (V01–X59, Y85–Y86)	8,100	42.2	35.4
2	Intentional self-harm (suicide) (ᶜU03, X60–X84, Y87.0)	3,245	16.9	14.2
3	Assault (homicide) (ᶜU01–ᶜU02, X85–Y09, Y87.1)	2,747	14.3	12.0
4	Malignant neoplasms (C00–C97)	957	5.0	4.2
5	Diseases of heart (I00–I09, I11, I13, I20–I51)	654	3.4	2.9
6	Congenital malformations, deformations and chromosomal abnormalities (Q00–Q99)	198	1.0	0.9
7	Diabetes mellitus (E10–E14)	147	0.8	0.6
8	Influenza and pneumonia (J09–J18)	135	0.7	0.6
9	Pregnancy, childbirth and the puerperium (O00–O99)	132	0.7	0.6
10	Chronic lower respiratory diseases (J40–J47)	123	0.6	0.5
—	All other causes (residual)	2,767	14.4	12.1

TABLE 7.13

Deaths and death rates for the 10 leading causes of death among young people, 2014 [CONTINUED]

[Rates are per 100,000 population in specified group. Data for races other than white and black should be interpreted with caution because of misreporting of race on death certificates.]

[a]Based on number of deaths.
[b]Figures for age not stated are included in "all ages" but not distributed among age groups.
[c]The code is not included in the *International Classification of Diseases, Tenth Revision* (ICD010).

SOURCE: Adapted from Melonie Heron, "Table 1. Deaths, Percentage of Total Deaths, and Death Rates for the 10 Leading Causes of Death in Selected Age Groups, by Race and Sex: United States, 2014," in "Deaths: Leading Causes for 2014," *National Vital Statistics Reports*, vol. 65, no. 5, National Center for Health Statistics, June 30, 2016, http://www.cdc.gov/nchs/data/nvsr/nvsr65/nvsr65_05.pdf (accessed September 10, 2016)

FIGURE 7.11

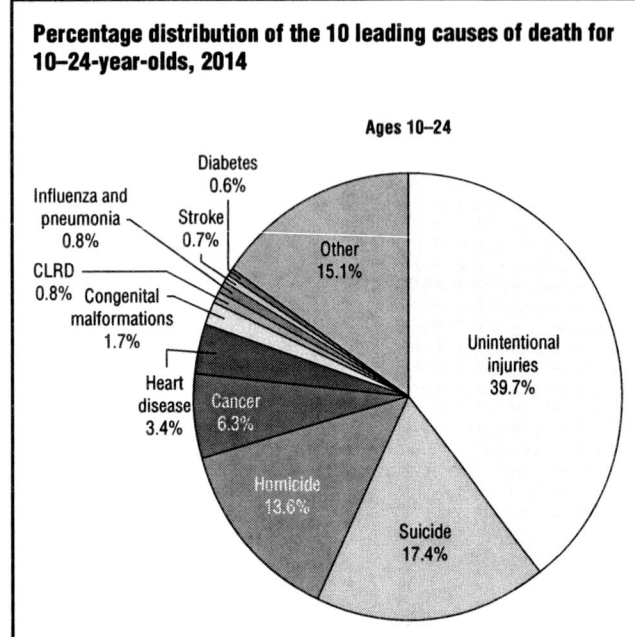

Percentage distribution of the 10 leading causes of death for 10–24-year-olds, 2014

Ages 10–24

- Diabetes 0.6%
- Influenza and pneumonia 0.8%
- Stroke 0.7%
- CLRD 0.8%
- Congenital malformations 1.7%
- Heart disease 3.4%
- Cancer 6.3%
- Homicide 13.6%
- Other 15.1%
- Unintentional injuries 39.7%
- Suicide 17.4%

CLRD = chronic lower respiratory disease.

SOURCE: Adapted from Melonie Heron, "Figure 2. Percent Distribution of the 10 Leading Causes of Death, by Age Group: United States, 2014," in "Deaths: Leading Causes for 2014," *National Vital Statistics Reports*, vol. 65, no. 5, National Center for Health Statistics, June 30, 2016, http://www.cdc.gov/nchs/data/nvsr/nvsr65/nvsr65_05.pdf (accessed September 10, 2016)

CHAPTER 8
PUBLIC ATTITUDES TOWARD GUN CONTROL

A DEEPLY PERSONAL ISSUE

Gun control first became a prominent issue during the 1960s, when the nation was rocked by the assassinations of President John F. Kennedy (1917–1963), his brother Senator Robert F. Kennedy (1925–1968; D-NY), and the civil rights leader Martin Luther King Jr. (1929–1968). Since that time, additional political assassination attempts, violent crime waves, and horrific mass shootings have fueled the ongoing controversy about the place of firearms in American life.

Political assassinations and mass shootings attract more media attention than other types of firearm violence, but they are, in fact, extremely rare, representing only a small fraction of the overall deaths and injuries caused by gun use annually. As Figure 8.1 illustrates, since 1980 mass shootings have consistently represented less than 1% of total annual gun homicides. It is the use of guns in everyday crime and suicide that most troubles public health professionals, if not ordinary Americans. As Table 6.7 in Chapter 6 shows, 320,575 people were killed by firearms in the United States between 2005 and 2014.

Nevertheless, as many gun rights advocates point out, it is difficult to prove that tighter gun ownership and purchase requirements would prevent many of these gun fatalities. Criminals often obtain weapons illegally, whereas law-abiding citizens are the primary purchasers of guns in stores that conduct background checks. Many people who use guns to commit suicide, including most young people, obtain those guns from their own home, following legal purchases by parents or other household members. Although gun control advocates point to numerous studies showing that the presence of a gun in the home positively correlates with higher rates of homicide and suicide, it is unclear that placing more restrictions on the purchase of firearms would address such deaths. Meanwhile, as is described in Chapter 1, the right to keep and bear arms is interwoven with the history and political foundations of the United States in such a way that many gun owners see their firearm possession as one of the basic features of their identity as an American. Thus, the gun control debate touches on some of the most emotionally charged territory in U.S. politics.

Some Americans are convinced that more federal regulation of firearms is necessary to reduce the number of gun-related fatalities and injuries and to ensure a safer, more civilized society. They see firearms as the cause of a national public health crisis and note that countries that restrict firearm ownership have much lower rates of firearm death. Their opponents on the other side of the debate insist that the right to bear arms is guaranteed by long-standing custom and by the Second Amendment to the U.S. Constitution. These gun rights advocates believe that no cyclical increase in crime, no mass killing, or any rash of political murders should lead the nation to violate the Constitution and the individual rights it guarantees. Moreover, many believe that more guns in the hands of law-abiding citizens will lead to a reduction of violent crime, as criminals will be less able to victimize people.

Both supporters and opponents of gun control agree that some means should be found to keep guns out of the hands of criminals. Not surprisingly, the two sides approach this issue differently. The two different strategies for gun control involve deterrence (discouraging by instilling fear of consequences) and interdiction (legally forbidding the use of). Advocates of deterrence, most notably the Second Amendment Foundation and the National Rifle Association of America (NRA), recommend consistent enforcement of current laws and instituting tougher penalties to discourage individuals from using firearms in crimes. They maintain that interdiction will not have any effect on crime but will strip away the constitutional rights and privileges of law-abiding Americans by taking away their right to own guns.

FIGURE 8.1

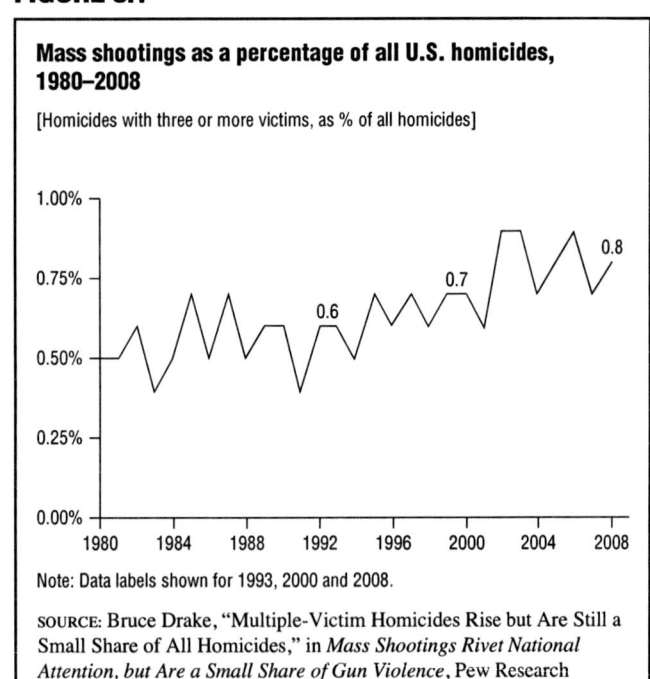

Mass shootings as a percentage of all U.S. homicides, 1980–2008

[Homicides with three or more victims, as % of all homicides]

Note: Data labels shown for 1993, 2000 and 2008.

SOURCE: Bruce Drake, "Multiple-Victim Homicides Rise but Are Still a Small Share of All Homicides," in *Mass Shootings Rivet National Attention, but Are a Small Share of Gun Violence*, Pew Research Center, September 17, 2013, http://www.pewresearch.org/fact-tank/2013/09/17/mass-shootings-rivet-national-attention-but-are-a-small-share-of-gun-violence/ (accessed September 15, 2016)

Advocates of interdiction, led by organizations such as the Brady Campaign to Prevent Gun Violence, the Law Center to Prevent Gun Violence, and the Violence Policy Center, believe controlling citizens' access to firearms will reduce crime. Therefore, they favor restrictions on public gun ownership.

EVALUATING PUBLIC OPINION POLLS

Public opinion polls, like all sources of information, must be used with care. Pollsters select sample populations because it is impossible to interview every American on a given question. The selection is usually performed randomly using computers. Most major pollsters interview between 1,000 and 2,000 people to establish a valid sample. Other pollsters may interview far fewer than 500, and the sample may be too small to fairly represent the opinions of all adult Americans. Generally, the larger the sample, the greater the chance for an adequate representation and a valid result.

The polling errors that concern most people are those caused by bias in the presentation of questions, which may influence the response. For example, is the question vague? Is it too long? Is it threatening? Is it leading? If the questions are asked in an in-person interview, was the interviewer too forceful or threatening? Did respondents provide answers they thought would please interviewers? Were respondents disqualified because of membership in gun control or gun rights organizations? What was the purpose of the poll? Who hired the polling organization, and what is its stance on the issue?

Respondents may be unwilling to candidly discuss their use of weapons. In addition, polling does not always determine how important a person considers an issue to be. The issue may be of absolutely no concern to the respondent, but when asked, the respondent then thinks about the topic and provides an answer. Five minutes after the question has been asked, the issue may completely disappear from his or her mind.

The polling organization might not include the number of times there was no response. If these "no replies" come predominantly from one group, they might influence the poll so that it does not truly represent the national opinion on a given issue. Pollsters are aware of these weaknesses, so they usually indicate how reliable they consider their polls to be. As a result, many polls provide a measurement along with a confidence interval (also called a standard error) of a "plus or minus" (+/−) value. An example might be a measurement and confidence interval of "37% +/−2%," which means that the margin of error of the measurement (37%) is estimated to be within 2 percentage points above (39%) or 2 percentage points below (35%).

PUBLIC OPINION ON GUN CONTROL

The Demographics of Gun Ownership

Gun ownership rates vary significantly by region, state, location, sex, race and ethnicity, and other demographic factors. Rich Morin explains in "The Demographics and Politics of Gun-Owning Households" (July 15, 2014, http://www.pewresearch.org/fact-tank/2014/07/15/the-demographics-and-politics-of-gun-owning-households) that the Pew Research Center, a nonpartisan research and polling firm, established the prevalence of gun ownership by demographic and political characteristics as part of its American Trends Panel survey conducted in April and May 2014.

By region, Pew found that the percentage of gun-owning households among all households in 2014 was similar in the West (34%), the Midwest (35%), and the South (38%). In contrast, the percentage of gun-owning households in the Northeast (27%) was much lower. Accordingly, residents of northeastern states, on average, favor stricter gun control laws than residents in other regions, with the notable exceptions of California, Hawaii, and Illinois, which have comparably strict gun control laws at the state level.

As part of its mid-2014 survey, Pew also assessed gun ownership rates by gender, race and ethnicity, age, location, and the presence of children in the home. It found a significant disparity by gender: 38% of American men reported the presence of a gun in their home, compared with 31% of women. (See Figure 8.2.) It also found that white Americans are much more likely to own guns than are African Americans and Hispanics. Whereas 41% of non-Hispanic whites reported the presence of a gun in their home, only 19% of African Americans and 20% of Hispanics reported a gun in the home. Gun ownership rates also increase with age. More than half (51%) of rural Americans reported the presence of a gun in the home in 2014, compared with 36% of suburban residents and 25% of urban residents. Gun ownership rates were not significantly different for households that included children under the age of 18 years (35%) than for the population at large (34%).

FIGURE 8.2

Prevalence of gun-ownership among selected groups, 2014

[Percent in each group who say they have a gun, rifle or pistol in their home]

Group	%
Gender	
Total	34%
Men	38%
Women	31%
Race/Ethnicity	
Non-Hispanic white	41%
Black	19%
Hispanic	20%
Age	
18–29	26%
30–49	32%
50–64	40%
65+	40%
Environment	
Urban	25%
Suburban	36%
Rural	51%
Children	
Children under 18 in home	35%
0–4 years old	33%
5–11	33%
12–17	37%
Party	
Republican	49%
Democrat	22%
Independent	37%
Ideology	
Conservative	41%
Moderate	36%
Liberal	23%

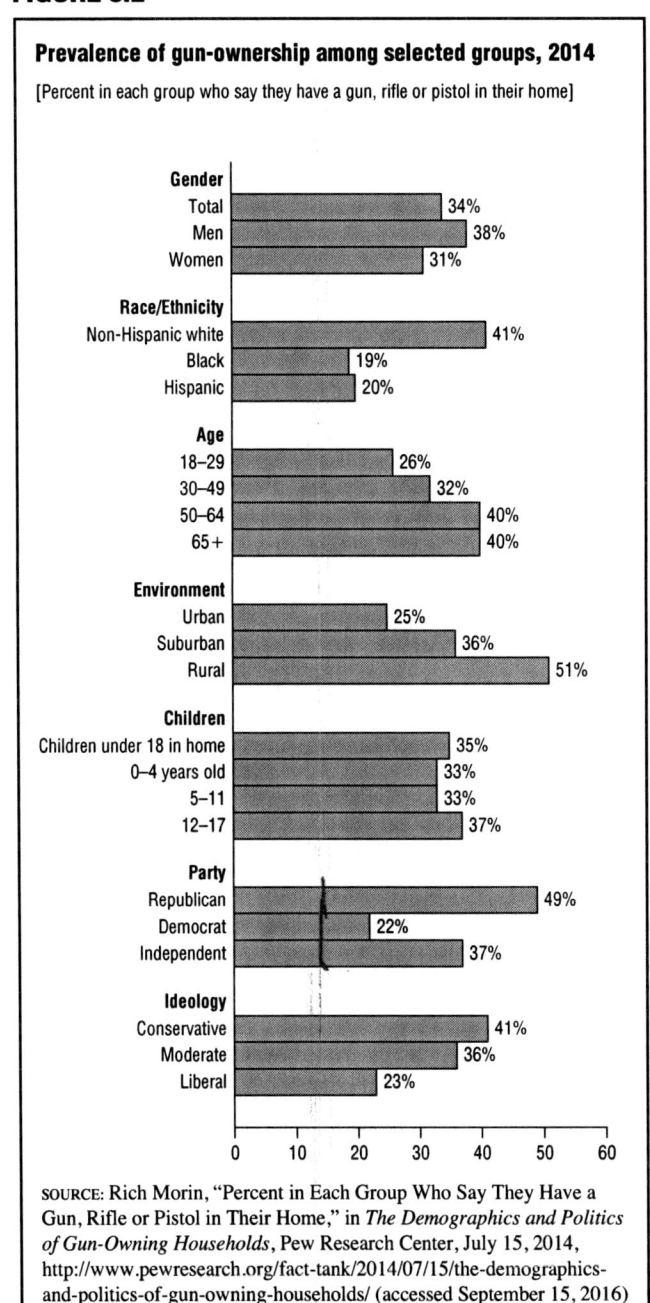

SOURCE: Rich Morin, "Percent in Each Group Who Say They Have a Gun, Rifle or Pistol in Their Home," in *The Demographics and Politics of Gun-Owning Households*, Pew Research Center, July 15, 2014, http://www.pewresearch.org/fact-tank/2014/07/15/the-demographics-and-politics-of-gun-owning-households/ (accessed September 15, 2016)

Gun ownership rates also track the division between the two major political parties in the United States, as well as divisions among broader ideological categories. Whereas almost half (49%) of Republicans reported the presence of a gun in their home in 2014, less than a quarter (22%) of Democrats did. (See Figure 8.2.) Independents (37%) were slightly closer to Republicans than to Democrats in their gun ownership status. Among those Americans who identify as conservative, 41% reported a gun in the home, compared with 36% who identify as moderate and 23% who identify as liberal.

Strictness of Gun Laws

Numerous polling organizations periodically conduct surveys assessing Americans' views on the strictness of gun control laws. In 2016, as in past years, Pew approached the question of the strictness of gun control laws through the lens of gun control versus gun rights, asking respondents whether they believe it is more important to control gun ownership or to protect the rights of gun owners. Figure 8.3 shows an overall trend in opinion toward increasing support for the rights of gun owners and decreasing support for gun control measures. In 2000, 66% of Americans favored gun control and only 29% favored the rights of gun owners. By 2016, 46% of Americans favored gun control and 52% favored the rights of gun owners.

Support for the rights of gun owners accelerated sharply in 2008. (See Figure 8.3.) This increase in support

FIGURE 8.3

Americans' views on controlling gun ownership vs. protecting the right of Americans to own guns, 1993–2016

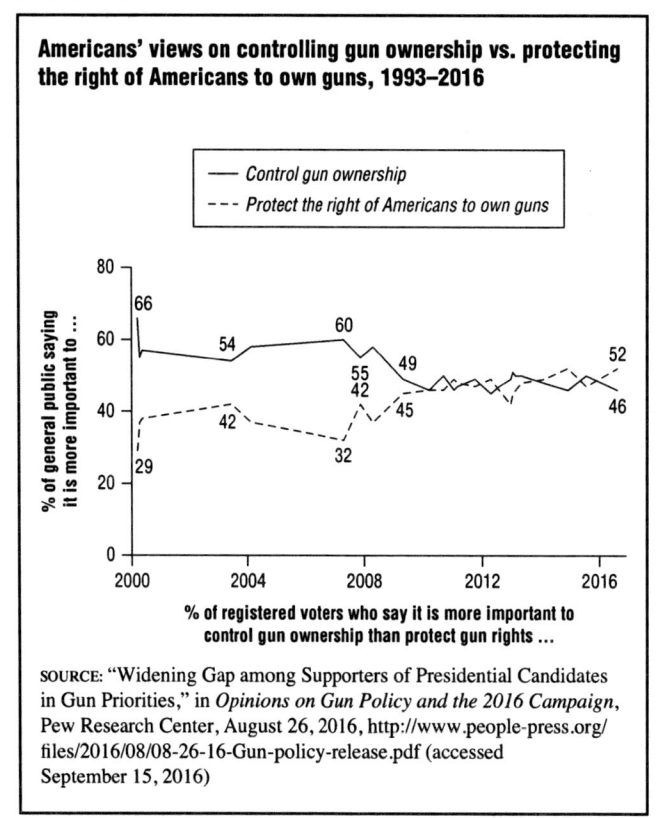

— Control gun ownership
--- Protect the right of Americans to own guns

% of general public saying it is more important to ...

Control gun ownership: 66 (2000), 54, 60, 55, 49, 46 (2016)
Protect right to own guns: 29 (2000), 42, 32, 42, 45, 52 (2016)

% of registered voters who say it is more important to control gun ownership than protect gun rights ...

SOURCE: "Widening Gap among Supporters of Presidential Candidates in Gun Priorities," in *Opinions on Gun Policy and the 2016 Campaign*, Pew Research Center, August 26, 2016, http://www.people-press.org/files/2016/08/08-26-16-Gun-policy-release.pdf (accessed September 15, 2016)

is widely believed to have been triggered by the election of Barack Obama (1961–) as president, who was believed by many gun rights advocates to be hostile to their position. The NRA saw its membership grow during the Obama years and the organization's rhetoric in matters relating to President Obama was notably urgent, casting the president and his administration as an unprecedented threat to Second Amendment rights. As correspondents from most major news outlets reported, ordinary gun owners tended to share the NRA's sense of alarm, and many responded by stockpiling weapons and ammunition in anticipation of new legislation that would outlaw them.

Increased demand for guns and ammunition persisted throughout Obama's first term, surging again in the wake of the December 2012 Sandy Hook shooting, which horrified the general public arguably more than any previous mass shooting, due to the fact that 20 of the 26 victims were young children. As Figure 8.3 indicates, support for the rights of gun owners fell sharply in early 2013, but this shift in public sentiment did not last long. As Obama and congressional Democrats called for increased gun control measures in the months and years that followed, the trends toward increased support for gun rights and increased demand for guns and ammunition continued.

Pew's election year poll, *Opinions on Gun Policy and the 2016 Campaign* (August 26, 2016, http://www.people-press.org/files/2016/08/08-26-16-Gun-policy-release.pdf), reveals an American electorate that was sharply divided along party lines on issues of gun policy. Among those who backed the Democratic candidate Hillary Rodham Clinton (1947–), 79% favored controlling gun ownership, compared with 19% who thought it was more important to protect gun rights. Among those who backed the Republican candidate Donald Trump (1946–), priorities were reversed and even more lopsided: 90% favored protecting gun rights, compared with just 9% who wanted gun control.

Overall, Pew finds a strong majority of the American public believed that gun ownership did more to protect people from becoming victims of crime (58%), compared with those who thought guns posed more of a safety risk (37%). (See Table 8.1.) Not surprisingly, an overwhelming majority of Republicans (or those who lean Republican) saw guns as a form of protection (82%), rather than a source of risk (14%). This view was also quite strong among men (64% protection versus 32% risk), whites (64% versus 33%), and those with a gun in their household (74% versus 22%). Among those with different levels of education, adults with postgraduate degrees were the only group where a majority believed that guns posed more of a safety risk (59%) than a means of protection (37%). Similarly, 57% of Democrats (or those who lean Democrat) saw guns as a safety risk, compared with 39% who saw them as a means of protection.

TABLE 8.1

Views of gun ownership as a matter of protection vs. a safety risk, by selected demographics, 2016

[% who say gun ownership in this country does more to…]

	Protect from becoming crime victims %	Put people's safety at risk %	Don't know
Total	58	37	4 = 100
Men	64	32	4 = 100
Women	53	43	4 = 100
White	64	33	3 = 100
Black	46	49	5 = 100
Hispanic	50	46	4 = 100
Postgrad	37	59	5 = 100
College grad	51	44	5 = 100
Some college	66	30	4 = 100
HS or less	61	35	4 = 100
Rep/lean Rep	82	14	4 = 100
Dem/lean Dem	39	57	4 = 100
Gun in household (44%)	74	22	4 = 100
No gun in household (51%)	43	52	4 = 100
Among whites			
College grad+	49	48	3 = 100
Non-college	71	26	3 = 100

Notes: Whites and blacks include only non-Hispanics; Hispanics can be of any race. Figures may not add to 100% because of rounding.

SOURCE: "Whites See Gun Ownership More as Protection, Non-Whites More Safety Risk," in *Opinions on Gun Policy and the 2016 Campaign*, Pew Research Center, August 26, 2016, http://www.people-press.org/files/2016/08/08-26-16-Gun-policy-release.pdf (accessed September 15, 2016)

In spite of the overarching polarization, however, there were areas of common ground. Perhaps most notably, public support for expanded background checks was high in both political parties, and indeed across nearly all demographic groups. As Table 8.2 shows, 86% of Democrats (or those who lean Democrat) and 78% of Republicans (or those who lean Republican) favored instituting background checks at gun shows. Support for this measure was similarly high (above 70%) regardless of sex, race or ethnicity, age, education level, and whether or not there was a gun in the household. Strong majorities of survey respondents also favored barring those on no-fly lists and those who are mentally ill from purchasing guns, although Hispanics were the least likely to support these two measures (53% and 54%, respectively). Those least likely to support creating a federal database to track gun sales were Republicans (50%) and those with a gun in their household (58%).

Polling conducted by the Gallup Organization in January 2016 reflects the American public's growing dissatisfaction toward existing gun laws. According to Art Swift of Gallup, in *Americans' Dissatisfaction with U.S. Gun Laws at New High* (January 14, 2016, http://www.gallup.com/poll/188219/americans-dissatisfaction-gun-laws-new-high.aspx), a greater percentage (62%) of Americans claimed to be dissatisfied with the nation's gun laws than at any point since 2001. (See Figure 8.4.)

TABLE 8.2

Opinions on expanded background checks among Americans in gun-owning households, 2016

[% who favor each policy proposal]

	Federal database of gun sales %	Background checks for gun shows %	Barring gun buys by people on no-fly lists %	Laws to prevent mentally ill from buying guns %
Total	68	81	71	76
Men	61	78	71	79
Women	74	84	70	73
White	64	84	76	83
Black	81	78	64	75
Hispanic	76	72	53	54
18–29	73	81	67	77
30–49	67	83	72	73
50–64	67	81	73	80
65+	65	79	70	77
College grad+	69	88	80	86
Some college	67	84	72	79
High school or less	68	74	63	67
Republican/lean Rep	50	78	69	79
Democrat/lean Dem	82	86	74	77
Gun in household (44%)	58	83	73	82
No gun in household (51%)	79	83	70	73

Note: Whites and blacks include only those who are not Hispanic; Hispanics are of any race.

SOURCE: "Broad Support for Expanded Background Checks among Those in Gun-Owning Households," in *Opinions on Gun Policy and the 2016 Campaign*, Pew Research Center, August 26, 2016, http://www.people-press.org/files/2016/08/08-26-16-Gun-policy-release.pdf (accessed September 15, 2016)

Dissatisfaction increased sharply between 2012 (42%) and 2013 (51%), following the Sandy Hook massacre. After remaining relatively stable for the next two years, dissatisfaction surged again, increasing 11 percentage points between 2015 (51%) and 2016 (62%)—another period during which the nation was rocked by high-profile mass shootings in Charleston, South Carolina, and San Bernardino, California. (See Table 5.9 in Chapter 5.) Accordingly, Swift notes, among people who reported dissatisfaction with existing gun laws, more than twice as many wanted stricter laws (38%) than wanted existing laws loosened (15%). As Table 8.3 shows, dissatisfaction was much higher among Democrats (75%) than among Republicans (54%).

In *American Public Opinion, Terrorism and Guns* (June 13, 2016, http://www.gallup.com/opinion/polling-matters/192695/american-public-opinion-terrorism-guns.aspx), Frank Newport reports on a Gallup survey of American attitudes toward gun control conducted in December 2015, shortly after the San Bernardino tragedy, which found that 55% of Americans favored stricter gun control laws. (See Figure 8.5.) This was higher than at any time during Obama's tenure in office, except in 2013 (after Sandy Hook), when 58% wanted stricter laws. Even after Sandy Hook, however, American support for stricter gun laws was much lower than during the early to mid-1990s, when, in the midst of the nation's worst homicide epidemic, support for stricter gun laws ranged between

FIGURE 8.4

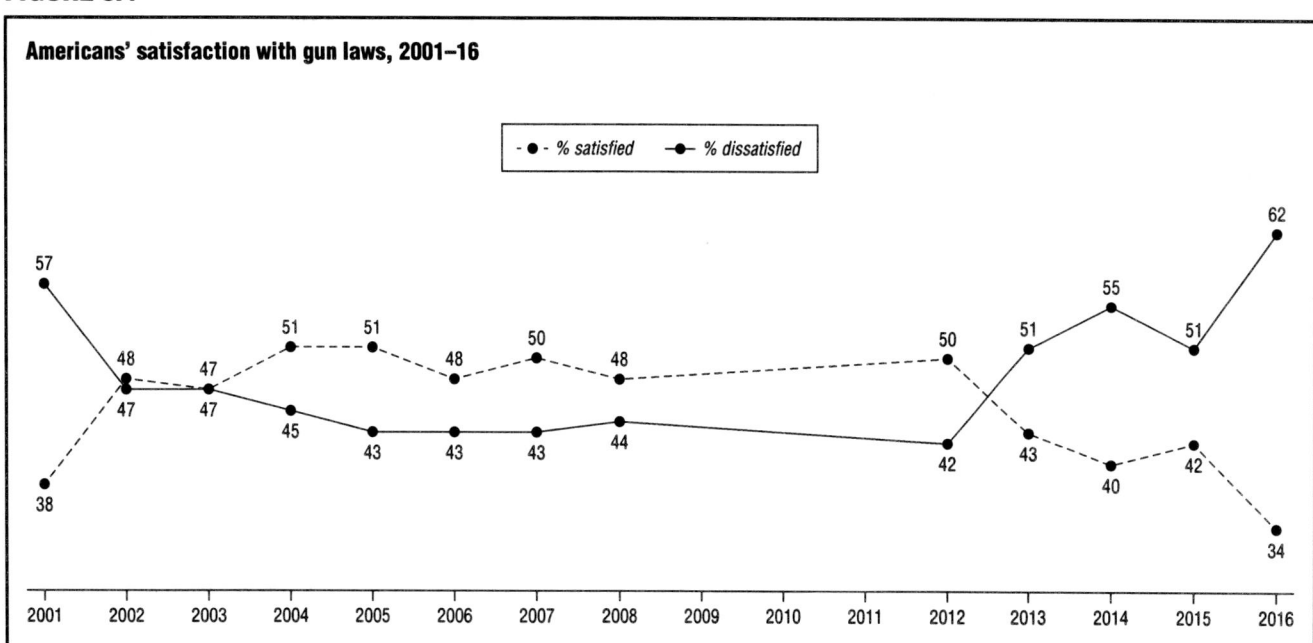

Americans' satisfaction with gun laws, 2001–16

Year	% satisfied	% dissatisfied
2001	38	57
2002	47	48
2003	47	47
2004	51	45
2005	51	43
2006	48	43
2007	50	43
2008	48	44
2012	50	42
2013	43	51
2014	40	55
2015	42	51
2016	34	62

SOURCE: Art Swift, "Americans' Satisfaction with the Nation's Laws or Policies on Guns," in *Americans' Dissatisfaction with U.S. Gun Laws at New High*, The Gallup Organization, January 14, 2016, http://www.gallup.com/poll/188219/americans-dissatisfaction-gun-laws-new-high.aspx (accessed September 15, 2016). Copyright © 2016 Gallup, Inc. All rights reserved. The content is used with permission; however, Gallup retains all rights of republication.

67% and 78%. Newport notes that 2015 survey respondents were "skeptical that gun control laws would be highly efficacious in controlling acts of gun violence and mass shootings."

Support for a ban on handguns, one of the strictest forms of gun control that has ever been proposed in the United States, has generally fallen over time since the middle of the 20th century. Such bans have been imposed at the municipal level but never at the state or federal level, and in 2008 the U.S. Supreme Court ruling in *District of Columbia v. Heller* (554 U.S. 570), which struck down the District of Columbia's ban on the possession of handguns, made it increasingly unlikely that such policies will be imposed in the future. As Figure 8.6 shows, support for a handgun ban fell steadily over the course of the late 20th and early 21st centuries, from 60% in 1959 to 27% in 2015.

Decreased support for a handgun ban seemed to correspond to the increase among Americans who saw firearms as a source of personal protection (as is discussed in Chapter 2). Newport reports in *Majority Say More Concealed Weapons Would Make U.S. Safer* (October 20, 2015, http://www.gallup.com/poll/186263/majority-say-concealed-weapons-safer.aspx) that 56% of American adults thought the country would be safer if more adults carried concealed weapons, provided they had passed a background check and completed a training course. (See Table 8.4.) The majority was much higher among gun owners, 74% of whom believed the country would be safer if more adults carried concealed weapons. By comparison, nongun owners were about evenly split on this issue: 48% believed the country would by more safe, while 49% believed it would be less safe. Attitudes were predictably polarized between Republicans and Democrats. Among Republicans, 82% thought more concealed carry weapons would make the country safer, compared with 16% who thought the country would be less safe. Among Democrats, only 31% thought more concealed carry weapons would make the country safer, compared with 67% who thought the country would be less safe.

TABLE 8.3

Americans' dissatisfaction with gun laws, by party, 2016

	% Dissatisfied
Republican	54
Independent	59
Democrat	75

SOURCE: Art Swift, "Dissatisfaction with Gun Laws in the U.S., by Party ID," in *Americans' Dissatisfaction with U.S. Gun Laws at New High*, The Gallup Organization, January 14, 2016, http://www.gallup.com/poll/188219/americans-dissatisfaction-gun-laws-new-high.aspx (accessed September 15, 2016). Copyright © 2016 Gallup, Inc. All rights reserved. The content is used with permission; however, Gallup retains all rights of republication.

FIGURE 8.5

Americans' views on the strictness of gun laws, 1990–2016

IN GENERAL, DO YOU FEEL THAT THE LAWS COVERING THE SALE OF FIREARMS SHOULD BE MADE MORE STRICT, LESS STRICT OR KEPT AS THEY ARE NOW?

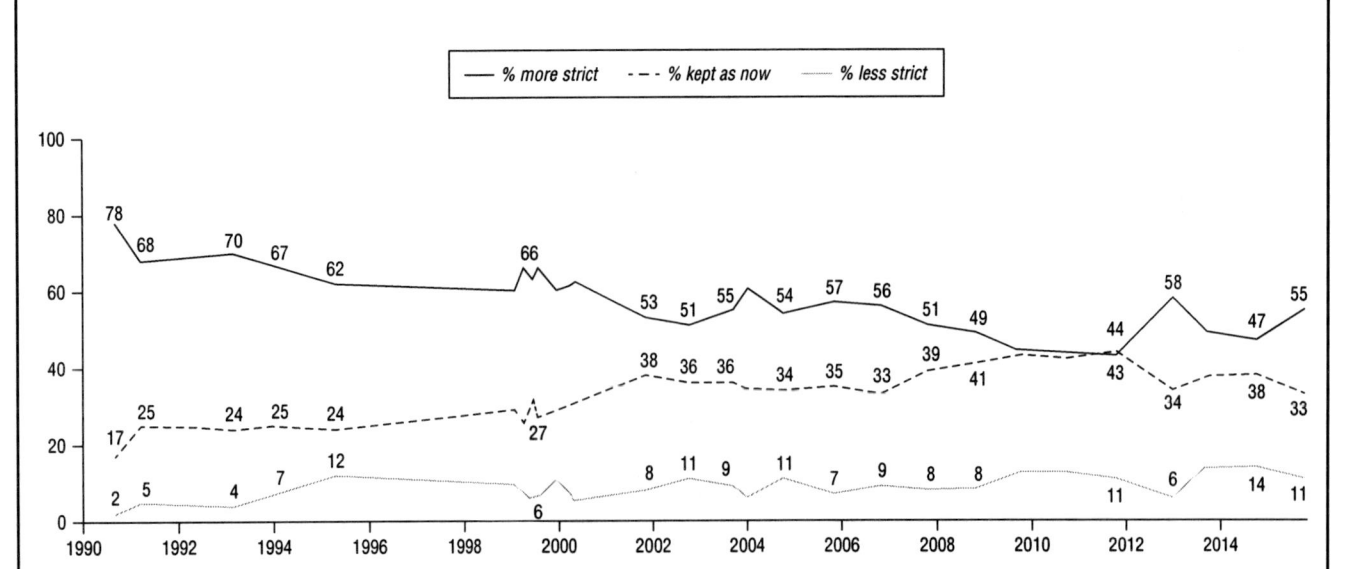

SOURCE: Frank Newport, "In General, Do You Feel That the Laws Covering the Sale of Firearms Should Be Made More Strict, Less Strict or Kept as They Are Now?" in *American Public Opinion, Terrorism and Guns*, The Gallup Organization, June 13, 2016, http://www.gallup.com/opinion/polling-matters/192695/american-public-opinion-terrorism-guns.aspx (accessed September 15, 2016). Copyright © 2016 Gallup, Inc. All rights reserved. The content is used with permission; however, Gallup retains all rights of republication.

FIGURE 8.6

Americans' views on a potential handgun ban, 1959–2015

DO YOU THINK THERE SHOULD OR SHOULD NOT BE A LAW THAT WOULD BAN THE POSSESSION OF HANDGUNS, EXCEPT BY THE POLICE AND OTHER AUTHORIZED PERSONS?

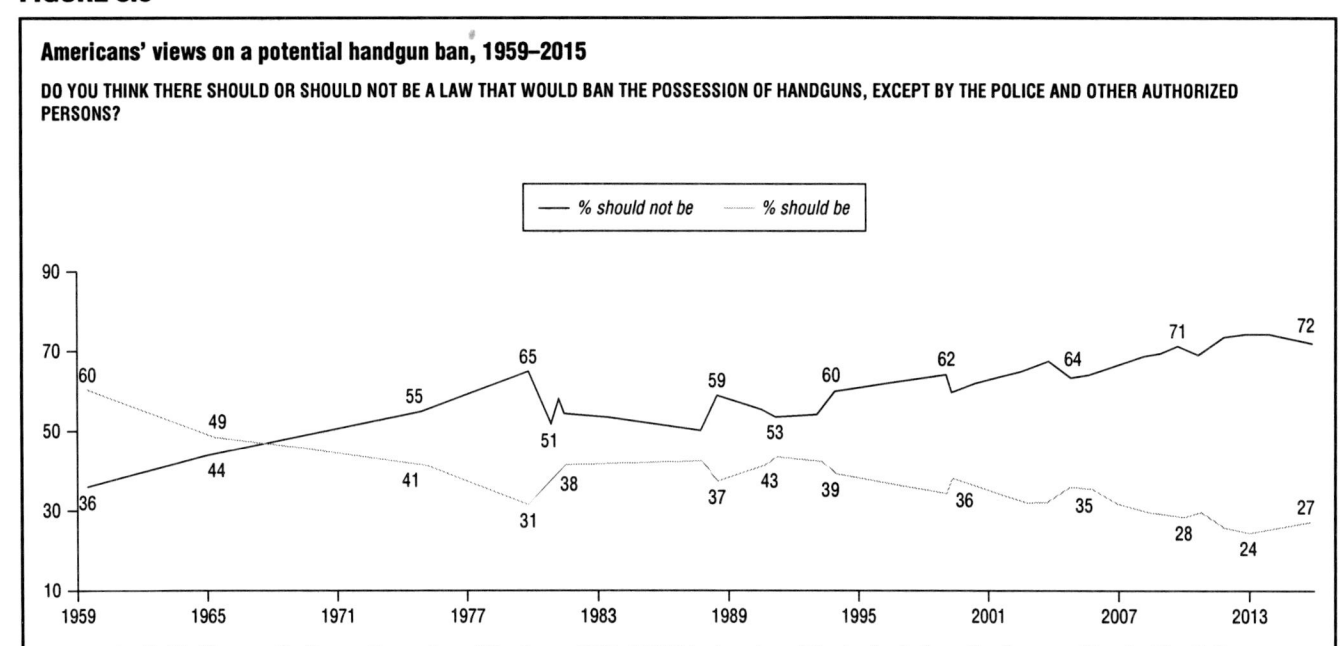

SOURCE: Art Swift, "Support for Ban on Possession of Handguns, 1959–2015," in *Americans' Desire for Stricter Gun Laws up Sharply*, The Gallup Organization, October 19, 2015, http://www.gallup.com/poll/186236/americans-desire-stricter-gun-laws-sharply.aspx?g_source=guns&g_medium=search&g_campaign=tiles (accessed September 15, 2016). Copyright © 2016 Gallup, Inc. All rights reserved. The content is used with permission; however, Gallup retains all rights of republication.

Gun Control and Mass Shootings

Support for increased gun control tends to spike in the aftermath of mass shootings, but such support has typically been fleeting, due in large part to a widespread belief that untreated mental illness is more to blame for mass shootings than is easy access to guns. In *Americans Fault Mental Health System Most for Gun Violence* (September 20, 2013, http://www.gallup.com/poll/164507/americans-fault-mental-health-system-gun-violence.aspx), Lydia Saad of Gallup notes that polling conducted in September 2013 indicated that only 61% of Americans believed access to guns had either a "great deal" or a "fair amount" to do with mass shootings, whereas 80% believed problems with the mental health system had either a "great deal" or a "fair amount" to do with such incidents.

Although there has been broad consensus that guns should be kept out of the hands of people with severe mental illness, many experts have sought to clarify that mental illness itself is not a significant predictor of violent behavior. For example, Jonathan M. Metzl and Kenneth T. MacLeish of Vanderbilt University state in "Mental Illness, Mass Shootings, and the Politics of American Firearms" (*American Journal of Public Health*, vol. 105, no. 2, February 2015) that "in the current political moment … relationships between shootings and mental illness often appear to be the only points upon which otherwise divergent voices in the contentious national gun debate agree." However, the researchers point out that only about 4% of violence in the United States can be attributed to people diagnosed with psychiatric disorders—including severe depression, bipolar disorder, and schizophrenia. Statistically, Metzl and MacLeish emphasize, "people are far more likely to be shot by relatives, friends, enemies, or acquaintances than they are by lone violent psychopaths."

In "Gun Violence Not a Mental Health Issue, Experts Say, Pointing to 'Anger,' Suicides" (CNN.com, January 25, 2016), Emanuella Grinberg quotes Jeffrey Swanson of Duke University, a professor in psychiatry and behavioral sciences who specializes in gun violence and mental illness, as saying that gun violence and mental illness should be understood as public health problems "that intersect at the edges" but are largely separate issues. Furthermore, Swanson contends that "fixing the mental health system is not a silver bullet for reducing gun violence." Rather, gun violence reduction would require multiple approaches, many of which are unrelated to mental illness. Among other recommendations, Swanson suggests that firearms background checks would be more effective if the criteria for denying a gun sale included indicators of aggressive, impulsive, or risky behavior, such as pending charges or convictions for violent assault, domestic violence restraining orders, or multiple citations for driving under the influence.

TABLE 8.4

Americans' views on carrying concealed weapons, by demographic group, 2015

SUPPOSE MORE AMERICANS WERE ALLOWED TO CARRY CONCEALED WEAPONS IF THEY PASSED A CRIMINAL BACKGROUND CHECK AND TRAINING COURSE. IF MORE AMERICANS CARRIED CONCEALED WEAPONS, WOULD THE UNITED STATES BE SAFER OR LESS SAFE?

[By demographic group]

	Safer %	Less safe %
National adults	56	41
Gender		
Men	62	37
Women	50	45
Age		
18 to 29	66	33
3o to 49	56	40
50 to 64	51	46
65+	50	45
Education		
High school or less	57	40
Some college	65	31
College graduate only	56	43
Postgraduate	35	64
Party ID		
Republicans	82	16
Independents	59	39
Democrats	31	67
Gun owner		
Yes	74	24
No	48	49
Area live in		
Big city	50	47
Suburb	52	47
Town/rural	63	33

SOURCE: Frank Newport, "Suppose more Americans were allowed to carry concealed weapons if they passed a criminal background check and training course. If more Americans carried concealed weapons, would the United States be safer or less safe? By demographic group," in *Majority Say More Concealed Weapons Would Make U.S. Safer*, The Gallup Organization, October 20, 2015, http://www.gallup.com/poll/186263/majority-say-concealed-weapons-safer.aspx (accessed September 15, 2016). Copyright © 2016 Gallup, Inc. All rights reserved. The content is used with permission; however, Gallup retains all rights of republication.

CHAPTER 9
THERE SHOULD BE STRICTER GUN CONTROL LAWS

This chapter presents a sample of the arguments used by the proponents of strong federal gun control to support their position. Chapter 10 provides arguments put forward by opponents of strong federal gun control.

EXCERPT OF TESTIMONY OF JACQUELYN CAMPBELL, PHD, RN, FAAN, ANNA D. WOLF CHAIR AND PROFESSOR, JOHNS HOPKINS UNIVERSITY SCHOOL OF NURSING, BEFORE THE U.S. SENATE COMMITTEE ON THE JUDICIARY FOR THE HEARING "VAWA NEXT STEPS: PROTECTING WOMEN FROM GUN VIOLENCE?" JULY 30, 2014

The United States has [a higher] homicide rate of women than all other westernized countries and among the highest rate in the world. This disparity is particularly pronounced for homicides of women committed with guns, in which the country's rate exceeds by 11 times the average rate in other comparable countries. . . .

Much of this fatal violence against women is committed by intimate partners. Although neither entirely complete and nor without coding errors, the FBI's Supplemental Homicide Reports are the most complete national database of homicide with information on the relationship of the perpetrator to the victim. In the most recent data available (2011), at least 45% of the murders of women were committed by a current husband or boyfriend or ex-husband. If we only examine the homicides where the perpetrator relationship to the victim was identified, more than half (54.2%) of the homicides of women are committed by a husband, boyfriend [or] former husband. . . . There were 10 times as many women killed by a current husband or boyfriend or ex-husband (926) as by a male stranger (92).

The majority of this violence is perpetrated with firearms. In the Violence Policy Center analysis of the 2011 murders of women, there were 1,707 females murdered by males in single victim/single offender incidents in 2011. Of those incidents of homicides in which the weapon could be determined (1,551), more of these homicides were committed with firearms (51 percent) than with any other weapon.

Women are also killed by partner or ex-partners when they are pregnant. In an important study of maternal mortality in the state of Maryland from 1993–2008, Dr. Diana Cheng and Dr. Isabelle Horon examined medical records of women who died during the pregnancy and the first postpartum year. Homicides (n=110) were the leading cause of death, and firearms were the most common (61.8%) method of death. A current or former intimate partner was the perpetrator in 54.5% (n=60) of the homicide deaths and nearly two-thirds of intimate partner homicide victims in their study were killed with guns. In a national study of pregnancy associated homicide, firearms again accounted for the majority of homicides (56.6%). . . .

Research my peers and I have conducted provides further insights into how firearm access and domestic abuse elevate the risk of homicide for American women, and explain why existing federal laws restrict certain convicted domestic abusers from buying or possessing guns.

Survey research of battered women indicates that when a firearm is present, a majority of abusers will use the gun to threaten or injure the victim. In a study Susan Sorenson and Douglas Weibe conducted with over 400 women in domestic violence shelters across California, two-thirds of abused women who reported a firearm in their home said their intimate partner used a gun against them, with 71.4% threatening to shoot/kill her and 5.1% actually shooting at her.

Among the most rigorous research available on factors that influence a woman's likelihood of homicide is the national, 12-city case-control study of intimate partner homicide (husband, boyfriend, ex-husband, ex-boyfriend) conducted by myself and colleagues. In the study we compared a group of abused women who were murdered by their partner or ex-partner to another group of abused women who were not. Controlling for other factors, we found that gun access/ownership increased the risk of homicide over and above prior domestic violence by 5.4 times. Gun access was the strongest risk factor for an abused woman to be killed by her partner or ex-partner. Among perpetrators who committed suicide after killing their abused wife, girlfriend or former wife or girlfriend, gun access was an even stronger risk factor. In those cases, access to a firearm increased the chances of the homicide-suicide by an adjusted odds ratio of 13.

Neither of these studies found evidence that women frequently use firearms to defend themselves against abuse or that access to a firearm reduces the risk of homicide for a woman. In the survey conducted by Sorenson and Wieber, fewer than 1 in 20 abused women that had access to a gun reported having ever used it in self-defense against their abuser (4.5% of the total). And our research showed that for the relatively few women in the study who owned a gun, firearm possession had no statistically significant effect on her risk of being killed by an intimate partner; it neither increased nor decreased her risk.

In leaving out abusive dating partners, current federal firearm prohibitions ignore the perpetrators of a large and growing share of intimate partner homicides. US Department of Justice data shows that the share of domestic violence homicides committed by dating partners has been rising for three decades, and boyfriends now commit more homicides than do spouses. And this data is likely an underestimate because it does not account for homicides committed by ex-boyfriends, which are seldom accurately coded in the Supplemental Homicide Reports. Our national case control data study of 12 cities found approximately 19% of intimate partner homicides are committed by an ex-boyfriend.... Estimating from that proportion, approximately 300–500 female intimate partner homicides each year should be added to the approximately 1000 already counted in the Supplemental Homicide Reports. S. 1290, the Protecting Domestic Violence and Stalking Victims Act, would expand our national domestic violence laws to include both former and current dating partners who together represented 48% of the male domestic violence homicide perpetrators in our study, and is therefore an extremely important way to keep women safe and save lives.

There is also evidence that state laws to strengthen firearm prohibitors against domestic abusers reduce intimate partner homicide. In two separate studies—one of 46 of the largest cities in the US ... and one of at the state level ... —researchers found that state statutes restricting those under domestic violence restraining orders from accessing firearms are associated with reductions in intimate partner homicides, driven by a reduction in those committed with firearms. [The second] study found that state laws prohibiting firearm possession by people under domestic violence restraining orders (along with entering state domestic violence restraining orders into the federal database) reduced intimate partner homicide of women by firearms by 12–13%, decreasing overall intimate partner homicide by 10%.

In conclusion, women who suffer abuse are among the most important for society to protect. Congress has an opportunity to do so by strengthening the laws to keep domestic abusers from getting guns. And ample scientific evidence shows that in doing so you will save lives. (http://www.judiciary.senate.gov/imo/media/doc/07-30-14 CampbellTestimony.pdf)

EXCERPT OF TESTIMONY OF MICHAEL A. NUTTER, MAYOR OF PHILADELPHIA AND PRESIDENT OF THE UNITED STATES CONFERENCE OF MAYORS, BEFORE THE U.S. SENATE COMMITTEE ON THE JUDICIARY ON THE TOPIC OF THE PROPOSED ASSAULT WEAPONS BAN OF 2013, FEBRUARY 27, 2013

The U.S. Conference of Mayors has been calling for sensible gun laws to protect the public for more than 40 years. Our call for a ban on assault weapons dates back to 1991. Mayors and police chiefs from cities of all sizes have worked together in this effort over the years.

Gun Violence in Cities

We have done that because of the tremendous toll gun violence takes on the American people day in and day out:

- Every year in America more than 100,000 people are shot, and 31,537 of them die, including 11,583 who are murdered.
- Every year, 18,000 children and teens are shot, and 2,829 of them die, including 1,888 who are murdered.
- Every day in America, 282 people are shot and 86 of them die, including 32 who are murdered.
- Every day 50 children and teens are shot and eight of them die, including five who are murdered.

Gun violence disproportionately affects urban areas. Our nation's 50 largest metro areas have 62 center cities, and these cities account for 15 percent of the population, but 39 percent of gun-related murders and 23 percent of total homicides.

Philadelphia, like many major cities, has struggled to control gun violence for years. However, despite our recent success at employing more effective policing techniques, deaths due to gun violence have not fallen. Let me use one set of statistics to illustrate this point:

Last year, the number of shooting victims in Philadelphia was 1,282. This is down considerably from the year before—and was the lowest number since we began tracking shooting victims in the year 2000. However, the number of homicides was up slightly—331, seven more than the previous year. How are these two statistics possible? The answer is that the homicide victims have more bullets killing them. Or, to put it another way, there are more rounds being fired and more intentional head shots. So despite better policing, when someone in Philadelphia is shot, they are now more likely to die.

I would note that Pennsylvania does not have stringent gun restrictions. When the City of Philadelphia adopted strict gun laws a few years ago, the state supreme court struck those laws down. This is why we need federal legislation. Cities alone cannot reduce gun violence. We are doing everything that we can, but we are still losing the battle thanks to the proliferation of guns in this nation.

Philadelphia's story is not unique. Mayors everywhere struggle with gun violence, using scarce city resources to fight it—resources which we should be using to educate our children, create jobs for our residents, and revitalize our cities.

I have with me this morning a letter originally sent just three days after the Newtown tragedy occurred and now signed by 212 mayors which calls on the President and Congress to take immediate action and make reasonable changes to our gun laws and regulations. Listed first among our recommended changes is enactment of legislation to ban assault weapons and high-capacity magazines.... I ask that you include this letter in the record of this hearing.

The Assault Weapons Ban Does Not Violate the Second Amendment

Since the shootings in Newtown, the question, "If not now, when?" has been raised often in news media columns, editorials, and other arguments for swift and meaningful action to combat gun violence. And well it should, because it's the right question. For too many in the gun rights community, however, the answer to the question is always "never," and the reason is always the Second Amendment's protection of the "right to bear arms."

Harvard University's Lawrence Tribe, one of the nation's most respected experts on Constitutional law, told this Committee just a few weeks ago that, after examining the various proposals being considered—including the ban on assault weapons and high-capacity magazines—"I am convinced that nothing under discussion in the Senate Judiciary Committee represents a threat to the Constitution or even comes close to violating the Second Amendment or the Constitution's structural limits either on congressional power or on executive authority."

The 1994 Assault Weapons Ban Worked

In addition we've been told by the gun rights community that the assault weapons ban didn't work before and it won't work now. Research shows that the 1994 ban did work and that since it expired the use of assault weapons by criminals has increased:

- A Justice Department study of the 1994 assault weapons ban found that it was responsible for a 6.7 percent decrease in total gun murders, holding all other factors equal. That study also found that "assault weapons are disproportionately involved in murders with multiple victims, multiple wounds per victim, and police officers as victims."

- An updated assessment of the federal ban found that the use of assault weapons in crime declined by more than two-thirds about nine years after 1994 Assault Weapons Ban took effect.

- A recent study by the Violence Policy Center found that, between 2005 and 2007, one in four law enforcement officers slain in the line of duty was killed with an assault weapon.

- The Police Executive Research Forum reports that 37 percent of police departments reported seeing a noticeable increase in criminals' use of assault weapons since the ban expired.

It's Time to Pass the Assault Weapons Ban

Mayors consider protecting the safety of their citizens and their cities their highest responsibility. We know that keeping our cities and our citizens safe requires more than passing sensible gun laws, including the assault weapons ban, but we also know that we cannot keep our cities safe unless we pass such laws. The Assault Weapons Ban of 2013 would:

- Prohibit the sale, manufacture, transfer and importation of 157 of the most commonly-owned military-style assault weapons and ban an additional group of assault weapons that can accept a detachable ammunition magazine and have one or more military characteristics;

- Ban large-capacity magazines and other ammunition feeding devices that hold more than 10 rounds of ammunition—devices which allow shooters to fire numerous rounds in rapid succession without having to stop and reload; and

- Protect the rights of law-abiding citizens who use guns for hunting, household defense or legitimate recreational purposes and exempt all assault weapons lawfully possessed at the date of enactment from the ban.

- Require background checks on all future transfers of assault weapons covered by the legislation, including sale, trade and gift;

- Require that grandfathered assault weapons be stored safely using a secure gun storage or safety device in order to keep them away from prohibited persons; and
- Prohibit the sale or transfer of high-capacity ammunition feeding devices currently in existence.

This is common sense legislation which will help us to reduce the number of people, including police officers, who are shot and killed in our cities and throughout our nation. This legislation deserves a vote. This legislation deserves to be passed, by this Committee, by the Senate, and by the House so that the President can sign it into law.

I know it will take an act of political courage for many Members of Congress to support the Assault Weapons Ban of 2013, but the time for such political courage is now. How many more children, how many more police officers do we have to lose for our elected representatives to do the right thing? The nation's mayors pledge to work with you to build a safer America for our children and all of our citizens. (http://www.judiciary.senate.gov/imo/media/doc/2-27-13NutterTestimony.pdf)

EXCERPT OF WRITTEN TESTIMONY SUBMITTED BY THE AMERICAN ACADEMY OF PEDIATRICS TO THE U.S. SENATE COMMITTEE ON THE JUDICIARY, SUBCOMMITTEE ON THE CONSTITUTION, CIVIL RIGHTS, AND HUMAN RIGHTS HEARING "PROPOSALS TO REDUCE GUN VIOLENCE: PROTECTING OUR COMMUNITIES WHILE RESPECTING THE SECOND AMENDMENT," FEBRUARY 12, 2013

The American Academy of Pediatrics (AAP) [is] a nonprofit professional organization of more than 60,000 primary care pediatricians, pediatric medical sub-specialists, and pediatric surgical specialists dedicated to the health, safety, and well-being of infants, children, adolescents, and young adults.

The AAP is committed to protecting children from the horrific consequences of gun violence and traumatic events, and ensuring children's safety within their homes, schools and communities. The tragedy at Sandy Hook Elementary School in Newtown, CT, serves as a stark reminder that gun violence affects communities nationwide. Unfortunately, while outbursts of mass violence like that at Sandy Hook are at least relatively rare, the scourge of gun violence is a phenomenon that our nation's children experience every single day. In 2008 and 2009, 5,740 children were killed by guns, meaning that 55 died each week during that period. The causes of gun violence are varied and complex but we must act to develop a comprehensive response centered on the rights [of] children and families to be safe and free from its harmful effect in their lives and within their communities.

A Public Health Approach to Reducing Gun Violence

Gun violence is a public health issue with particularly pernicious effects on children. Firearm related deaths continue to be one of the top three causes of death among American youth, causing twice as many deaths as cancer, five times as many as heart disease, and 15 times as many as infections. In 2009, 84.5 percent of all homicides of people 15 to 19 years of age were firearm-related. The United States has the highest rates of firearm-related death (including homicide, suicide and unintentional deaths) among high income countries. For youth ages 15 to 24 years of age, [firearm] homicide rates were 35.7 times higher than in other high income countries. For over 20 years, the AAP has supported stronger gun violence prevention policies because of the public health implications of this problem. Reducing its impact must be consistent with other initiatives that have reduced injury and mortality through evidence-based prevention efforts.

Policy of the AAP, based on extensive research, is that absence of guns from children's homes and communities is the most reliable and effective measure to prevent firearm-related injuries in children and adolescents. Access to a firearm increases the risk of unintended injury or death among all children. A gun stored in the home is associated with a threefold increase in the risk of homicide and a fivefold increase in the risk of suicide. Individuals possessing a firearm are more than four times more likely to be shot during an assault than those who do not own one. The association of a gun in the home and increased risk of suicide among adolescents is well-documented, even among teens with no underlying psychiatric diagnosis. These health risks associated with gun violence point toward the need for long term research investments on effective strategies to protect children and adolescents, particularly those within at-risk communities.

As part of its engagement with the White House Taskforce on Gun Violence Prevention, the AAP recommended federal support for gun violence research, and is pleased the President's plan recommended $10 million to support Centers for Disease Control and Prevention (CDC) research into the causes and prevention of gun violence; $10 million for CDC to conduct further research regarding the relationship between video games, media images and violence; and $20 million to expand the National Violent Death Reporting System from 18 to 50 states. The AAP urges Congress to support these efforts within the annual appropriations process and to eliminate any restrictive language that may discourage gun violence research.

The AAP supports policies aimed at protecting children and adolescents from the destructive effect[s] of guns through strong gun safety legislation that bans assault weapons and high-capacity magazines, requires universal background checks, and mandates safe firearm storage. Consistent with this policy, the AAP has endorsed the *Assault Weapons Regulatory Act of 2013* (S. 150/H.R.

437). According to a recent analysis by the Violence Policy Center (VPC), the five states (Alabama, Alaska, Louisiana, Montana, and Wyoming) with the least restrictive gun laws and high gun ownership rates also had the highest per capita gun death rates. States with strong gun laws and low rates of gun ownership had far lower rates of firearm-related death. The AAP is encouraged that bipartisan efforts are underway to strengthen gun laws, and looks forward to the opportunity to review those plans as they materialize.

Responsible Gun Ownership

Policies to support safe and responsible ownership can go a long way toward keeping firearms out of the hands of children and adolescents who may harm themselves or others. An estimated 57 million Americans own[ed] 283 million firearms in 2004. Among gun owners with a child 18 years old or younger, 31 percent store their guns unlocked, 21 percent store them loaded, and 8.3 percent store them unlocked and loaded. Safe gun storage can reduce the risk of youth injury and suicide by more than 70 percent; therefore, efforts to educate families and require responsible practices through child access prevention (CAP) laws should be supported as important, but common sense, interventions.

CAP laws impose criminal liability on adults who negligently leave firearms accessible to children or otherwise allow children access to firearms. One study found that in twelve states where such laws had been in effect for at least one year, unintentional firearm deaths fell by 23% from 1990–94 among children under 15 years of age. Laws reducing child access also are associated with lower overall adolescent suicide.

The AAP commends the Obama Administration's safe gun storage campaign proposal and urges Congress to support this initiative. Medical professionals and law enforcement officials should play an important role in implementing this campaign. The AAP's *Bright Futures* clinical guidance recommends that pediatricians ask about guns in the home and that they provide age-appropriate safety counseling, similar to the guidance they provide on other injury risks, like drowning and parental tobacco use. Physician counseling of parents about firearm safety, particularly when combined with the distribution of gun locks, has been demonstrated to be an effective prevention measure and shown to increase compliance with safe storage principles.

At the federal level, the Affordable Care Act includes language barring the Secretary of the Department of Health and Human Services and health plans participating in the exchanges from collecting and housing information regarding the presence of firearms in the home. The AAP welcomes the president's guidance that the Affordable Care Act does not prohibit physicians from counseling patients regarding firearms.

The AAP remains concerned about state efforts to infringe upon physicians' rights to provide this crucial counsel such as the Firearm Owners' Privacy Act, enacted in Florida, which prevented physicians from providing such counsel under threat of financial penalty and potential loss of licensure. The law has been blocked from implementation by a U.S. District court but similar policies have been introduced in seven other states: Alabama, Minnesota, North Carolina, Oklahoma, South Carolina, Tennessee, and West Virginia. This right must be protected to mitigate risk of injury to children in the environments in which they live and play. (http://www.judiciary.senate.gov/imo/media/doc/021213RecordSubmission-Durbin.pdf)

WRITTEN TESTIMONY OF CHIEF JIM BUEERMANN (RET.), PRESIDENT, POLICE FOUNDATION, WASHINGTON, D.C., BEFORE THE U.S. SENATE COMMITTEE ON THE JUDICIARY FOR THE HEARING "WHAT SHOULD AMERICA DO ABOUT GUN VIOLENCE?" JANUARY 30, 2013

I write to you in my capacity as both President of the Police Foundation and the former Chief of Police of the Redlands, CA Police Department. The Police Foundation, established in 1970 by the Ford Foundation, is a nonpartisan, non-constituency research organization. Our mission is to advance policing through innovation and scientific research. The Foundation is committed to disseminating science and evidence-based practices to the field. My written testimony reflects these principles and my personal experience after 33 years as a police officer during which time I witnessed countless acts of violence. I urge the passage of the Assault Weapons Ban Act of 2013 and ask Congress to consider funding additional scientific research to help this country implement evidence-based approaches to reducing gun violence in our communities and schools.

The most recent available data reveal this alarming picture of America's experience with gun-related violence: in 2011, of the 32,163 deaths from firearms, 19,766 were suicides and 11,101 were homicides. Additionally, there were 467,321 non-fatal violent crimes committed with a firearm. These numbers all reflect the unique position of the United States in relation to other high-income nations: our homicide rate is 6.9 times higher than the combined homicide rate of 22 other high-income countries. We all know that gun violence must be stemmed. The Police Foundation supports a comprehensive and holistic approach to preventing and reducing gun violence that includes:

- Legislation that bans assault weapons, requires universal background checks for all firearm purchases and limits high capacity ammunition feeding devices to ten rounds;

- Enhanced funding for research on the availability of firearms, the causes and prevention of gun violence and the connection between mental health and gun violence;
- Specific funding to replicate the 1996 US DOJ, National Institute of Justice study *Guns in America* that provided a comprehensive view of guns in our society;
- Increased funding to states for community-based mental health treatment; and,
- Sustained funding and support of the Justice and Mental Health Collaboration Program Act, which allows for collaborative efforts between law enforcement, criminal justice and mental health professionals.

Gun violence, especially violence that is mental health-related, is a complex social, cultural, health and safety issue. It is one that we do not know enough about. As the leader of a research organization that focuses on policing crime and disorder, I stress the need for scientific research and an evidence-based approach to understanding important societal issues. As a country, we need a robust and rigorous agenda on the causes of gun violence, effective, community-based prevention and intervention strategies and the link between mental illness and gun violence. Lifting the freeze on gun violence research at the Centers for Disease Control [and Prevention] is heartening, and I hope Congress will support additional funding for interdisciplinary, scientific research and collaboration across government agencies, including the Department of Justice and the Department of Health and Human Services.

Mental health-related gun violence has been brought to the fore with the shootings in Newtown, CT, Aurora, CO, and Tucson, AZ. While these tragic incidents are statistically rare, when combined with the number of gun-related suicides each year, the necessity of addressing the mental health needs of individuals, and the availability of firearms in our communities, is paramount.

We do not want to stigmatize individuals with mental illness nor solely focus the current dialogue on gun violence on the role of mental illness. The best available data on violence attributable to mental illness shows that 3–5% of violent acts are committed by individuals with mental illness and most of these acts do not involve guns. Yet, we cannot ignore the number of gun-involved suicides each year and the connection between mass shootings and mental illness. Increased scientific research across the fields of medicine, public health, criminal justice and law will help us understand how to prevent mental health-related gun violence. This requires both robust funding and time.

As a former chief of police, I recognize that local law enforcement agencies require immediate strategies to prevent another incident of mass violence. Earlier this month, the Police Foundation convened a roundtable meeting of expert researchers and practitioners from the fields of law enforcement, mental health, public health, criminal justice and policy. The group discussed how available interdisciplinary research might be used to develop practical strategies for law enforcement that prevent mental health-related gun violence. Existing research establishes the difficulty in predicting a violent act, but the group committed to three strategies that law enforcement can adopt now. Based on innovative practices defined in the literature, the group proposed that law enforcement executives:

- Create local partnerships with mental health service providers, school officials and appropriate community groups to develop a mental health crisis response capacity;
- Advocate for increased mental health services in their communities. Law enforcement executives should convene local service providers and community members to assess local mental health services and community needs and increase community members' knowledge of the exiting science on mental health and gun violence;
- Adopt specific policies and practices that reduce the availability of guns to people in mental health crisis, institutionalize mental health training for their officers and facilitate community-wide "mental health first aid" training for all community members.

Clearly, more work needs to be done in this area so police departments can effectively operationalize these ideas. With additional Congressional support, strategies like these can be supported by legislation such as the Justice and Mental Health Collaboration Act or through an enhancement of programs at the Department of Justice and the Departments of Health and Human Services and Education. The JMHC Act has bipartisan support across the House of Representatives and Senate, and I ask that Congress sustain funding for these important ideas as part of a targeted approach to specifically reducing gun violence.

Charting a path to respond to gun violence will not be easy, but I encourage Congress to rely on the police, community leaders and science to guide that path. The Police Foundation, along with law enforcement leaders across the country, support reducing the availability of assault weapons and high capacity ammunition feeding device[s] as a first step to reducing gun violence. However, to effectively reduce gun violence, there must be more comprehensive action. Congress should prioritize funding to better understand guns in America, research on the causes and prevention of gun violence and the connection between mental illness and gun violence. It should also enhance the funding and availability of mental health services in communities, and support programs that increase local collaboration between law enforcement, criminal justice and mental health professionals. (https://www.judiciary.senate.gov/imo/media/doc/013013 RecordSubmission-Feinstein.pdf)

CHAPTER 10
THERE SHOULD NOT BE STRICTER GUN CONTROL LAWS

This chapter presents a sample of the arguments used by the opponents of strong federal gun control to support their position. Chapter 9 provides arguments put forward by supporters of strong federal gun control.

EXCERPT OF TESTIMONY OF JOYCE LEE MALCOLM, PATRICK HENRY CHAIR OF CONSTITUTIONAL LAW AND THE SECOND AMENDMENT, GEORGE MASON UNIVERSITY SCHOOL OF LAW, BEFORE THE U.S. SENATE COMMITTEE ON THE JUDICIARY FOR THE HEARING "VAWA NEXT STEPS: PROTECTING WOMEN FROM GUN VIOLENCE?" JULY 30, 2014

We are all concerned about protection of victims of domestic violence and more generally public safety. While current law is not perfect, it has the great virtue of according with the long-standing traditions of American law by protecting the rights of all concerned, rights recognized by the Supreme Court as "deeply rooted in the Nation's history and tradition" and "fundamental to our scheme of ordered liberty." The two bills under consideration, by contrast, ride rough shod over those key rights, over our Second Amendment right to keep and bear arms, our Fourth Amendment right against unreasonable search and seizure, and the all important right of Due Process. To put the problem these bills address, protection of women, in perspective, I will begin with a few statistics then address the way in which these bills impact the basic rights in question.

First, a fact we should celebrate rarely mentioned by those desiring more gun control, from 1992 to 2010 the homicide rate in this country has declined sharply, falling by nearly half. Looking at homicide by sex, while female victims were more likely than men to be killed by an intimate, the rate of family violence fell between 1993–2002 and throughout the period accounted for only one in ten violent victimizations and only 11% of reported and unreported violence between 1998 and 2002. In those years the most frequent type of family violence was simple assault, while murder of a family member was less than half of 1%. For women in particular, after 1980 the proportion of female intimate homicide victims killed by guns decreased while the proportion killed by other weapons increased. More generally how were these victims killed[?] Between 1980–2008 17.4% of female homicide victims were killed by guns, 45.3% by arson and 43.9% by poison.

Although trends are in the right direction, people are still being assaulted and some of them killed, so it is understandable that better procedures and laws are sought. But the way in which the bills in question plan to accomplish this would do violence to our system of individual rights, fairness, and due process.

The Klobuchar Bill and Lautenberg Amendment greatly expand the sorts of individuals who fall within its reach by adding the crime of stalking, a misdemeanor crime, and including a series of individuals who were or are non-co-habiting, those dating, formerly dating or known to the potential victim. The opening sentence of the fact sheet, "Women Under the Gun," from the Center for American Progress, demonstrates how broad the new standard would be, referring to sexual assault on college campuses and military violence by a partner.

In addition to netting large numbers of innocent individuals, the bill would also make the loss of the right to be armed retroactive. Thousands of individuals have made plea bargains. In federal court felony convictions 79% make a guilty plea. These individuals would not have known that part of the guilty plea would be to be deprived of the ability to own firearms forever.

The Blumenthal Bill goes several steps further in its attack on traditional rights. It would permit the seizure of firearms from anyone subject to a temporary restraining order upon the filing [of] such a complaint, and before the individual complained of had an opportunity to be heard. The police would be sent to search for, and forcibly seize any firearms found in his possession. This is a serious infringement of due process, an Alice in Wonderland world in which, like the Red Queen, the new rule is "Sentence first! Verdict afterwards." After his property is taken, the defendant would have to prove himself innocent to get his property back. Although half of those who have been subject to temporary restraining orders are subsequently found not guilty, the bill tilts sharply in favor of the plaintiff. Since the drafters favor taking guns away from civilians, this works to accomplish that goal. Individuals who have committed violent offences are already prohibited from possessing firearms. The new names are to be added quickly to the [National Instant Criminal Background Check System] list but that would take some time.

Law-abiding people in this country have a right to keep firearms for their [defense] and other lawful purposes. Taking these guns without due process violates that fundamental right. Law-abiding people in this country are to be free from unreasonable search and seizure. Yet the bills would permit their homes to be ransacked for weapons on the mere allegation that they pose a danger to someone else. The taking of property, the violation of rights without due process on the mere assumption it will protect someone else is deeply offensive to our legal tradition. Justice Antonin Scalia, in *District of Columbia v. Heller* is emphatic: "the enshrinement of constitutional rights necessarily takes certain policy choices off the table." The policy choices in these two bills must be taken off the table. (https://www.judiciary.senate.gov/imo/media/doc/07-30-14Malcolm Testimony.pdf)

TESTIMONY OF WAYNE LAPIERRE, EXECUTIVE VICE PRESIDENT OF THE NATIONAL RIFLE ASSOCIATION OF AMERICA, BEFORE THE U.S. SENATE COMMITTEE ON THE JUDICIARY FOR THE HEARING "WHAT SHOULD AMERICA DO ABOUT GUN VIOLENCE?" JANUARY 30, 2013

It's an honor to be here today on behalf of more than 4.5 million moms and dads and sons and daughters, in every state across our nation, who make up the National Rifle Association of America. Those 4.5 million active members are joined by tens of millions of NRA supporters.

And it's on behalf of those millions of decent, hardworking, law-abiding citizens...to give voice to their concerns...that I'm here today.

The title of today's hearing is "What should America do about gun violence?"

We believe the answer to that question is to be honest about what works—and what doesn't work.

Teaching safe and responsible gun ownership works—and the NRA has a long and proud history of teaching it.

Our "Eddie Eagle" children's safety program has taught over 25 million young children that if they see a gun, they should do four things: "Stop. Don't touch. Leave the area. Tell an adult." As a result of this and other private sector programs, fatal firearm accidents are at the lowest levels in more than 100 years.

The NRA has over 80,000 certified instructors who teach our military personnel, law enforcement officers and hundreds of thousands of other American men and women how to safely use firearms. We do more—and spend more—than anyone else on teaching safe and responsible gun ownership.

We joined the nation in sorrow over the tragedy that occurred in Newtown, Connecticut. There is nothing more precious than our children. We have no more sacred duty than to protect our children and keep them safe. That's why we asked former Congressman and Undersecretary of Homeland Security, Asa Hutchinson, to bring in every expert available to develop a model School Shield Program—one that can be individually tailored to make our schools as safe as possible.

It's time to throw an immediate blanket of security around our children. About a third of our schools have armed security already—because it works. And that number is growing. Right now, state officials, local authorities and school districts in all 50 states are considering their own plans to protect children in their schools.

In addition, we need to enforce the thousands of gun laws that are currently on the books. Prosecuting criminals who misuse firearms works. Unfortunately, we've seen a dramatic collapse in federal gun prosecutions in recent years. Overall in 2011, federal weapons prosecutions per capita were down 35 percent from their peak in the previous administration. That means violent felons, gang members and the mentally ill who possess firearms are not being prosecuted. And that's unacceptable.

And out of more than 76,000 firearms purchases denied by the federal instant check system, only 62 were referred for prosecution and only 44 were actually prosecuted. Proposing more gun control laws—while failing to enforce the thousands we already have—is not a serious solution to reducing crime.

I think we can also agree that our mental health system is broken. We need to look at the full range of mental health issues, from early detection and treatment, to civil commitment laws, to privacy laws that needlessly prevent mental health records from being included in the National Instant Criminal Background Check System.

While we're ready to participate in a meaningful effort to solve these pressing problems, we must respectfully—but honestly and firmly—disagree with some members of this committee, many in the media, and all of the gun control groups on what will keep our kids and our streets safe.

Law-abiding gun owners will not accept blame for the acts of violent or deranged criminals. Nor do we believe the government should dictate what we can lawfully own and use to protect our families.

As I said earlier, we need to be honest about what works and what does not work. Proposals that would only serve to burden the law-abiding have failed in the past and will fail in the future.

Semi-automatic firearms have been around for over 100 years. They are among the most popular guns made for hunting, target shooting and self-defense. Despite this fact, Congress banned the manufacture and sale of hundreds of semi-automatic firearms and magazines from 1994 to 2004. Independent studies, including a study from the Clinton Justice Department, proved that ban had no impact on lowering crime.

And when it comes to the issue of background checks, let's be honest—background checks will never be "universal"—because criminals will never submit to them.

But there are things that can be done and we ask you to join with us. The NRA is made up of millions of Americans who support what works … the immediate protection for all—not just some—of our school children; swift, certain prosecution of criminals with guns; and fixing our broken mental health system.

We love our families and our country. We believe in our freedom. We're the millions of Americans from all walks of life who take responsibility for our own safety and protection as a God-given, fundamental right. (http://www.judiciary.senate.gov/imo/media/doc/1-30-13LaPierre Testimony.pdf)

EXCERPT OF WRITTEN TESTIMONY OF DAVID B. KOPEL, RESEARCH DIRECTOR OF THE INDEPENDENCE INSTITUTE AND ASSOCIATE POLICY ANALYST AT THE CATO INSTITUTE, BEFORE THE U.S. SENATE COMMITTEE ON THE JUDICIARY FOR THE HEARING "WHAT SHOULD AMERICA DO ABOUT GUN VIOLENCE?" JANUARY 30, 2013

Gun prohibition advocates have been pushing the "assault weapon" issue for a quarter century. Their political successes on the matter have always depended on public confusion. The guns are *not* machine guns. They do *not* fire automatically. They fire only one bullet each time the trigger is pressed, just like every other ordinary firearm. They are *not* more powerful than other firearms; to the contrary, their ammunition is typically intermediate in power, less powerful than guns and ammunition made for big game hunting.

The Difference between Automatic and Semi-automatic

For an automatic firearm (commonly called a "machine gun"), if the shooter presses the trigger and holds it, the gun will fire continuously, automatically, until the ammunition runs out. Ever since the National Firearms Act of 1934, automatics have been very strictly regulated by federal law: Every person who wishes to possess one must pay a $200 federal transfer tax, must be fingerprinted and photographed, and must complete a months-long registration process with the federal Bureau of Alcohol, Tobacco, Firearms, and Explosives (BATFE). In addition, the transferee must be granted written permission by local law enforcement, via ATF Form 4. Once registered, the gun may not be taken out of state without advance written permission from BATFE.

Since 1986, the manufacture of new automatics for sale to persons other than government agents has been forbidden by federal law. As a result, automatics in [the] U.S. are rare (there are about a hundred thousand legally registered ones), and expensive, with the least expensive ones costing nearly ten thousand dollars.

The automatic firearm was invented in 1883 by Hiram Maxim. The early Maxim Guns were heavy and bulky, and required a two-man crew to operate. In 1943, a new type of automatic was invented, the "assault rifle." The assault rifle is light enough for a soldier to carry for long periods of time. Soon, the assault rifle became the ubiquitous infantry weapon. Examples include the U.S. Army M-16, the Soviet AK-47, and the Swiss militia SIG SG 550. The AK-47 (and its various updates, such as the AK-74 and AKM) can be found all over the Third World, but there are only a few hundred in the United States, mostly belonging to firearms museums and wealthy collectors.

The precise definition of "assault rifle" is supplied by the Defense Intelligence Agency. If you use the term "assault rifle," persons who are knowledgeable about firearms will know precisely what kinds of guns you are talking about. The definition of "assault rifle" has never changed, because the definition describes a particular type of thing in the real world—just like the definitions of "apricot" or "Minnesota."

In contrast, the definition of "assault weapon" has never been stable. The phrase is merely an epithet. It has been applied to things which are not even firearms (namely, air guns). It has been applied to double-barreled shotguns, to single-shot guns (guns whose ammunition capacity is only a single round), and to many other sorts of ordinary handguns, shotguns, and rifles.

The first "assault weapon" ban in the United States, in California in 1989, was created by legislative staffers thumbing through a picture book of guns, and deciding which guns looked bad. The result was an incoherent law which, among other things, outlawed certain firearms that do not exist, since the staffers just copied the typographical errors from the book, or associated a model by one manufacturer with another manufacturer whose name appeared on the same page.

Over the last quarter century, the definition has always kept shifting. One recent version is Sen. Dianne Feinstein's new bill. Another is the pair of bills defeated in the January 2013 lame duck session of the Illinois legislature, which would have outlawed most handguns (and many long guns as well) by dubbing them "assault weapons."

While the definitions of what to ban keep changing, a few things remain consistent: The definitions do *not* cover automatic firearms, such as assault rifles. The definitions do *not* ban guns based on how fast they fire, or how powerful they are. Instead, the definitions are based on the name of a gun, or on whether a firearm has certain superficial accessories (such as a bayonet lug, or a grip in the "wrong" place).

Most, but not all, of the guns which have been labeled "assault weapons" are semi-automatics. Many people think that a gun which is "semi-automatic" must be essentially the same as an automatic. This is incorrect.

Semi-automatic firearms were invented in the 1890s, and have been common in the United States ever since. Today, about three-quarters of new handguns are semi-automatics. A large share of rifles and shotguns are also semi-automatics. Among the most popular semi-automatic firearms in the United States today are the Colt 1911 pistol (named for the year it was invented, and still considered one of the best self-defense handguns), the Ruger 10/22 rifle (which fires the low-powered .22 Long Rifle cartridge, popular for small game hunting or for target shooting at distances less than a hundred yards), the Remington 1100 shotgun (very popular for bird hunting and home defense), and the AR-15 rifle (popular for hunting game no larger than deer, for target shooting, and for defense). All of these guns were invented in the mid-1960s or earlier. All of them have, at various times, been characterized as "assault weapons."

Unlike an automatic firearm, a semi-automatic fires only one round of ammunition when the trigger is pressed. (A "round" is one unit of ammunition. For a rifle or handgun, a round has one bullet. For a shotgun, a single round contains several pellets).

In some other countries, a semi-automatic is usually called a "self-loading" gun. This accurately describes what makes the gun "semi"-automatic. When the gun is fired, the bullet (or shot pellets) travel[s] from the firing chamber, down the barrel, and out the muzzle. Left behind in the firing chamber is the now empty case or shell that contained the bullets (or pellets) and the gunpowder.

In a semi-automatic, some of the energy from firing is used to eject the empty shell from the firing chamber, and then load a fresh round of ammunition into the firing chamber. Then, the gun is ready to shoot again, when the user is ready to press the trigger.

In some other types of firearms, the user must perform some action in order to eject the empty shell and load the next round. This could be moving a bolt back and forth (bolt action rifles), moving a lever down and then up (lever action rifles), or pulling and then pushing a pump or slide (pump action and slide action rifles and shotguns). A revolver (the second-most popular type of handgun) does not require the user to take any additional action in order to fire the next round.

The semi-automatic has two principle advantages over lever action, bolt action, slide action, and pump action guns. First, many hunters prefer it because the semi-automatic mechanism allows a faster second shot. The difference may be less than a second, but for a hunter, this can make all the difference.

Second, and more importantly, the semi-automatic's use of gunpowder energy to eject the empty case and then to load the next round substantially reduces how much recoil is felt by the shooter. This makes the gun much more comfortable to shoot, especially for beginners, or for persons without substantial upper body strength and bulk.

The reduced recoil also make[s] the gun easier to keep on target for the next shot, which is important for hunting and target shooting, and extremely important for self-defense.

Semi-automatics also have their disadvantages. They are much more prone to misfeeds and jams than are simpler, older types of firearms, such as revolvers or lever action.

Contrary to the hype of anti-gun advocates and less-responsible journalists, there is no rate of fire difference between a so-called "assault" semi-automatic gun and any other semi-automatic gun.

How Fast Does a Semi-automatic Fire?

Here is a report on the test-firing of a new rifle:

> 187 shots were fired in three minutes and thirty seconds and one full fifteen shot magazine was fired in only 10.8 seconds.

Does that sound like a machine gun? A "semi-automatic assault weapon"? Actually it is an 1862 test report of the then-new lever-action Henry rifle, manufactured by Winchester. If you have ever seen a Henry rifle, it was probably in the hands of someone at a cowboy re-enactment, using historic firearms from 150 years ago.

The Winchester Henry is a lever-action, meaning that after each shot, the user must pull out a lever, and then push it back in, in order to eject the empty shell casing, and then load a new round into the firing chamber.

The lever-action Winchester is not an automatic. It is not a semi-automatic. It was invented decades before either of those types of firearms. And yet that old-fashioned Henry lever action rifle can fire one bullet per second.

By comparison, the murderer at Sandy Hook fired 150 shots over a 20 minute period, before the police arrived. In other words, a rate of fewer than 8 shots per minute. This is a rate of fire far slower than the capabilities of a lever-action Henry Rifle from 1862, or a semi-automatic AR-15 rifle from 2010. Indeed, his rate of fire could have been far exceeded by a competent person using very old technology, such as a break-open double-barreled shotgun.

Are Semi-automatics More Powerful Than Other Guns?

The power of a firearm is measured by the kinetic energy it delivers. Kinetic energy is based on the mass (the weight) of the projectile, and its velocity. So a heavier bullet will deliver more kinetic energy than a lighter one. A faster bullet will deliver more kinetic energy than a slower bullet.

How much kinetic energy a gun will deliver has nothing to do with whether it is a semi-automatic, a lever action, a bolt action, a revolver, or whatever. What matter[s] is, first of all, the weight of the bullet, how much gunpowder is in the particular round of ammunition, and the length of the barrel.

None of this has anything to do with whether the gun is or is not a semi-automatic. Manufacturers typically produce the same gun in several different calibers, sometimes in more than a dozen calibers.

Regarding the rifles which some people call "assault weapons," they tend to be intermediate in power, as far as rifles go. Consider the AR-15 rifle in its most common caliber, the .223. The bullet is only a little bit wider than the puny .22 bullet, but it is longer, and thus heavier.

Using typical ammunition, an AR-15 in .223 would have 1,395 foot-pounds of kinetic energy. That's more than a tiny rifle cartridge like the .17 Remington, which might carry 801 foot-pounds of kinetic energy. In contrast, a big-game cartridge, like the .444 Marlin, might have 3,040 [foot-pounds of kinetic energy]. This is why rifles like the AR-15 are suitable and often used for hunting small to medium animals (such as rabbits or deer), but are not suitable for the largest animals, such as elk or moose.

Many (but not all) of the ever-changing group of guns which are labeled "assault weapons" use detachable magazines (a box with an internal spring) to hold their ammunition. But this is a characteristic shared by many other firearms, including many non-semiautomatic rifles (particularly, bolt-actions), and by the large majority of handguns. Whatever the merits of restricting magazine size..., the size of the magazine depends on the size of the magazine. If you want to control magazine size, there is no point in banning certain guns which can take detachable magazines, while not banning other guns which also take detachable magazines. (http://www.judiciary.senate.gov/imo/media/doc/1-30-13KopelTestimony.pdf)

EXCERPT OF TESTIMONY OF GAYLE S. TROTTER, SENIOR FELLOW AT THE INDEPENDENT WOMEN'S FORUM, BEFORE THE U.S. SENATE COMMITTEE ON THE JUDICIARY FOR THE HEARING "WHAT SHOULD AMERICA DO ABOUT GUN VIOLENCE?" JANUARY 30, 2013

I would like to begin with the compelling story of Sarah McKinley. Home alone with her baby, she called 911 when two violent intruders began to break down her front door. The men wanted to force their way into her home so they could steal the prescription medication of her deceased husband, who had recently died of cancer. Before the police could arrive, while Ms. McKinley was on the line with the 911 operator, these violent intruders broke down her door. One of the men brandished a foot-long hunting knife. As the intruders forced their way into her home, Ms. McKinley fired her weapon, fatally wounding one of the violent attackers and causing the other to flee the scene. Later, Ms. McKinley reflected on the incident: "It was either going to be him or my son," she said. "And it wasn't going to be my son."

Guns make women safer. Most violent offenders actually do not use firearms, which makes guns the great equalizer. In fact, over 90 percent of violent crimes occur without a firearm. Over the most recent decade, from 2001 to 2010, "about 6 percent to 9 percent of all violent victimizations were committed with firearms," according to a federal study. Violent criminals rarely use a gun to threaten or attack women. Attackers use their size and physical strength, preying on women who are at a severe disadvantage.

Guns reverse that balance of power in a violent confrontation. Armed with a gun, a woman can even have the advantage over a violent attacker. How do guns give women the advantage? An armed woman does not need superior strength or the proximity of a hand-to-hand struggle. She can protect her children, elderly relatives, herself or others who are vulnerable to an assailant. Using a firearm with a magazine holding more than 10 rounds of ammunition, a woman would have a fighting chance even against multiple attackers....

Concealed-carry laws reverse that balance of power even before a violent confrontation occurs. In this way, armed women indirectly benefit those who choose not to carry. For a would-be criminal, concealed-carry laws dramatically increase the cost of committing a crime, paying safety dividends to those who do not carry. All women in these jurisdictions reap the benefits of concealed-carry laws because potential assailants face a much higher risk when they attempt to threaten or harm a potential victim. As a result, in jurisdictions with concealed-carry laws, women are less likely to be raped, maimed or murdered than they are in states with stricter gun ownership laws.

Research has shown that states with nondiscretionary concealed handgun laws have 25 percent fewer rapes than states that restrict or forbid women from carrying concealed handguns. The most thorough analysis of concealed-carry laws and crime rates indicates that "there are large drops in overall violent crime, murder, rape, and aggravated assault that begin right after the right-to-carry laws have gone into effect" and that "in all those crime categories, the crime rates consistently stay much lower than they were before the law." Among the ten states that adopted concealed-carry laws over a fifteen-year span, there were 0.89 shooting deaths and injuries per 100,000 people, representing less than half the rate of 2.09 per 100,000 experienced in states without these laws....

We need sensible enforcement of the gun laws that are already on the books. Currently, we have more than 20,000 under-enforced or selectively enforced gun laws. Gun regulation affects only the guns of the law-abiding. Criminals will not be bound by such gestures, especially as we continually fail to prosecute serious gun violations or provide meaningful and consistent penalties for violent felonies involving firearms....

In lieu of empty, self-defeating gestures, we should address gun violence by doing what works. By safeguarding our Second Amendment rights, we preserve meaningful protection for women. Our nation made significant progress in that regard when, in recent memory, the United States Supreme Court held that the Second Amendment protects an individual's right to possess a firearm for traditionally lawful purposes, such as self-defense within the home.

For those who believe in safeguarding the civil liberties enshrined in our Bill of Rights, you might consider this an unremarkable conclusion. The constitutional text expressly guarantees the right "to keep and bear Arms," and that right is specifically enumerated—not implied—and guaranteed to "the people." In other words, unlike many of the individual rights that the Supreme Court has recognized—some would say invented—you can actually find the right to bear arms in the literal text of the Second Amendment. Moreover, the Constitution guarantees a "right of the people" only two other times, both of which clearly describe individual rights: The First Amendment protects the "right of the people" to assemble and to petition the government, and the Fourth Amendment protects the "right of the people" against "unreasonable searches and seizures"....

In lieu of empty gestures, we should address gun violence based on what works. Guns make women safer. The Supreme Court has recognized that lawful self-defense is a central component of the Second Amendment's guarantee of the right to keep and bear arms. For women, the ability to arm ourselves for our protection is even more consequential than for men because guns are the great equalizer in a violent confrontation. As a result, we preserve meaningful protection for women by safeguarding our Second Amendment rights. Every woman deserves a fighting chance. (http://www.judiciary.senate.gov/imo/media/doc/1-30-13TrotterTestimony.pdf)

APPENDIX: STATE CONSTITUTION ARTICLES CONCERNING WEAPONS

The Second Amendment of the Bill of Rights of the U.S. Constitution reads: "A well regulated militia, being necessary to the security of a free state, the right of the people to keep and bear arms, shall not be infringed."

Alabama: That every citizen has a right to bear arms in defense of himself and the state. Art. I, § 26 (enacted 1819).

Alaska: A well-regulated militia being necessary to the security of a free state, the right of the people to keep and bear arms shall not be infringed. The individual right to keep and bear arms shall not be denied or infringed by the State or a political subdivision of the State. Art. I, § 19 (first sentence was enacted in 1959; second sentence was added in 1994).

Arizona: The right of the individual citizen to bear arms in defense of himself or the State shall not be impaired, but nothing in this section shall be construed as authorizing individuals or corporations to organize, maintain, or employ an armed body of men. Art. II, § 26 (enacted in 1912).

Arkansas: The citizens of this State shall have the right to keep and bear arms for their common defense. Art. II, § 5 (enacted in 1874).

California: No constitutional provision.

Colorado: The right of no person to keep and bear arms in defense of his home, person and property, or in aid of the civil power when thereto legally summoned, shall be called in question; but nothing herein contained shall be construed to justify the practice of carrying concealed weapons. Art. II, § 13 (enacted in 1876).

Connecticut: Every citizen has a right to bear arms in defense of himself and the state. Art. I, § 15 (enacted in 1818).

Delaware: A person has the right to keep and bear arms for the defense of self, family, home and State, and for hunting and recreational use. Art. I, § 20 (enacted in 1987).

Florida: (a) The right of the people to keep and bear arms in defense of themselves and of the lawful authority of the state shall not be infringed, except that the manner of bearing arms may be regulated by law. (b) There shall be a mandatory period of three days, excluding weekends and legal holidays, between the purchase and delivery at retail of any handgun. For the purposes of this section, "purchase" means the transfer of money or other valuable consideration to the retailer, and "handgun" means a firearm capable of being carried and used by one hand, such as a pistol or revolver. Holders of a concealed weapon permit as prescribed in Florida law shall not be subject to the provisions of this paragraph. (c) The legislature shall enact legislation implementing subsection (b) of this section, effective no later than December 31, 1991, which shall provide that anyone violating the provisions of subsection (b) shall be guilty of a felony. (d) This restriction shall not apply to a trade in of another handgun. Art. I, § 8 (section (a) adopted in 1968, sections (b) to (d) adopted in 1990).

Georgia: The right of the people to keep and bear arms shall not be infringed, but the General Assembly shall have power to prescribe the manner in which arms may be borne. Art. I, § 1, Paragraph VIII (enacted in 1877).

Hawaii: A well regulated militia being necessary to the security of a free state, the right of the people to keep and bear arms shall not be infringed. Art. I, § 17 (enacted in 1959).

Idaho: The people have the right to keep and bear arms, which right shall not be abridged; but this provision shall not prevent the passage of laws to govern the carrying of weapons concealed on the person nor prevent passage of legislation providing minimum sentences for crimes committed while in possession of a firearm, nor prevent the passage of legislation providing penalties for the possession of firearms by a convicted felon, nor prevent the passage of any legislation punishing the use of a firearm. No law shall impose licensure, registration or special taxation on the ownership or possession of firearms or ammunition. Nor shall any law permit the confiscation of firearms, except those actually used in the commission of a felony. Art. I, § 11 (enacted in 1978).

Illinois: Subject only to the police power, the right of the individual citizen to keep and bear arms shall not be infringed. Art. I, § 22 (enacted in 1970).

Indiana: The people shall have a right to bear arms, for the defense of themselves and the State. Art. I, § 32 (enacted in 1851).

Iowa: No constitutional provision.

Kansas: The people have the right to bear arms for their defense and security; but standing armies, in time of peace, are dangerous to liberty, and shall not be tolerated, and the military shall be in strict subordination to the civil power. Bill of Rights, § 4 (enacted in 1859).

Kentucky: All men are, by nature, free and equal, and have certain inherent and inalienable rights, among which may be reckoned:... The right to bear arms in defense of themselves and of the State, subject to the power of the General Assembly to enact laws to prevent persons from carrying concealed weapons. Bill of Rights, § 1 (enacted in 1891).

Louisiana: The right of each citizen to keep and bear arms shall not be abridged, but this provision shall not prevent the passage of laws to prohibit the carrying of weapons concealed on the person. Art. I, § 11 (enacted in 1974).

Maine: Every citizen has a right to keep and bear arms and this right shall never be questioned. Art. I, § 16 (enacted in 1987).

Maryland: No constitutional provision.

Massachusetts: The people have a right to keep and to bear arms for the common defence [*sic*]. And as, in time of peace, armies are dangerous to liberty, they ought not to be maintained without the consent of the legislature; and the military power shall always be held in an exact subordination to the civil authority, and be governed by it. Pt. I, Art. XVII (enacted in 1780).

Michigan: Every person has a right to keep and bear arms for the defense of himself and the state. Art. I, § 6 (enacted in 1963).

Minnesota: No constitutional provision.

Mississippi: The right of every citizen to keep and bear arms in defense of his home, person, or property, or in aid of the civil power when thereto legally summoned, shall not be called in question, but the legislature may regulate or forbid carrying concealed weapons. Art. III, § 12 (enacted in 1890).

Missouri: That the right of every citizen to keep and bear arms in defense of his home, person and property, or when lawfully summoned in aid of the civil power, shall not be questioned; but this shall not justify the wearing of concealed weapons. Art. I, § 23 (enacted in 1945).

Montana: The right of any person to keep or bear arms in defense of his own home, person, and property, or in aid of the civil power when thereto legally summoned, shall not be called in question, but nothing herein contained shall be held to permit the carrying of concealed weapons. Art. II, § 12 (enacted in 1889).

Nebraska: All persons are by nature free and independent, and have certain inherent and inalienable rights; among these are life, liberty, the pursuit of happiness, and the right to keep and bear arms for security or defense of self, family, home, and others, and for lawful common defense, hunting, recreational use, and all other lawful purposes, and such rights shall not be denied or infringed by the state or any subdivision thereof. Art. I, § 1 (right to keep and bear arms enacted in 1988).

Nevada: Every citizen has the right to keep and bear arms for security and defense, for lawful hunting and recreational use and for other lawful purposes. Art. I, § 11(1) (enacted in 1982).

New Hampshire: All persons have the right to keep and bear arms in defense of themselves, their families, their property and the state. Pt. I, Art. IIa (enacted in 1982).

New Jersey: No constitutional provision.

New Mexico: No law shall abridge the right of the citizen to keep and bear arms for security and defense, for lawful hunting and recreational use and for other lawful purposes, but nothing herein shall be held to permit the carrying of concealed weapons. No municipality or county shall regulate, in any way, an incident of the right to keep and bear arms. Art. II, § 6 (first sentence enacted in 1971; second sentence added in 1986).

New York: No constitutional provision.

North Carolina: A well regulated militia being necessary to the security of a free State, the right of the people to keep and bear arms shall not be infringed; and, as standing armies in time of peace are dangerous to liberty, they shall not be maintained, and the military shall be kept under strict subordination to, and governed by, the civil power. Nothing herein shall justify the practice of carrying concealed weapons, or prevent the General Assembly from enacting penal statutes against that practice. Art. I, § 30 (enacted in 1971).

North Dakota: All individuals are by nature equally free and independent and have certain inalienable rights, among which are those of enjoying and defending life and liberty; acquiring, possessing and protecting property and reputation; pursuing and obtaining safety and happiness; and to keep and bear arms for the defense of their person, family, property, and the state, and for lawful hunting, recreational, and other lawful purposes, which shall not be infringed. Art. I, § 1 (right to keep and bear arms enacted in 1984).

Ohio: The people have the right to bear arms for their defense and security; but standing armies, in time of peace, are dangerous to liberty, and shall not be kept up; and the military shall be in strict subordination to the civil power. Art. I, § 4 (enacted in 1851).

Oklahoma: The right of a citizen to keep and bear arms in defense of his home, person, or property, or in aid of the civil power, when thereunto legally summoned, shall never be prohibited; but nothing herein contained shall prevent the Legislature from regulating the carrying of weapons. Art. II, § 26 (enacted in 1907).

Oregon: The people shall have the right to bear arms for the defence [sic] of themselves, and the State, but the Military shall be kept in strict subordination to the civil power. Art. I, § 27 (enacted in 1857).

Pennsylvania: The right of the citizens to bear arms in defence [sic] of themselves and the State shall not be questioned. Art. I, § 21 (enacted in 1790).

Rhode Island: The right of the people to keep and bear arms shall not be infringed. Art. I, § 22 (enacted in 1842).

South Carolina: A well regulated militia being necessary to the security of a free State, the right of the people to keep and bear arms shall not be infringed. As, in times of peace, armies are dangerous to liberty, they shall not be maintained without the consent of the General Assembly. The military power of the State shall always be held in subordination to the civil authority and be governed by it. Art. I, § 20 (enacted in 1895).

South Dakota: The right of the citizens to bear arms in defense of themselves and the state shall not be denied. Art. VI, § 24 (enacted in 1889).

Tennessee: That the citizens of this State have a right to keep and to bear arms for their common defense; but the Legislature shall have power, by law, to regulate the wearing of arms with a view to prevent crime. Art. I, § 26 (enacted in 1870).

Texas: Every citizen shall have the right to keep and bear arms in the lawful defense of himself or the State; but the Legislature shall have power, by law, to regulate the wearing of arms, with a view to prevent crime. Art. I, § 23 (enacted in 1876).

Utah: The individual right of the people to keep and bear arms for security and defense of self, family, others, property, or the state, as well as for other lawful purposes shall not be infringed; but nothing herein shall prevent the Legislature from defining the lawful use of arms. Art. I, § 6 (enacted in 1984).

Vermont: That the people have a right to bear arms for the defence [sic] of themselves and the State—and as standing armies in time of peace are dangerous to liberty, they ought not to be kept up; and that the military should be kept under strict subordination to and governed by the civil power. Ch. I, Art. XVI (enacted in 1777).

Virginia: That a well regulated militia, composed of the body of the people, trained to arms, is the proper, natural, and safe defense of a free state, therefore, the right of the people to keep and bear arms shall not be infringed; that standing armies, in time of peace, should be avoided as dangerous to liberty; and that in all cases the military should be under strict subordination to, and governed by, the civil power. Art. I, § 13 (enacted in 1776 without explicit right to keep and bear arms; "therefore, the right of the people to keep and bear arms shall not be infringed" added in 1971).

Washington: The right of the individual citizen to bear arms in defense of himself, or the state, shall not be impaired, but nothing in this section shall be construed as authorizing individuals or corporations to organize, maintain or employ an armed body of men. Art. I, § 24 (enacted in 1889).

West Virginia: A person has the right to keep and bear arms for the defense of self, family, home and state, and for lawful hunting and recreational use. Art. III, § 22 (enacted in 1986).

Wisconsin: The people have the right to keep and bear arms for security, defense, hunting, recreation or any other lawful purpose. Art. I, § 25 (enacted in 1998).

Wyoming: The right of citizens to bear arms in defense of themselves and of the state shall not be denied. Art. I, § 24 (enacted in 1889).

IMPORTANT NAMES AND ADDRESSES

Americans for Responsible Solutions
PO Box 15642
Washington, DC 20003
(571) 295-7807
E-mail: info@americansforresponsible
solutions.org
URL: http://www.americansforresponsible
solutions.org/

Brady Campaign to Prevent Gun Violence
840 First St. NE, Ste. 400
Washington, DC 20002
(202) 370-8100
URL: http://www.bradycampaign.org/

Bureau of Alcohol, Tobacco, Firearms, and Explosives
Office of Public and Governmental Affairs
99 New York Ave. NE
Washington, DC 20226
(202) 648-8500
URL: http://www.atf.gov/

Centers for Disease Control and Prevention
1600 Clifton Rd.
Atlanta, GA 30329-4027
1-800-232-4636
URL: http://www.cdc.gov/

Coalition to Stop Gun Violence
805 15th St. NW, Ste. 502
Washington, DC 20005

(202) 408-0061
E-mail: csgv@csgv.org
URL: http://www.csgv.org/

Federal Bureau of Investigation
935 Pennsylvania Ave. NW
Washington, DC 20535-0001
(202) 324-3000
URL: http://www.fbi.gov/

Gun Owners of America
8001 Forbes Pl., Ste. 102
Springfield, VA 22151
(703) 321-8585
FAX: (703) 321-8408
URL: http://www.gunowners.org/

Johns Hopkins Center for Gun Policy and Research
Johns Hopkins Bloomberg School of Public Health
615 N. Wolfe St.
Baltimore, MD 21205
(410) 955-3543
URL: http://www.jhsph.edu/research/
centers-and-institutes/johns-hopkins-center-
for-gun-policy-and-research/

Law Center to Prevent Gun Violence
268 Bush St., Ste. 555
San Francisco, CA 94104
(415) 433-2062
FAX: (415) 433-3357
URL: http://smartgunlaws.org/

National Institute of Justice
810 Seventh St. NW
Washington, DC 20531
(202) 307-2942
URL: https://www.nij.gov/Pages/welcome
.aspx

National Rifle Association of America
11250 Waples Mill Rd.
Fairfax, VA 22030
1-800-672-3888
URL: http://www.nra.org/

Office of Juvenile Justice and Delinquency Prevention
810 Seventh St. NW
Washington, DC 20531
(202) 307-5911
URL: http://www.ojjdp.gov/

Second Amendment Foundation
12500 NE 10th Pl.
Bellevue, WA 98005
(425) 454-7012
E-mail: info@saf.org
URL: http://www.saf.org/

Violence Policy Center
1730 Rhode Island Ave. NW, Ste. 1014
Washington, DC 20036
(202) 822-8200
URL: http://www.vpc.org/

RESOURCES

The Bureau of Alcohol, Tobacco, Firearms, and Explosives (ATF) monitors the production and regulation of alcohol, tobacco, firearms, and explosives and is the major source for statistical and technical information on these categories. The agency publishes the annual report *Firearms Commerce in the United States*, which supplies authoritative data on the manufacturing, import/export, and sales of firearms. The ATF also publishes annual *Firearms Trace Data*, which provides insight into firearms recoveries.

The Federal Bureau of Investigation (FBI) and the Bureau of Justice Statistics (BJS) maintain statistics on crime in the United States. The FBI's annual *Crime in the United States* is based on crime statistics reported by law enforcement agencies through its Uniform Crime Reports program. The FBI also publishes the annual *Law Enforcement Officers Killed and Assaulted*. The BJS supplements this information with data derived from surveys of victims and other supplementary sources. Among the most useful BJS publications in the preparation of this book were *Criminal Victimization, 2014* (Jennifer L. Truman and Lynn Langton, August 2015), *A Study of Active Shooter Incidents, 2000–2013* (J. Pete Blair and Katherine W. Schweit, 2014), *Homicide in the U.S. Known to Law Enforcement, 2011* (Erica L. Smith and Alexia Cooper, December 2013), *Firearm Violence, 1993–2011* (Michael Planty and Jennifer L. Truman, May 2013), *Criminal Victimization, 2014* (Jennifer L. Truman and Lynn Langton, August 2015), and *Homicide Trends in the United States, 1980–2008* (Alexia Cooper and Erica L. Smith, November 2011). Information on the National Instant Criminal Background Check System is available online at https://www.fbi.gov/services/cjis/nics.

Other important government data came from the Centers for Disease Control and Prevention, which operates the Web-based Injury Statistics Query and Reporting System (http://www.cdc.gov/injury/wisqars/). It also publishes the annual *Health, United States*, which provides data on mortality and injuries in addition to many other health-related topics, and periodic reports relevant to the topic of firearm violence as part of its National Violent Death Reporting System in the *Morbidity and Mortality Weekly Report*.

Also valuable were *Juvenile Offenders and Victims: 2014 National Report* (Melissa Sickmund and Charles Puzzanchera, December 2014), published by the National Center for Juvenile Justice, and *Indicators of School Crime and Safety: 2015* (Anlan Zhang, Lauren Musu-Gillette, and Barbara A. Oudekerk, May 2016), published by the National Center for Education Statistics and Bureau of Justice Statistics.

Gale, Cengage Learning would like to express its continuing appreciation to the Gallup Organization for its kind permission to publish its surveys and to the National Opinion Research Center of the University of Chicago for use of its material. Thanks also goes to the Pew Research Center and Harris Interactive for permission to publish material from their surveys. Gale, Cengage Learning would also like to thank the following organizations for permission to use information and tables from their web postings, reports, and journals: the National Rifle Association of America, the Brady Campaign to Prevent Gun Violence, Child Trends, the Violence Policy Center, the Law Center to Prevent Gun Violence, and Every Town for Gun Safety.

INDEX

Page references in italics refer to photographs. References with the letter t following them indicate the presence of a table. The letter f indicates a figure. If more than one table or figure appears on a particular page, the exact item number for the table or figure being referenced is provided.

A

AAP (American Academy of Pediatrics), 16, 134–135
Abraham, Peter, 91, 93
Access Denied: How the Gun Lobby Is Depriving Police, Policy Makers, and the Public of the Data We Need to Prevent Gun Violence (Mayors against Illegal Guns), 78–79
Active shooter incidents
    FBI study of, 65
    increase in number of, 70
    number of, annually, 70f
    number of casualties (killed or wounded) by, 71f
    See also Mass shootings
"Active Shooter Incidents Continue to Rise, New FBI Data Show" (Date, Thomas, & Levine), 70
Affordable Care Act, 135
African Americans
    arrests for weapons offenses by, 80
    firearm fatalities among, 87–89
    gun ownership by, 125
    gun use for self-defense, 93–94
    homicide rate for, 58, 60
    weapon-carrying by students, 110, 111
    youth homicide offenders, 102, 104
    youth homicide victims, 107–108
Age
    arrest trends, by offense/age, 81t
    arrests for weapons offenses by, 80
    cost of firearm injuries and, 93

    death rates for suicide, by sex/age, 90t–91t
    firearm fatalities by, 84, 86–89
    firearm homicides/suicides by, 91
    gun ownership by, 16, 125
    of homicide victims, 58
    homicides involving guns by age of victim, 106(f7.7)
    injury deaths, 10 leading causes of, by age group, 92t
    juveniles murdered by age/relationship to offender, 105t
    nonfatal gunshot injuries by, 84
    nonfatal injuries treated in hospital emergency departments, 10 leading causes of, by age group, 85t
    youth homicide victimization by, 100–101
    youth suicide by, 120
    of youthful homicide offenders, 102
    of youthful homicide victims, 104, 105, 107–108
Aggravated assault
    definition of, 57
    nonfatal firearm violence used in, 74, 76
    robberies/aggravated assaults, by type of weapon used, 77t
    by type of weapon used, percentage distribution, by region, 78(t5.18)
Alabama
    gun control laws of, 33
    state constitution article concerning weapons, 143
Alaska
    concealed weapons law of, 35–36
    gun control laws of, 33
    state constitution article concerning weapons, 143
Alito, Samuel A., 53
American Academy of Pediatrics (AAP), 16, 134–135
American health-wealth paradox, 15–16

American Public Opinion, Terrorism and Guns (Newport), 127–128
American Revolutionary War, 3
American Trends Panel survey (Pew Research Center), 124–125
Americans' Dissatisfaction with U.S. Gun Laws at New High (Swift), 126–127
Americans Fault Mental Health System Most for Gun Violence (Saad), 129
Ammunition
    armor-piercing ammunition, ban on, 21–22
    assault weapons ban, 29, 133
    juveniles banned from possession of, 29–30
Aneja, Abhay, 36
Aristotle, 2
Arizona
    concealed weapons law of, 35–36
    gun control laws of, 33
    state constitution article concerning weapons, 143
Arkansas
    gun control laws of, 33
    state constitution article concerning weapons, 143
"Armed Citizen" (NRA website), 95
Armed: New Perspectives on Gun Control (Kleck & Kates), 6–7, 16
"Armed Resistance to Crime: The Prevalence and Nature of Self-Defense with a Gun" (Kleck & Gertz), 6
Armijo v. Ex Cam Inc. (New Mexico), 56
Arming Pilots against Terrorism Act, 31
Armor-piercing ammunition, 21–22
Arms Export Control Act, 11
Armslist, 28
Aron, Laudan, 15–16
Aronsen, Gavin, 65
Arrests
    juvenile arrests, 102, 104t
    by race, 82t

trends, by offense/age, 81*t*
for weapons offenses, 80
Ashcroft, John D., 27
Asian Americans/Pacific Islanders
firearm fatalities among, 87
weapon-carrying by students, 110, 111
Assassinations
gun control issue and, 123
prompting gun control, 1
Assault rifle, 139
Assault weapons
ban on, 29
Connecticut ban on, 49–50
New Jersey's ban on, challenge to, 50
overview of, 12–14
Assault Weapons Ban of 1994, 29, 133
Assault Weapons Ban of 2013
arguments against, 139–141
arguments for, 132–136
failure of, 13
Assault Weapons Ban of 2015
AAP endorsement of, 134–135
disposition of, 14
Assaults
on law enforcement officers, 72
among youth, 99–102
See also Aggravated assaults
Assize of Arms (England), 2
Association of Corporate Counsel, 50
"At Least 28 People Holding BB or Pellet Guns Were Killed by Police in US in 2015" (McCarthy & Swaine), 22
ATF. *See* Bureau of Alcohol, Tobacco, Firearms, and Explosives
Aurora, Colorado, 77
Automatic firearms, 12–14
Azrael, Deborah, 120

B

Background checks
exemption from, 28
expanded, public opinion on, 126, 127*t*
five-day waiting period of interim Brady, 23
gun show/online regulation, 28–29
on handgun purchases, 5
LaPierre, Wayne on, 138, 139
NICS, components of, 25*f*
NICS checks, total, 24*t*
NICS denial process per 100 applicants, 27*f*
NICS firearms denials, by reason, 27(*t*3.2)
NICS Index, active records in, 27(*t*3.3)
NICS participation map, 26*f*
NICS records, retention of, 27–28
NICS system, operation of, 24–25
for prevention of gun violence, 129
state firearms control laws on, 33

states' right-to-carry weapons laws and, 36
waiting period, elimination of, 26–27
"Background Checks for Guns" (Institute for Legislative Action), 28
Bailey, Benjamin v., 49–50
Bailey, John M., 49–50
Bailey, Roland J., 45
Bailey v. United States, 45
Barron v. City of Baltimore, 41
Baton Rouge, Louisiana, 71
BB gun
BB/pellet gun injuries among children aged 0–19 years, rates per 100,000, 86(*t*6.4)
rate of injuries from, 84
Beemiller, Williams v. (New York), 56
Beland, Louis-Philippe, 120
Benjamin, DeForest H., Jr., 49–50
Benjamin v. Bailey, 49–50
Berman, Mark, 71
Bill of Rights, 4
Bill of Rights (England), 2
BJS. *See* Bureau of Justice Statistics
Black market, 78, 79–80
Blackstone, William, 2–3
Blocker, State v. (Oregon), 51
Bloomberg, Michael, 22
Blumenthal Bill, 137–138
Bostic, James Nigel, 56
Boston Massacre, 3
Boyce, Michael, 51
Boyce, State v. (Oregon), 51
Bradlees, 22
Brady, James, 5, 23
The Brady Campaign State Scorecard (Brady Campaign to Prevent Gun Violence), 33
Brady Campaign to Prevent Gun Violence
as interdiction advocate, 124
state "scorecard" rankings for gun control laws, 33
on unregulated gun sales, 28
Brady Handgun Violence Prevention Act
controversies over provisions of, 26–29
five-day waiting period, 23
impact on gun violence, 25–26
NICS, 24–25
overview of, 4–5
passage of, 23
Permanent Brady permit chart, 37*t*
Printz v. United States, 46–47
prohibited people under, 23
states' right-to-carry weapons laws and, 36
Braga, Anthony A., 108
Bridge, Jeffrey A., 95–96
Bright Futures (AAP), 135
Brown, Charles, 56
Bueermann, Jim, 135–136

Bureau of Alcohol, Tobacco, Firearms, and Explosives (ATF)
assault weapons ban of, 13
automatic firearm registration with, 139
creation of, 20
enforcement of federal gun laws, 20
estimates on gun ownership, 9–10
FFLs, new regulations to obtain, 30, 31
gun trace data for youth offenders, 108–110
on gun trafficking, 79–80
Stewart, Robert Wilson, Jr., investigation of, 47
Bureau of Justice Statistics (BJS)
on criminals' acquisition of firearms, 79
on gun-related homicides, numbers of, 1
on homicide rate, 58
NCVS crime statistics, 57
on nonfatal violent crimes involving firearms, 73
Bush, George W.
NICS Improvement Amendments Act, 31
Protection of Lawful Commerce in Arms Act, 56
Virginia Tech shooting report, 118
Butterfield, Fox, 56
Byrne, Sklar v. (Illinois), 53

C

Caldwell, Cornell, 56
California
FFLs in, 31
firearms control laws of, 32–33
firearms used in homicides in, 60
gun law reform by, 33
gun violations, gun dealers liability for in, 54
local ordinances in, 38, 39
as non-RTC state, 36
Oikos University, shooting at, 118
open carrying prohibited in, 38
San Francisco handgun ban, 50–51
Saturday Night Specials ban in West Hollywood, 53–54
state constitution article concerning weapons, 143
youth crime guns recovered in, 110
Camden, New Jersey, mass shooting in, 63, 65
Campbell, Jacquelyn, 131–132
CAP laws. *See* Child access prevention (CAP) laws
Cases v. United States, 43
Castile, Philando, 71
Casualties
from active shooter incidents, 70
active shooters, number of casualties (killed or wounded) by, 71*f*
See also Deaths

"Causes of Law Enforcement Deaths" (NLEOMF), 70–71
CDC. *See* Centers for Disease Control and Prevention
CDF (Children's Defense Fund), 120
Centers for Disease Control and Prevention (CDC)
    on CAP laws, 95
    cost of firearm injuries, 91
    data on gun violence, 83–84
    freeze on gun violence research, 136
    funding for gun violence research, 134
    on nonfatal gunshot injuries, 84
    on students and guns, 110–111
    on youth acquisition of guns, 110
    on youth suicide prevention, 120
"Changes in Suicide Rates by Hanging and/or Suffocation and Firearms among Young Persons Aged 10–24 Years in the United States: 1992–2006," 95–96
Chappell, Bill, 33
Chardon High School, Chardon, Ohio, 117
Charles II, King of England, 2
Cheng, Diana, 131
Chicago, Illinois, 52–53
"Child Access Prevention" (Law Center to Prevent Gun Violence), 95
Child access prevention (CAP) laws
    AAP on, 135
    influence of, 95–96
    by state, 96*f*
"Child Access Prevention Policy Summary" (Law Center to Prevent Gun Violence), 50
Children
    BB/pellet gun injuries among children aged 0–19 years, rates per 100,000, 86(*t*6.4)
    child access prevention laws, by state, 96*f*
    firearm deaths of children aged 0–19 years, 101*t*
    firearm homicide deaths among, 105–106
    firearm injury/death, susceptibility to, 16
    Florida child access prevention law, 50
    gun control laws and, 134–135
    gun safety programs to protect, 96–97
    homicide firearm deaths of children aged 12/under, 107*t*
    injured/killed by gunfire, 120
    LaPierre, Wayne, testimony of on gun violence, 138–139
    nonfatal firearm injuries among children aged 0–19 years, rates per 100,000, 86(*t*6.3)
    nonfatal gunshot injuries among, 84
    in Ohio, security against juvenile theft of firearms by, 54
    safe storage of guns in home and, 95–96
    See also Youth

Children's Defense Fund (CDF), 120
Cho, Seung-Hui
    description of shooting at Virginia Tech, 118
    shooting at Virginia Tech by, 31–32, 65
Cicero, Marcus Tullius, 2
Cities
    gun violence in, 132–133
    youth gangs in, 108
City and County of San Francisco, Fiscal v., 50–51
City of Baltimore, Barron v., 41
City of Chicago, McDonald v., 48
City of Chicago, National Rifle Association of America v., 48
Clinton, Hillary Rodham
    gun control and, 126
    public opinion on background checks, 29
Coalition of New Jersey Sportsmen v. Whitman, 50
Coker, Billy Wayne, 55
Coker, Wal-Mart Stores, Inc. v. (Florida), 55
Collective rights, 5, 41–42
College shootings, 118–119
Colorado
    Columbine High School shooting, 65, 115–116, 117
    Deer Creek Middle School shooting, 119
    movie theater shooting in Aurora, 77
    state constitution article concerning weapons, 143
Columbine High School, Littleton, Colorado
    shooting at, 65, 117
    as turning point, 115–116
Commentaries (Blackstone), 2–3
Commerce environments, 70
Concealed weapons
    Permanent Brady permit chart, 37*t*
    public opinion on, 128, 130*t*
    right-to-carry weapons laws, 36*t*
    state laws on, 35–36
    Trotter, Gayle S., on, 142
"Concord Hymn" (Emerson), 3
Connecticut
    child access prevention challenge in, 50
    firearms control laws of, 32, 33
    Sandy Hook Elementary School shooting, 1, 65, 119–120, 141
    semiautomatic firearms ban in, 49–50
    state constitution article concerning weapons, 143
Constitution. *See* U.S. Constitution
Constitution Society, 3
Constitutions, state, 143–145
Cook, Philip J., 25–26
Cooper, Alexia
    on homicide trends, 58
    on youth homicide trends, 99
    on youth homicide victims, 107–108

Cop killer bullets, 21–22
Cost, of firearm injuries, 91, 93
"The Counted" project (*Guardian*), 22
Court cases
    Armijo v. Ex Cam Inc. (New Mexico), 56
    Bailey v. United States, 45
    Barron v. City of Baltimore, 41
    Benjamin v. Bailey, 49–50
    Cases v. United States, 43
    Coalition of New Jersey Sportsmen v. Whitman, 50
    District of Columbia v. Heller, 5, 42, 47, 128, 138
    Doe v. Portland Housing Authority, 49
    Drake v. Jerejian, 48
    Farmer v. Higgins, 21
    Fiscal v. City and County of San Francisco, 50–51
    Gonzales v. Raich, 47
    Hoosier v. Randa (California), 54
    Kalodimos v. Village of Morton Grove (Illinois), 52
    Kelley v. R.G. Industries Inc. (Maryland), 55
    McDonald v. City of Chicago, 5, 48
    Muscarello v. United States, 45
    National Rifle Association of America v. City of Chicago, 48
    NRA v. Reno, 27
    Printz v. United States, 5, 46–47
    Quilici v. Village of Morton Grove (Illinois), 52
    Ramsey Winch Inc. v. Henry, 50
    Schubert v. DeBard, 48–49
    Second Amendment Foundation v. Renton (Washington), 51–52
    Silveira v. Lockyer, 5, 44
    Sklar v. Byrne (Illinois), 53
    Smith v. United States, 44
    State v. Blocker (Oregon), 51
    State v. Boyce (Oregon), 51
    State v. Kessler (Oregon), 51
    State v. Owenby (Oregon), 49
    State v. Wilchinski (Florida), 50
    United States v. Cruikshank, 42
    United States v. Edwards, 46
    United States v. Emerson, 5, 43
    United States v. Harris, 44–45
    United States v. Hayes, 20
    United States v. Lopez, 23, 45–46
    United States v. Miller, 41–42
    United States v. Phelps, 44
    United States v. Robinson, 45
    United States v. Stewart, 47
    United States v. Synnes, 44
    United States v. Tot, 43
    United States v. Warin, 44
    Wal-Mart Stores, Inc. v. Coker (Florida), 55
    Williams v. Beemiller (New York), 56

Court rulings
- federal court cases, 42–48
- gun industry, liability for gunshot injuries, 55–56
- handgun deaths, responsibility for, 54–55
- local, 51–54
- NICS, firearm prohibitive criteria under, 42t
- overview of, 41
- Second Amendment interpretations, 41–42
- state laws, 48–51
- states with laws protecting right to use deadly force in self-defense, 48t

Crime
- active shooter incidents, 65, 70
- active shooter incidents, number of, annually, 70f
- active shooters, number of casualties (killed or wounded) by, 71f
- aggravated assault, by type of weapon used, percentage distribution by region, 78(t5.18)
- arrest trends, by offense/age, 81t
- arrests, by race, 82t
- convicted felons, gun possession by, federal court cases on, 44
- crime trends, by population group, 59t
- criminally active nongang groups, 108
- federal officers, assaults on, by extent of injury/type of weapon, 75(t5.13)
- firearm violence, 76t
- firearms, acquisition of, 77–80
- firearms used in, types of, 4t
- gun control debate and, 123–124
- handguns, use of in, 6
- homicide circumstances, by relationship of victim to offender, 64t
- homicide circumstances, by weapon, 62t
- homicide victims by age, sex, race, ethnicity, 61t
- homicide/murder, 58, 60–61, 63, 65
- homicides, by number of victims/weapon type, 65t
- homicides, by state/weapon type, 63t
- homicides, number/rate of, by victim demographics, 60t
- justifiable homicides committed by law enforcement officers, by weapon used, 78(t5.19)
- justifiable homicides committed by private citizens, by weapon used, 79t
- law enforcement officers accidentally killed, by circumstance at scene of incident, 74t
- law enforcement officers assaulted, by circumstance at scene of incident/type of weapon, 75(t5.14)
- law enforcement officers killed, by circumstance at scene of incident, 73t
- law enforcement officers killed, by type of weapon/region, 72t
- mass shootings, notable, 1949–2008, 66t–67t
- mass shootings, notable, 2009–2016, 68t–69t
- nonfatal firearm violence, 76f
- nonfatal firearm violence, by race/Hispanic origin, 77f
- nonfatal violent crimes involving firearms, 72–74, 76–77
- police deaths/injuries, 70–72
- robberies/aggravated assaults, by type of weapon used, 77t
- robbery, by type of weapon used, percentage distribution by region, 78(t5.17)
- statistics on, source of, 57
- use of firearms in drug crimes, federal court cases on, 44–45
- violent crime, trends in, 57–58
- violent crime offenses reported to law enforcement, 58f
- weapons offenses, 80
- youth gangs, 108

"Crime, Deterrence, and Right-to-Carry Concealed Handguns" (Lott & Mustard), 36

Crime in the United States (FBI)
- on justifiable homicide, 76
- on nonfatal firearm violence used in robbery/aggravated assault, 74, 76
- as source of crime statistics, 57

Cruikshank, United States v., 42

D

Date, Jack, 70
De Officiis (Cicero), 2
Deaths
- from active shooter incidents, 65, 70
- active shooters, number of casualties (killed or wounded) by, 71f
- in America from gun violence, 132
- of children from gun violence, 120, 134
- death rates for firearm-related injuries, by sex, race, Hispanic origin, age, 87t–89t
- death rates for suicide, by sex/age, 90t–91t
- death/death rates for 10 leading causes of death among young people, 121t–122t
- fatal gunshot (homicide and suicide) deaths/rates per 100,000 people, 91(t6.7)
- firearm deaths of children aged 0–19 years, 101t
- firearm deaths per 100,000 people, 86f
- firearm fatalities, 84, 86–91
- firearm-related deaths among youths, 99
- gun control issue and, 123
- gun death rates, gun ownership rates and, 7
- gun homicides, number of justifiable, compared to criminal, 94(t6.10)
- handgun deaths, responsibility for, 54
- homicide firearm deaths of children aged 12/under, 107t
- homicide/suicide/gun-related deaths among youths aged 15–19, 100f
- injury deaths, 10 leading causes of, by age group, 92t
- from justifiable firearms homicides, 93–94
- justifiable gun homicides, race of shooter/person killed in, 94(t6.11)
- justifiable homicides, firearms used in, by type, 95t
- juveniles murdered 1980–2010, by age/weapon type, 100t
- law enforcement officers accidentally killed, by circumstance at scene of incident, 74t
- law enforcement officers killed, by circumstance at scene of incident, 73t
- law enforcement officers killed, by type of weapon/region, 72t
- leading causes of death for 10–24-year-olds, percentage distribution of, 122f
- from mass shootings, 63, 65
- police deaths/injuries, 70–72
- rates of, gun ownership rates and, 7
- reduction of with CAP laws, 95
- unintentional firearm deaths/rates per 100,000 people, 91(t6.8)

Deaths: Leading Causes for 2014 (Heron), 95
DeBard, Robert L., 48
DeBard, Schubert v., 48–49
Deer Creek Middle School, Littleton, Colorado, 119
Default process transactions, 25
Delaware
- as non-RTC state, 36
- state constitution article concerning weapons, 143

Democratic Party
- Americans' dissatisfaction with gun laws, by party, 128t
- gun control, position on, 1
- gun ownership by, 125
- public opinion on concealed weapons, 128
- public opinion on strictness of gun laws, 126

"The Demographics and Politics of Gun-Owning Households" (Morin), 124
Department of Defense Appropriations Act, 46
Designated point of contact, 24
Deterrence, 123
Directorate of Defense Trade Controls, 11
Discourse (Machiavelli), 2

District of Columbia
  handgun ban, constitutionality of, 47
  manufacturer liability in, 56
  open carrying prohibited in, 38
District of Columbia v. Heller
  constitutionality of handgun ban, 47
  handgun ban ruling, 128
  on individual rights to bear arms, 42
  Scalia, Antonin, statement in, 138
  on Second Amendment guarantee to keep/bear arms, 5
"Doctors Condemn the NRA-Fueled Ban on Gun Violence Research" (Schumaker), 83
Documents on the First Congress Debate on Arms and Militia (Constitution Society), 3
Doe, Jane, 49
Doe, John, 49
Doe v. Portland Housing Authority, 49
Domestic violence
  Brady Handgun Violence Prevention Act and, 25
  gun control laws, argument for stricter, 131–132
  guns in home and, 15
  Lautenberg Amendment and, 20, 43
  state gun control laws and, 33
  United States v. Emerson, 43
Domestic Violence Offender Gun Ban, 20
Donohue, John J., III, 36
Drake v. Jerejian, 48
Drugs
  gun use and, 108
  use of firearms in drug crimes, federal court cases on, 44–45
Due process, 137–138

E
Eastwood, Bruco Strongeagle, 119
Economy, 91, 93
Eddie Eagle GunSafe Program
  LaPierre, Wayne, testimony of on, 138
  overview of, 96–97
Edmond, Oklahoma, 65
Educational environments, 70
Edward III, King of England, 2
Edwards, Ray Harold, III, 46
Edwards, United States v., 46
"The Effect of Child Access Prevention Laws on Unintentional Child Firearm Fatalities, 1979–2000" (Hepburn et al.), 95
"The Effect of High School Shootings on Schools and Student Performance" (Beland & Kim), 120
Eighteenth Amendment, 19
Emergency department, 91, 93
Emerson, Ralph Waldo, 3
Emerson, Timothy Joe, 43
Emerson, United States v., 5, 43

England, early gun control laws in, 2
"The Epidemiology of Case Fatality Rates for Suicide in the Northeast" (Miller, Azrael, & Hemenway), 120
"An Evaluation of Two Procedures for Training Skills to Prevent Gun Play in Children" (Himle et al.), 97
Ex Cam Inc., Armijo v. (New Mexico), 56
Exports
  of firearms, numbers of, 12t
  of firearms, overview of, 11–12

F
Family, 60
Farmer v. Higgins, 21
"Fatal Texas Shooting Highlights Struggle to Regulate Replica Guns" (Ferriss), 22
Federal Bureau of Investigation (FBI)
  on active shooter incidents, 65, 70
  crime statistics collected by, 57
  on justifiable homicide, 76
  NICS, 5
  NICS background checks, 24–25
  NICS records, destruction of, 28
  on nonfatal firearm violence used in robbery/aggravated assault, 74, 76
  on police deaths/injuries, 70
  Supplemental Homicide Reports, 131
  on violent crime trends, 57–58
Federal court cases
  on Brady law, constitutionality of, 46–47
  on convicted felons, gun possession by, 44
  on District of Columbia's handgun ban, 47
  on drug crimes, use of firearms in, 44–45
  on gangster-type weapons, possession of, 42–44
  on homemade machine guns, possession of, 47
  on machine guns, possession of, 44
  on right to bear arms, 42
  on schools, possession of firearms near, 45–46
  on Second Amendment rules, states' rights and, 48
  "use," new interpretations of, 45
Federal crimes, 21
Federal Energy Management Improvement Act, 22
Federal Firearms Act
  Cases v. United States, 43
  requirements of, 19
Federal firearms licensee (FFL)
  active firearms licensees between 1975 and 2015, 30t
  description of, 5
  federal firearms licenses, by state, 31t
  firearms/ammunition purchase regulations, 20–21
  juveniles, state laws prohibiting FFLs for, 38

  new regulations for obtaining, 30–31
  NICS system and, 24–25
  prohibited people, 20
Federal Firearms Licensees: Various Factors Have Contributed to the Decline in the Number of Dealers (GAO), 31
Federal government
  federal firearms licensees, 30t
  gun violence research, defunding of, 83
  regulation of guns, 19
The Federalist, 3
Fee, for FFL, 30
Feinstein, Dianne, 13, 140
Felons
  felony as reason for NICS denial, 25
  gun possession by, federal court cases on, 44
  as prohibited people, 20
  right to possess firearms, 21
Females
  firearm fatalities among, 84, 86
  gun ownership by gender, 125
  homicide rate for, 58
  relationship of homicide victim to murderer, 60–61
  violence against, as argument for gun control laws, 131–132
  See also Gender; Women
Ferriss, Susan, 22
FFL. *See* Federal firearms licensee
Firearm Justifiable Homicides and Non-fatal Self-Defense Gun Use (Violence Policy Center), 93–94
"Firearm Legislation and Firearm-Related Fatalities in the United States" (Fleegler et al.), 7
Firearm Owners' Privacy Act (Florida), 135
"The Firearm Owners' Protection Act of 1986: A Historical and Legal Perspective" (Hardy), 20
Firearm transferee, 24
Firearm Use by Offenders (Harlow), 79
Firearm Violence, 1993–2011 (Planty & Truman), 73, 84
"Firearm-Related Injuries Affecting the Pediatric Population" (AAP), 16
Firearms
  automatic/semiautomatic, 12–14
  criminals' acquisition of, 77–80
  exports, 11–12, 12t
  firearm violence, 76t
  in the home, 14–16
  homicides, firearms used in, 60–61
  imported, by country, 15(t2.5)
  imported, numbers of, 11t
  imports, 11
  justifiable homicide involving, 76–77
  manufactured, numbers of, 10t
  nonfatal firearm violence, 76f
  nonfatal firearm violence, by race/Hispanic origin, 77f

nonfatal violent crimes involving, 72–74, 76–77
owners, characteristics of, 16–17
police deaths/injuries from, 70–72
state constitution articles concerning weapons, 143–145
U.S. manufacturing of, 10–11
used in assaults on law enforcement officers, 72
used in crime, types of, 4t
in violent crimes, 57
youth suicide and, 120
See also Guns

Firearms and Violence: A Critical Review (National Research Council of the National Academies), 36

Firearms Commerce in the United States—2001/2002 (ATF)
on decline in number of FFLs, 31
on numbers of guns in U.S., 9–10

Firearms Owners' Protection Act
changes implemented with, 20–21
machine gun ownership ban of, 12

Firearms sales
acquisition of firearms by criminals, 77–80
acquisition of guns by youth, 108–110
Brady law on gun show/online regulation, 25–29
firearms/ammunition purchase regulations, 20–21
five-day waiting period of interim Brady, 23
Gun Control Act of 1968 and, 19–20
NICS system, operation of, 24–25
people who may not purchase firearms under Brady law, 23
See also Gun dealers

"Firearms Trace Data—2015" (ATF), 109–110

Fiscal v. City and County of San Francisco, 50–51

Five-day waiting period
elimination of, 26–27
for handgun purchases, 5
of interim Brady, 23

Fleegler, Eric W., 7

Florida
FFLs in, 31
Firearm Owners' Privacy Act, 135
gun control laws of, 33
juvenile purchase of ammunition, store responsibility for, 54–55
mass shooting at Pulse nightclub in, 1, 33, 65
open carrying prohibited in, 38
state constitution article concerning weapons, 143

Follman, Mark
on cost of firearm injuries, 93
on mass shootings, 65

Forcible rape, 57
Forfeiture, 21
Fourth Amendment, 137
Fryberg, Jaylen, 117–118
Funding, for gun violence research, 136

G

Gallup Organization
on guns in homes, 14
public opinion on strictness of gun laws, 126–128

Game Act (England), 2

Gangs, youth
firearm use among youths, increase in, 99
overview of, 108

Gangster-type weapons
National Firearms Act and, 19
possession of, federal court cases on, 42–44

Gender
cost of firearm injuries and, 93
death rates for suicide, by sex/age, 90t–91t
firearm fatalities by, 84, 86
gun ownership by, 124–125
of homicide victims, 58
nonfatal gunshot injuries by, 84
relationship of homicide victim to murderer, 60
student reports of threats/injuries by, 111
students grades 9–12 who carried weapon/gun, by sex/race/ethnicity/grade, 113t
students grades 9–12 who reported carrying weapon by location/sex, 111f
students who carried weapon/gun, by sex/state/urban area, 114t
youth homicide victimization by, 100–101
of youthful homicide offenders, 102, 104
of youthful homicide victims, 105, 106–108

General Social Survey Final Report: Trends in Gun Ownership in the United States, 1972–2014 (Smith & Son), 14

General Social Surveys (National Opinion Research Center), 14

Georgia, 143
Gertz, Marc, 6
Goh, One L., 118
Gonzales v. Raich, 47
Gorski, Gary, 44
Grinberg, Emanuella, 129
Guardian (newspaper), 22

"A Guide to Mass Shootings in America" (Follman, Aronsen, & Pan), 65

Gun ban
Americans' views on potential handgun ban, 129f

Assault Weapons Ban of 2013, arguments against, 139–141
Assault Weapons Ban of 2013, arguments for, 132–136
handgun ban, public support for, 128

"Gun Carrying and Drug Selling among Young Incarcerated Men and Women" (Kacanek & Hemenway), 108

Gun control
arguments for/against, 5–6
constitutionality of, 5
expert opinions on, 6–7
mass shootings and, 129
overview of, 1–2
as personal issue, 123–124
public opinion on, 124–129
views of gun ownership as matter of protection vs. safety risk, 126t

Gun Control Act of 1968
gun imports stipulations of, 11
passage of, 4
passage of/requirements of, 19–20

Gun Control Act of 1986. See Firearms Owners' Protection Act

Gun control advocates
arguments for stricter gun control laws, 131–136
arguments of, 19
Brady law, controversies over, 26–29
on CAP laws, 95
Gun Control Act of 1968 and, 20
gun control as personal issue, 123–124
gun use in self-defense and, 93
on Gun-Free School Zones Act, 23
public opinion on strictness of gun laws, 125–128
on state preemption statutes, 39

Gun control debate
Americans' views on controlling gun ownership vs. protecting right of Americans to own guns, 125(f8.3)
arguments for/against, 19
about assault weapons ban, 29
gun control as personal issue, 123–124
gun use in self-defense and, 93
right to bear arms, 4–7
youth at center of, 99

"Gun Control Explained" (Pérez-Peña), 1

Gun control laws
Americans' dissatisfaction with gun laws, by party, 128t
Americans' satisfaction with gun laws, 127f
Americans' views on strictness of gun laws, 128f
arguments for stricter, 131–136
early, 2
public opinion on strictness of gun laws, 125–128
state constitution articles concerning weapons, 143–145

Gun control laws, arguments against stricter
   Kopel, David B., 139–141
   LaPierre, Wayne, 138–139
   Malcolm, Joyce Lee, 137–138
   Trotter, Gayle S., 141–142
Gun control laws, arguments for stricter
   American Academy of Pediatrics, 134–135
   Bueermann, Jim, 135–136
   Campbell, Jacquelyn/Anna D. Wolf, 131–132
   Nutter, Michael A., 132–134
Gun Control: Potential Effects of Next-Day Destruction of NICS Background Check Records (GAO), 27–28
Gun dealers
   acquisition of firearms by criminals, 77
   Brady law, controversies over, 26–29
   in California, liability for gun violations of others, 54
   FFLs, new regulations to obtain, 30–31
   firearms/ammunition purchase regulations, 20–21
   five-day waiting period of interim Brady, 23
   NICS system, operation of, 24–25
   straw purchasing by criminals, 79
   See also Firearms sales
"Gun Homicides Steady after Decline in '90s; Suicide Rate Edges Up" (Krogstad), 90–91
Gun industry, 55–56
"Gun Industry Is Gaining Immunity from Suits" (Butterfield), 56
"Gun Law Trendwatch" (Law Center to Prevent Gun Violence), 33, 34
"Gun Market Is Wide Open in America" (Sugarmann), 30
Gun ownership
   AAP on, 135
   Americans' views on controlling gun ownership *vs.* protecting right of Americans to own guns, 125(*f*8.3)
   background checks, expanded, public opinion on, 127*t*
   demographics of, 124–125
   firearms, automatic/semiautomatic, 12–14
   firearms exported, numbers of, 12*t*
   firearms exports, 11–12
   firearms imported, by country, 15(*t*2.5)
   firearms imported, numbers of, 11*t*
   firearms imports, 11
   firearms in the home, 14–16
   firearms manufactured, numbers of, 10*t*
   firearms manufacturing, U.S., 10–11
   firearms owners, characteristics of, 16–17
   gun owners by selected demographic characteristics, 16*t*

guns at home, non-gun households attitudes toward having, 17(*t*2.9)
   National Firearms Act registered weapons, by state, 13*t*–14*t*
   numbers of guns, 9
   numbers of guns, ATF estimates on, 9–10
   prevalence of, among selected groups, 125(*f*8.2)
   public opinion on strictness of gun laws, 125–128
   reasons for owning a gun, 17, 17(*t*2.8)
   U.S. households with guns, percentage of, 15(*t*2.6)
   views of gun ownership as matter of protection *vs.*. safety risk, 126*t*
Gun rights advocates
   arguments of, 19
   Brady law, controversies over, 26–29
   Gun Control Act of 1968 and, 20
   gun control as personal issue, 123–124
   gun control laws, arguments against stricter, 137–142
   gun use in self-defense and, 93, 94–95
   gun violence research and, 83
   on Gun-Free School Zones Act, 23
   public opinion on strictness of gun laws, 125–128
   on right-to-carry laws, 36
   on state preemption statutes, 38–39
Gun safety. *See* Safety
Gun shows
   Brady law, controversies over, 28–29
   in Ohio, security against juvenile theft by, 54
   public opinion on background checks at, 126
Gun trace data
   acquisition of guns by youth, 108–110
   guns recovered after use in crimes by state/age of possessor, 109*t*
   Tiahrt Amendment and, 9
Gun trafficking, 79–80
Gun violence
   gun control laws, arguments against stricter, 137–142
   gun control laws, arguments for stricter, 131–136
   impact of Brady law on, 25–26
   research, controversy over, 83
   youth, trends in, 99
Gun Violence among Serious Young Offenders (Braga & Wiener), 108
"Gun Violence Not a Mental Health Issue, Experts Say, Pointing to 'Anger,' Suicides" (Grinberg), 129
"Gun Violence Research: History of the Federal Funding Freeze" (Jamieson), 83
Gun-Free School Zones Act, 23, 45–46
Gun-Free Schools Act, 115
Gun-related injuries/fatalities

   BB/pellet gun injuries among children aged 0–19 years, rates per 100,000, 86(*t*6.4)
   CDC data on, 83–84
   child access prevention laws, by state, 96*f*
   cost of firearm injuries, 91, 93
   death rates for firearm-related injuries, by sex, race, Hispanic origin, age, 87*t*–89*t*
   death rates for suicide, by sex/age, 90*t*–91*t*
   fatal gunshot (homicide and suicide) deaths/rates per 100,000 people, 91(*t*6.7)
   firearm deaths per 100,000 people, 86*f*
   firearm fatalities, 84, 86–91
   gun homicides, number of justifiable, compared to criminal, 94(*t*6.10)
   gun safety, efforts to promote, 95–97
   gun violence research, controversy over, 83
   injury deaths, leading causes of, by age group, 92*t*
   justifiable gun homicides, race of shooter/person killed in, 94(*t*6.11)
   justifiable homicides, firearms used in, by type, 95*t*
   nonfatal firearm injuries among children aged 0–19, rates per 100,000, 86(*t*6.3)
   nonfatal gunshot injuries, 84
   nonfatal gunshot injuries/rates per 100,000 people, 84*t*
   nonfatal injuries treated in hospital emergency departments, leading causes of, by age group, 85*t*
   self-defense, guns and, 93–95
   unintentional firearm deaths/rates per 100,000 people, 91(*t*6.8)
Guns
   child access prevention laws, by state, 96*f*
   criminal advantages of, 61
   criminals' acquisition of, 77–80
   gun safety, efforts to promote, 95–97
   guns recovered after use in crimes by state/age of possessor, 109*t*
   numbers of, ATF estimates on, 9–10
   self-defense and, 93–95
   state constitution articles concerning weapons, 143–145
   youth and, 99
   youthful offenders, acquisition of, 108–110
   See also Firearms
"Guns, Fear, the Constitution, and the Public's Health" (Wintemute), 15

H
Hamilton, Alexander, 3
Handgun ban
   Chicago, Illinois, 48
   District of Columbia, constitutionality of, 47

Handguns
- Americans' views on potential handgun ban, 129f
- ban, public support for, 128
- control of, debate over, 6
- deaths from, 54–55
- justifiable homicides involving, 76–77
- juveniles banned from possession of, 29–30
- manufactured in U.S., 10
- nonfatal violent crimes involving, 73
- Trotter, Gayle S., on, 142
- used in homicides, 60
- used in justifiable homicides, 94

Hanging, 96
Hardy, David T., 20
Harlow, Caroline Wolf, 79
Harper-Mercer, Chris, 118–119
Harris, Eric, 65, 117
Harris, United States v., 44–45
Harvard Injury Control Research Center
- on firearm ownership/use, 6–7
- on gun use for self-defense, 94

Hatalsky, Lanae Erickson, 28
Hawaii
- firearms control laws of, 32–33
- no preemption statute in, 38
- as non-RTC state, 36
- state constitution article concerning weapons, 143

Hayes, United States v., 20
Health, public, 7, 134–135
Heath, Brad, 80
Heller, Dick Anthony, 47
Heller, District of Columbia v.
- constitutionality of handgun ban, 47
- handgun ban ruling, 128
- on individual rights to bear arms, 42
- Scalia, Antonin, statement in, 138
- on Second Amendment guarantee to keep/bear arms, 5

Hemenway, David
- on drug dealing/guns, 108
- on firearm ownership/use as public health issue, 7
- on Gary Kleck's research, 6–7
- on gun use for self-defense, 93, 94
- on youth suicide, 120

Hennard, George, 65
Henry II, King of England, 2
Henry, Ramsey Winch Inc. v., 50
Hepburn, Lisa, 95
Heron, Melonie, 95
Heyward, Nicholas, Jr., 22
Higgins, Farmer v., 21
High-risk individual, 20
Himle, Michael B., 97
Hinckley, John W., Jr., 23

Hispanics
- firearm fatalities among, 87
- gun ownership by, 16, 125
- homicide rate for, 58
- weapon-carrying by students, 110

Holmes, James Eagan, 77
Home Depot, 37
Homes, guns in
- adults having, percentages of, 14
- non-gun households attitudes toward having, 17(t2.9)
- safety and, 14–16
- U.S. households with guns, percentage of, 15(t2.6)

Homicide in the U.S. Known to Law Enforcement, 2011 (Cooper & Smith)
- on homicide rate, 58
- on youthful homicide victims, 107–108

Homicide Trends in the United States (BJS), 1
Homicide Trends in the United States, 1980–2008 (Cooper & Smith)
- on homicide victimization rate, 58
- on youth homicide trends, 99

Homicide/murder
- AAP on gun violence, 134
- among children, CAP laws and, 95
- circumstances, by relationship of victim to offender, 64t
- circumstances, by weapon, 62t
- committed by juveniles, by firearm involvement, 103(f7.4)
- criminal advantages of guns, 61
- definition of, 57
- demographics of, 58, 60
- fatal gunshot (homicide and suicide) deaths/rates per 100,000 people, 91(t6.7)
- firearm deaths of children aged 12/under, 107t
- firearm fatalities, 84, 86–90
- firearm suicides *vs.*, 90–91
- firearms used in, 60–61
- gun control debate and, 123
- gun control laws, argument for stricter, 131–134
- gun control laws, arguments against stricter, 137
- gun homicides, number of justifiable, compared to criminal, 94(t6.10)
- guns used in, numbers of, 15
- homicide/suicide/gun-related deaths among youths aged 15–19, 100f
- involving guns by age of victim, 106(f7.7)
- justifiable gun homicides, race of shooter/person killed in, 94(t6.11)
- justifiable homicide, 76–77, 93–94
- justifiable homicides, firearms used in, by type, 95t
- juvenile homicide offenders, by race, 105f
- juvenile homicide offenders, by relationship to victim, 103(f7.3)
- juvenile homicide offenders, demographics, 102t
- juvenile homicide victims by race, 107f
- juvenile homicide victims killed with/without firearms, 106(f7.6)
- juveniles murdered 1980–2010, by age/weapon type, 100t
- juveniles murdered by age/relationship to offender, 105t
- mass shooting in U.S., 61, 63, 65
- mass shootings as percentage of all U.S. homicides, 124f
- by number of victims/weapon type, 65t
- number/rate of, by victim demographics, 60t
- school shootings, 115–120
- by state/weapon type, 63t
- trends in, 58
- by victim/offender relationship/weapon use, 101f
- victims by age, sex, race, ethnicity, 61t
- among youth, trends in, 99–102
- by youth gangs, 108
- youthful offenders, 102, 104
- youthful victims, 104–108

Hoosier, Bryan, 54
Hoosier v. Randa (California), 54
Horon, Isabelle, 131
The Hospital Costs of Firearm Assaults (Howell & Abraham), 91, 93
Hospitals, cost of firearm injuries to, 91, 93
"How Delinquent Youths Acquire Guns: Initial versus Most Recent Gun Acquisitions" (Webster et al.), 110
"How Open Carry Forced Businesses in Texas to Take a Side in the Culture War" (Solomon), 37–38
Howell, Embry M., 91, 93
Hutchinson, Asa, 138

I

Idaho
- gun control laws of, 33
- state constitution article concerning weapons, 143–144

ILA. *See* Institute for Legislative Action
Illinois
- Chicago, handgun possession, limits on, 52–53
- FFLs in, 31
- firearms control laws of, 33
- Morton Grove, handgun ban in, 52
- open carrying prohibited in, 38
- shooting at Northern Illinois University, 118
- state constitution article concerning weapons, 144

The Impact of Right to Carry Laws and the NRC Report: The Latest Lessons for the Empirical Evaluation of Law and Policy (Aneja, Donohue, & Zhang), 36

Imports
  of firearms, 11
  of firearms, by country, 15(*t*2.5)
  of firearms, numbers of, 11*t*
  Gun Control Act of 1968 and, 20
Independents
  Americans' dissatisfaction with gun laws, by party, 128*t*
  gun ownership by, 125
Indiana
  right to possess handgun in, 48–49
  state constitution article concerning weapons, 144
Indicators of School Crime and Safety: 2015 (Zhang, Musu-Gillette, & Oudekerk), 116
Individual rights
  collective rights *vs.*, 5
  collective rights *vs.*, interpretations of, 41–42
Ingraham, Christopher, 63
Injuries
  active shooters, number of casualties (killed or wounded) by, 71*f*
  BB/pellet gun injuries among children aged 0–19 years, rates per 100,000, 86(*t*6.4)
  as causes of deaths among children, 95
  cost of firearm injuries, 91, 93
  death rates for firearm-related injuries, by sex, race, Hispanic origin, age, 87*t*–89*t*
  federal officers, assaults on, by extent of injury/type of weapon, 75(*t*5.13)
  gun industry liability for, 55–56
  injury deaths, 10 leading causes of, by age group, 92t
  nonfatal firearm injuries among children aged 0–19, rates per 100,000, 86(*t*6.3)
  nonfatal gunshot injuries, 84
  nonfatal gunshot injuries/rates per 100,000 people, 84*t*
  nonfatal injuries treated in hospital emergency departments, leading causes of, by age group, 85*t*
  police deaths/injuries, 70–72
  student reports of threats/injuries, 111–113, 115
"Injuries and Deaths Due to Firearms in the Home" (Kellerman et al.), 15
Inside Straw Purchasing: How Criminals Get Guns Illegally (MAIG), 79
Institute for Legislative Action (ILA)
  on gun shows/online sales of guns, 28
  on RTC states, 35–36
Interdiction, 123–124
Internet, online gun sales, 28–29

Iowa
  firearms control laws of, 32, 33
  state constitution article concerning weapons, 144

J

Jamieson, Christine, 83
Jefferson, Thomas, 3
Jerejian, Drake v., 48
"Joe Camel with Feathers: How the NRA with Gun and Tobacco Industry Dollars Uses Its Eddie Eagle Program to Market Guns to Kids" (Violence Policy Center), 96–97
Johnson, Jamiel, 22
Jostad, Candace M., 97
Jourdain, Louis, 117
Justice and Mental Health Collaboration Program Act, 136
"Justices Decline New York Gun Suit" (Stout), 56
"Justices Extend Gun Owner Rights Nationwide; McKenna Statement Inside" (KHQ.com), 54
Justifiable homicide
  committed by law enforcement officers, by weapon used, 78(*t*5.19)
  committed by private citizens, by weapon used, 79*t*
  firearms used in, by type, 95*t*
  gun homicides, number of justifiable, compared to criminal, 94(*t*6.10)
  gun use for self-defense, demographics of, 76–77, 93–94
  race of shooter/person killed in, 94(*t*6.11)
"Juvenile Homicide" (Virginia Youth Violence Project), 102
Juvenile Offenders and Victims: 2014 National Report (Sickmund & Puzzanchera)
  on youth homicide, 99–102
  on youthful homicide victims, 104–107
Juveniles
  ban from possession of handguns/ammunition, 29–30
  state laws prohibiting FFLs for, 38
  See also Children; Youth

K

Kacanek, Deborah, 108
Kalodimos v. Village of Morton Grove (Illinois), 52
Kansas, 144
Kates, Don B.
  NRA reliance on research of, 16
  on use of firearms for self-defense, 6–7
Kay-Bee Stores, 22
Kazmierczak, Steven Phillip, 118
Kellerman, Arthur L., 15
Kelley, Olen J., 55

Kelley v. R.G. Industries Inc. (Maryland), 55
Kennedy, John F.
  assassination of, 1, 123
  Gun Control Act of 1968 and, 19
Kennedy, Robert F.
  assassination of, 1, 123
  Gun Control Act of 1968 and, 4, 19
Kentucky
  gun control laws of, 33
  state constitution article concerning weapons, 144
Kern, Terence C., 50
Kessler, State v. (Oregon), 51
KHQ.com, 54
Killeen, Texas, mass shooting in, 65
The Killer at Thurston High (PBS program), 117
Kim, Dong-woo, 120
King, Martin Luther, Jr.
  assassination of, 1, 123
  Gun Control Act of 1968 and, 4, 19
Kinkel, Faith, 116
Kinkel, Kip, 115–116
Klebold, Dylan, 65, 117
Kleck, Gary
  on gun use for self-defense, 93, 94
  NRA reliance on research of, 16
  on use of firearms for self-defense, 6–7
Klobuchar Bill, 137
Kopel, David B., 139–141
Krogstad, Jens Manuel, 90–91
Kutner, Max, 118

L

Laine, Christine, 83
Lane, Thomas "T.J.," 117
Lanza, Adam, 65, 119
LaPierre, Wayne
  argument of, 34
  on gun control laws, 138–139
  response to Sandy Hook Elementary School shooting, 119
Lautenberg Amendment
  Malcolm, Joyce Lee, on, 137
  prohibited people under, 20
  United States v. Emerson, 43
Law Center to Prevent Gun Violence
  on background checks, 29
  on child access prevention laws, 95
  "Child Access Prevention Policy Summary," 50
  database of federal/state firearms laws, 32
  "Gun Law Trendwatch," 34
  as interdiction advocate, 124
  on local firearms ordinances, 38
  "Open Carrying," 38
  on state gun laws, 33
  on state preemption statutes, 39

Law enforcement
  active shooter incidents and, 65, 70
  ban on armor-piercing ammunition and, 21–22
  federal officers, assaults on, by extent of injury/type of weapon, 75(t5.13)
  justifiable homicide, 76–77
  justifiable homicides committed by law enforcement officers, by weapon used, 78(t5.19)
  officers accidentally killed, by circumstance at scene of incident, 74t
  officers assaulted, by circumstance at scene of incident/type of weapon, 75(t5.14)
  officers killed, by circumstance at scene of incident, 73t
  officers killed, by type of weapon/region, 72t
  police deaths/injuries, 70–72
  violent crime offenses reported to, 58f
Law Enforcement Officers Killed and Assaulted, 2014 (FBI), 70
Law Enforcement Officers' Protection Act, 21–22
Law Enforcement Officers Safety Act, 31
Law Enforcement Steering Committee (LESC), 22
Laws, regulations, ordinances
  Brady Handgun Violence Prevention Act, 23
  federal firearms licensees, 30t
  federal firearms licenses, by state, 31t
  federal government regulation of guns, 19
  Firearms Owners' Protection Act, 20–21
  firearms purchase laws, by state, 32t
  Gun Control Act of 1968, 19–20
  Gun-Free School Zones Act, 23
  Law Enforcement Officers' Protection Act, 21–22
  local ordinances, 38–39
  NICS, components of, 25f
  NICS checks, total, 24t
  NICS denial process per 100 applicants, 27f
  NICS firearms denials, by reason, 27(t3.2)
  NICS Improvement Amendments Act, 31–32
  NICS Index, active records in, 27(t3.3)
  NICS participation map, 26f
  permanent Brady permit chart, 37t
  public opinion on strictness of gun laws, 125–128
  right-to-carry weapons laws, 36t
  September 11, 2001, gun laws passed in response to, 31
  state firearms control laws, 32–38
  states with loosest gun laws, 34f
  states with strictest gun laws, 35f

"Toy Gun Law," 22–23
Undetectable Firearms Act, 22
Violent Crime Control and Law Enforcement Act, 29–31
See also Court rulings; Gun control laws
Lee, Richard Henry, 3
Legislation and international treaties
  Affordable Care Act, 135
  Arming Pilots against Terrorism Act, 31
  Arms Export Control Act, 11
  Assault Weapons Ban of 1994, 29, 133
  Assault Weapons Ban of 2013, 13, 132–136, 139–141
  Bill of Rights (England), 2
  Brady Handgun Violence Prevention Act, 4–5, 23–29, 36, 46–47
  Department of Defense Appropriations Act, 46
  Domestic Violence Offender Gun Ban, 20
  Federal Energy Management Improvement Act, 22
  Federal Firearms Act, 19, 43
  Firearm Owners' Privacy Act (Florida), 135
  Firearms Owners' Protection Act, 12, 20–21
  Game Act (England), 2
  Gun Control Act of 1968, 4, 19–20, 43
  Gun-Free School Zones Act, 23, 45–46
  Gun-Free Schools Act, 115
  Justice and Mental Health Collaboration Program Act, 136
  Law Enforcement Officers' Protection Act, 21–22
  Law Enforcement Officers Safety Act, 31
  National Firearms Act, 4, 19, 42
  NICS Improvement Amendments Act, 7, 31–32
  Occupational Safety and Health Act, 50
  Oklahoma Firearms Act, 50
  Oklahoma Self-Defense Act, 50
  Omnibus Consolidated and Emergency Supplemental Appropriations Act, 20
  "Permanent Brady," NICS, 24–29
  Protection of Lawful Commerce in Arms Act, 56
  September 11, 2001, gun laws passed in response to, 31
  Tiahrt Amendment, 9
  "Toy Gun Law," 22–23
  Treasury and General Government Appropriations Act, 21
  Undetectable Firearms Act, 22
  Violent Crime Control and Law Enforcement Act, 13, 21, 29–31
LESC (Law Enforcement Steering Committee), 22
Levine, Mike, 70
Levy, Robert A., 47
License. *See* Federal firearms licensee

"The Life Cycle of Crime Guns: A Description Based on Guns Recovered from Young People in California" (Wintemute et al.), 110
"The Limited Impact of the Brady Act: Evaluation and Implications" (Webster & Vernick), 25–26
"Local Authority to Regulate Firearms" (Law Center to Prevent Gun Violence), 38
Local laws
  Chicago, Illinois, handgun possession limits in, 52–53
  on guns, 38–39
  Morton Grove, Illinois, handgun ban of, 52
  Portland, Oregon, gun possession regulation in, 51
  Renton, Washington, guns not permitted where alcohol is served, 51–52
  Seattle, Washington, city gun ban, 54
  West Hollywood, California, Saturday Night Specials ban of, 53–54
Lockyer, Silveira v., 5, 44
Look-alike toy guns, 22–23
Loopholes, 25, 28–29
Lopez, Alfonso, Jr., 45–46
Lopez, United States v., 23, 45–46
Lott, John R., Jr., 36
Louisiana
  ambush on police in Baton Rouge, 71
  gun control laws of, 33
  state constitution article concerning weapons, 144
Ludwig, Jens, 25–26
Luo, Michael, 28

M
MAC-10, 44–45
Machiavelli, Niccolò, 2
Machine guns
  Assault Weapons Ban of 2013, arguments against, 139–141
  ban on, 29
  freeze on, 21
  manufacture of, 10
  National Firearms Act and, 19
  possession of, federal court cases on, 44, 47
Mack, Richard, 46–47
MacLeish, Kenneth T., 129
Madison, James, 3
Magazines, assault weapon, 12–13
MAIG (Mayors against Illegal Guns), 79–80
Mailing, of firearms, 19
Maine
  concealed weapons law of, 35–36
  gun control laws of, 33
  Portland Housing Authority provision against gun possession, 49
  state constitution article concerning weapons, 144

Majority Say More Concealed Weapons Would Make U.S. Safer (Newport), 128
Malcolm, Joyce Lee, 137–138
Males
  cost of firearm injuries and, 93
  firearm fatalities among, 84, 86
  gun ownership by gender, 124–125
  homicide rate for, 58
  relationship of homicide victim to murderer, 60–61
  See also Gender; Men
Manes, Mark E., 117
Manufacturing, firearms manufactured in U.S., 10*t*
Mary II, Queen of England, 2
Maryland
  firearms control laws of, 32, 33
  gun injuries, gun industry liability for, 55
  maternal mortality study in, 131
  as non-RTC state, 36
  state constitution article concerning weapons, 144
Marysville-Pilchuck High School, Marysville, Washington, 117–118
Mass shootings
  dissatisfaction with gun control laws and, 127–128
  notable, 1949–2008, 66*t*–67*t*
  notable, 2009–2016, 68*t*–69*t*
  percentage of total gun homicides, 123, 124*f*
  public opinion on gun control/mass shootings, 129
  Sandy Hook shooting, 126, 134
  state gun control laws in response to, 33–34
  in U.S., 61, 63, 65
  See also Active shooter incidents; School shootings
Massachusetts
  firearms control laws of, 33
  no preemption statute in, 38
  as non-RTC state, 36
  open carry law in, 38
  state constitution article concerning weapons, 144
Mateen, Omar, 65
Matthew, Miller, 7
Maxim, Hiram, 139
Mayors against Illegal Guns (MAIG), 79–80
McCarthy, Ciara, 22
McDonald, Otis, 48
McDonald v. City of Chicago, 5, 48
McIntire, Mike, 28
McKenna, Rob, 54
McKinley, Sarah, 141
Medicaid/Medicare, 93
Men
  gun ownership among, 16
  on guns in homes, 16
  See also Gender; Males

Mental illness
  Cho, Seung-Hui, 118
  mass shootings and, 129
  reduction of gun violence and, 136
"Mental Illness, Mass Shootings, and the Politics of American Firearms" (Metzl & MacLeish), 129
Metler, Robert, 52
Metzl, Jonathan M., 129
Michigan
  firearms control laws of, 33
  state constitution article concerning weapons, 144
Midwest
  gun ownership in, 124
  robberies involving firearms in, 76
Militias
  American Revolutionary War era, 3
  early, ownership of weapons and, 2
  U.S. Constitution and, 3–4
Miller, Matthew, 120
Miller, Ted, 93
Miller, United States v., 41–43
Minnesota
  ambush on police in Minneapolis, 71
  firearms control laws of, 32, 33
  open carry law in, 38
  Red Lake High School shooting, 117
  state constitution article concerning weapons, 144
Minors. *See* Children; Juveniles; Youth
Mississippi
  concealed weapons law of, 35–36
  gun control laws of, 33
  state constitution article concerning weapons, 144
Missouri, 144
Molde, Jade, 10–11
Montana
  gun control laws of, 33
  state constitution article concerning weapons, 144
More Guns, Less Crime: Understanding Crime and Gun Control Laws (Lott), 36
Morin, Rich, 124
Morton Grove, Illinois, 51–52
Mother Jones, 93
Mufflers, 44
Murder. *See* Homicide/murder
Muscarello v. United States, 45
Mustard, David B., 36
Musu-Gillette, Lauren, 116

N
National Center for Injury Prevention and Control, 83–84
National Crime Victimization Survey (NCVS), 57
National Firearms Act
  automatic firearms regulation with, 139

  gangster-type weapons, ban of, 42
  machine gun freeze, 21
  reenaction of 1934 National Firearms Act, 20
  registered weapons by state, 13*t*–14*t*
  requirements of, 19
  Title II weapons, tax on, 4
  weapons prohibited by, 10
National Instant Criminal Background Check System (NICS)
  Brady law, controversies over, 26–29
  checks, total, 24*t*
  components of, 25*f*
  denial process per 100 applicants, 27*f*
  description of, 5
  firearm prohibitive criteria under, 42*t*
  firearms denials, by reason, 27(*t*3.2)
  how it works, 24–25
  Improvement Amendments Act, 31–32
  Index, active records in, 27(*t*3.3)
  participation map, 26*f*
  replacement of five-day waiting period, 23
  states' right-to-carry weapons laws and, 36
National Institute of Justice (NIJ)
  defunding of firearms research, 79
  Guns in America, funding for, 136
National Law Enforcement Officers Memorial Fund (NLEOMF), 70–71
National Opinion Research Center, 14
National Research Council of the National Academies, 36
National Rifle Association (NRA)
  ban on armor-piercing ammunition and, 21–22
  Connecticut semiautomatic firearms ban, challenge to, 49
  as deterrence advocate, 16, 123
  Eddie Eagle GunSafe Program, 96–97
  on gun shows/online sales of guns, 28
  gun violence research and, 83
  LaPierre, Wayne, testimony of, 138–139
  membership growth, 126
  opposition to gun control studies, 78–79
  San Francisco firearms ban, challenge to, 51
  Sandy Hook Elementary School shooting and, 119
  self-defense narratives of, 94–95
  Undetectable Firearms Act and, 22
National Rifle Association of America v. City of Chicago, 48
National Violent Death Reporting System
  funding to expand, 134
  gun violence data of, 84
National Youth Gang Center, 108
National Youth Gang Surveys (NYGS), 108
Native Americans and Alaskan Natives
  firearm fatalities among, 87

Marysville-Pilchuck High School shooting, 117–118
Red Lake High School shooting, 117
weapon-carrying by students, 110
NCVS (National Crime Victimization Survey), 57
Nebraska
  firearms control laws of, 33
  local ordinances in, 38
  state constitution article concerning weapons, 144
Nevada, 144
New Hampshire, 144
New Jersey
  assault weapons ban challenge in, 50
  firearms control laws of, 32, 33
  mass shooting in Camden, 63, 65
  as non-RTC state, 36
  open carry law in, 38
  state constitution article concerning weapons, 144
New Mexico
  gun control laws of, 33
  state constitution article concerning weapons, 144
New York
  FFLs in, 31
  firearms control laws of, 32, 33
  no preemption statute in, 38
  as non-RTC state, 36
  open carry law in, 38
  state constitution article concerning weapons, 144
New York City
  gun control laws of, 38
  toy guns, ending sales of, 22
New York Police Department, 22
Newport, Frank, 127–128
Newtown, Connecticut. *See* Sandy Hook Elementary School, Newtown, Connecticut
NICS. *See* National Instant Criminal Background Check System
NICS Improvement Amendments Act, 31–32
NIJ. *See* National Institute of Justice
Niles Gun Show Inc., 54
NLEOMF (National Law Enforcement Officers Memorial Fund), 70–71
Nonfatal gunshot injuries. *See* Injuries
Nonnegligent manslaughter, 57
North Carolina, 33, 144
North Dakota, 145
Northeast
  gun ownership in, 124
  robberies involving firearms in, 76
Northern Illinois University, DeKalb, 118
NRA. *See* National Rifle Association
"NRA: 'Only Thing That Stops a Bad Guy with a Gun Is a Good Guy with a Gun'" (Overby), 33–34

NRA v. Reno, 27
"Number of Law Enforcement Officers Fatally Shot this Year up Significantly after Ambush Attacks, Report Says" (Berman), 71
Nutter, Michael A., 132–134
NYGS (National Youth Gang Surveys), 108

O

Oates, Dan, 77
Obama, Barack
  gun rights advocates and, 126
  gun violence research and, 83, 134
  safe gun storage campaign, 135
  Sandy Hook Elementary School shooting and, 119, 127
Occupational Safety and Health Act, 50
O'Connor, Sandra Day, 44–45
Offenders
  homicide circumstances, by relationship of victim to offender, 64*t*
  homicides, by victim/offender relationship/weapon use, 101*f*
  juvenile homicide offenders, by race, 105*f*
  juvenile homicide offenders, by relationship to victim, 103(*f*7.3)
  juvenile homicide offenders, demographics, 102*t*
  juveniles murdered by age/relationship to offender, 105*t*
  youth homicide trends, 99
  youthful homicide offenders, 102, 104
Offenses
  violent crime offenses reported to law enforcement, 58*f*
  weapons offenses, 80
Ohio
  Chardon High School, shooting at, 117
  juvenile gun theft, gun show promoters security against, 54
  state constitution article concerning weapons, 145
Oikos University, Oakland, California, 118
Oklahoma
  guns on corporate property challenge in, 50
  shooting at Edmond post office, 65
  state constitution article concerning weapons, 145
Oklahoma Firearms Act, 50
Oklahoma Self-Defense Act, 50
Omnibus Consolidated and Emergency Supplemental Appropriations Act, 20
Online gun sales, 28–29
Open carry, 36–38
"Open Carry Group Holds Rally at Home Depot in North Richland Hills" (Swanson), 37
"Open Carrying" (Law Center to Prevent Gun Violence), 38

Opinions on Gun Policy and the 2016 Campaign (Pew Research Center), 28–29, 126
Ordinances. *See* Laws, regulations, ordinances
Oregon
  mental illness limits on right to bear arms, 49
  Portland, gun possession regulation in, 51
  school shooting at Thurston High School, 115–116
  state constitution article concerning weapons, 145
  Umpqua Community College shooting, 118–119
Orlando, Florida, mass shooting in, 1, 33, 65
Oudekerk, Barbara A., 116
Overby, Peter, 33–34
Owenby, Patrick, 49
Owenby, State v., 49
Ownership. *See* Gun ownership

P

Palmer, Griff, 28
Pan, Deanna, 65
Parker, Clifton B., 36
Pavlides, Greg L., 54
Pechman, Marsha J., 54
"Peer Tutoring to Prevent Firearm Play: Acquisition, Generalization, and Long-Term Maintenance of Safety Skills" (Jostad et al.), 97
Pellet gun
  BB/pellet gun injuries among children aged 0–19 years, rates per 100,000, 86(*t*6.4)
  rate of injuries from, 84
Pennsylvania
  state constitution article concerning weapons, 145
  West Nickel Mines Amish School, shooting at, 119
Peradotto, Erin, 56
Pérez-Peña, Richard, 1
Permits
  for concealed weapon, 35–36
  exemption from background checks with, 28
  Permanent Brady permit chart, 37*t*
  state firearms control laws, 33
  See also Federal firearms licensee
Pew Research Center
  on gun ownership demographics, 124–125
  on public opinion on background checks, 28–29
  public opinion on strictness of gun laws, 125–126
  Why Own a Gun? Protection Is Now Top Reason, 16
Phelps, United States v., 44
Philadelphia, Pennsylvania, 132–133

Pistols, 10
Planty, Michael, 73, 84
Plastic firearms, 22
Plato, 2
Point of contact (POC), 24
Poisoning, 120
Police. *See* Law enforcement
Police Executive Research Forum, 133
Police Foundation, 135–136
Political parties
    Americans' dissatisfaction with gun laws, by party, 128*t*
    gun ownership by, 125
    public opinion on concealed weapons, 128
    public opinion on strictness of gun laws, 126
Politics (Aristotle), 2
The Politics of Gun Control (Spitzer), 5, 6
Polls, 124
Pollsters, 124
Population
    crime trends, by population group, 59*t*
    firearm fatalities by, 84
    violent crime, distribution of, 57–58
    youth gang activity and, 108
Porter, Caroline, 119–120
Portland, Oregon, gun possession regulation in, 51
Portland Housing Authority, Doe v., 49
Preemption, 38–39
Pregnancy, 131
Printz, Jay, 46–47
Printz v. United States, 5, 46–47
Private citizens
    justifiable homicides by, 76–77
    justifiable homicides committed by, by weapon used, 79*t*
Private sale loophole, 25
Prohibited people
    under Brady law, 23
    under Firearms Owners' Protection Act, 20
    in NICS records, 25
    online purchases of guns by, 28
Prohibition, 19
Protection of Lawful Commerce in Arms Act, 56
Public health, 7, 134–135
"Public Health Approach to the Prevention of Gun Violence" (Hemenway & Miller), 7
Public opinion
    background checks, expanded, opinions on, among Americans in gun-owning households, 127*t*
    on carrying concealed weapons, by demographic group, 130*t*
    on controlling gun ownership *vs.* protecting right of Americans to own guns, 125(*f*8.3)

dissatisfaction with gun laws, by party, 128*t*
gun control as personal issue, 123–124
on gun control/mass shootings, 129
gun ownership, demographics of, 124–125
gun ownership, prevalence of, among selected groups, 125(*f*8.2)
mass shootings as percentage of all U.S. homicides, 124*f*
on potential handgun ban, 129*f*
public opinion polls, evaluation of, 124
satisfaction with gun laws, 127*f*
on strictness of gun laws, 125–128, 128*f*
views of gun ownership as matter of protection *vs.* safety risk, 126*t*
Pulse nightclub, Orlando, Florida, 1, 33, 65
Purchase laws, 32*t*
Purdy, Patrick E., 29
Puzzanchera, Charles
    on youth gangs, 108
    on youth homicide, 99–102, 104
    on youthful homicide victims, 104–107

Q

Quilici, Victor, 52
Quilici v. Village of Morton Grove (Illinois), 52

R

Race/ethnicity
    arrests by, 82*t*
    arrests for weapons offenses by, 80
    firearm fatalities by, 87–89
    of gun owners, 16, 125
    gun use for self-defense, demographics of, 93–94
    homicide rate by, 58, 60
    justifiable gun homicides, race of shooter/person killed in, 94(*t*6.11)
    juvenile homicide offenders by, 105*f*
    juvenile homicide victims by, 107*f*
    nonfatal firearm violence by, 77*f*
    police, ambush attacks on, 71
    student reports of threats/injuries by, 111–112
    weapon-carrying by students by, 110–111, 113*f*, 113*t*
    of youthful homicide offenders, 102, 104
    of youthful homicide victims, 106–108
"Racial Gap in U.S. Arrest Rates: 'Staggering Disparity'" (Heath), 80
Raich, Gonzales v., 47
Ramsey Winch Inc. v. Henry, 50
Randa, Hoosier v. (California), 54
Randa, Jeff, 54
Rape, 57
Reagan, Ronald, 5, 23
Records, 27–28

Red Lake High School, Red Lake, Minnesota, 117
"Reducing Firearm-Related Harms: Time for Us to Study and Speak Out" (Taichman & Laine), 83
Reducing Gun Violence in America: Informing Policy with Evidence and Analysis (Cook & Ludwig), 25–26
Region
    gun ownership by, 124
    law enforcement officers killed, by type of weapon/region, 72*t*
    robberies involving firearms by, 76
Registration, 33
Regulation. *See* Laws, regulations, ordinances
Reichert, George, 52
Relationship
    homicide circumstances, by relationship of victim to offender, 64*t*
    of homicide victim to perpetrator, 60–61, 131–132
    homicides, by victim/offender relationship/weapon use, 101*f*
    juvenile homicide offenders, by relationship to victim, 103(*f*7.3)
    juveniles murdered by age/relationship to offender, 105*t*
    of youthful homicide offenders to victims, 101–102
    of youthful homicide victims to offenders, 104
Remington Arms, 11
Reno, NRA v., 27
Renton, Second Amendment Foundation v. (Washington), 51–52
Renton, Washington, 51–52
Replica guns, 22–23
Report of the February 14, 2008 Shootings at Northern Illinois University (Northern Illinois University's Department of Safety), 118
Report on the Implementation of the Gun-Free Schools Act of 1994 in the States and Outlying Areas: School Years 2005–06 and 2006–07 (U.S. Department of Education), 115
Report to the President on Issues Raised by the Virginia Tech Tragedy (U.S. Departments of Health and Human Services, Education, and Justice), 118
Republic (Plato), 2
Republican Party
    Americans' dissatisfaction with gun laws, by party, 128*t*
    gun control, position on, 1
    gun ownership by, 125
    public opinion on concealed weapons, 128
    public opinion on strictness of gun laws, 126
R.G. Industries Inc., Kelley v. (Maryland), 55

Rhode Island
  firearms control laws of, 33
  as non-RTC state, 36
  state constitution article concerning weapons, 145
Rice, Tamir, 22
Rifles, 10
Right to bear arms
  debate over, 4–7
  federal court cases on, 42
  firearms used in crime, types of, 4t
  gun control, overview of, 1–2
  gun control laws, early, 2
  militia, American, 2–3
  militias/ownership of weapons, early, 2
  overview of, 1
  Second Amendment, poll respondents on meaning of, 2t
  U.S. Constitution and, 3–4
Right to carry (RTC)
  Americans' views on carrying concealed weapons, by demographic group, 130t
  state concealed weapons laws, 35–36
  states' right-to-carry weapons laws, 36t
  weapons laws by state, 36t
"Right-to-Carry Gun Laws Linked to Increase in Violent Crime, Stanford Research Shows" (Parker), 36
Robbery
  definition of, 57
  nonfatal firearm violence used in, 74, 76
  robberies/aggravated assaults, by type of weapon used, 77t
  by type of weapon used, percentage distribution by region, 78(t5.17)
Roberts, Charles C., IV, 119
Robinson, Candisha, 45
Robinson, United States v., 45
Rohm Gesellschaft, 55
Roosevelt, Franklin D., 4
RTC. See Right to carry
Rucker, Philip, 28
Ruger, 10–11

S

Saad, Lydia, 129
Safety
  AAP on gun violence, 134–135
  child access prevention laws, by state, 96f
  Florida's safe-storage law, 50
  guns in homes, research on, 14–16
  programs to protect young children, 96–97
  safe storage of guns in home, 95–96
  school safety efforts post-Newtown, 119–120
  views of gun ownership as matter of protection vs.. safety risk, 126t
Sales. See Firearms sales; Gun dealers

San Bernardino, California, 127
Sandy Hook Elementary School, Newtown, Connecticut
  AAP on gun control laws after, 134
  assault weapons ban, effect on, 13
  demand for guns/ammunition after, 126
  dissatisfaction with gun control laws after massacre, 127
  effect on gun control debate, 1
  gun violence research and, 83
  LaPierre, Wayne, on, 138
  school safety efforts post-Newtown, 119–120
  shooting at, 65, 119
  shots fired by shooter at, 141
Sawed-off shotguns, 19
Scalia, Antonin
  District of Columbia v. Heller, 47, 138
  on gun control, constitutionality of, 5
School shootings
  increase of, 115–116
  notable mass shootings, 65
  postsecondary, 118–119
  schoolchildren shot by adults, 119–120
  secondary, 116–118
Schools
  barring guns from, 115
  Gun-Free School Zones Act, 23
  possession of firearms near, federal court cases on, 45–46
  student reports of threats/injuries, 111–113, 115
  students and guns, 110–111
  students grades 9–12 who carried weapon/gun, by sex/race/ethnicity/grade, 113t
  students grades 9–12 who reported carrying weapon by location/sex, 111f
  students grades 9–12 who reported carrying weapon by location/student characteristics, 112t
  students grades 9–12 who reported carrying weapon by race/ethnicity/location, 113f
  students who carried weapon/gun, by sex/state/urban area, 114t
  students who reported being threatened/injured with weapon on school property, 115t
  students who reported being threatened/injured with weapon on school property, by state, 116t
Schubert, Joseph L., Jr., 48–49
Schubert v. DeBard, 48–49
Schumaker, Erin, 83
Seattle, Washington, 53–54
Second Amendment
  adoption of, 4
  Assault Weapons Ban of 2013 and, 133
  federal vs. state protections, 41
  gun control debate and, 123

gun control laws, arguments against stricter, 137, 142
  in gun rights argument, 19
  individual vs. collective rights, interpretations of, 41–42
  interpretation of, 1–2
  poll respondents on meaning, 2t
  public opinion on strictness of gun laws, 126
Second Amendment Foundation
  as deterrence advocate, 123
  on modern weapons, control of, 6
  Renton, Washington, gun ban challenge, 51–52
  San Francisco firearms ban, challenge to, 51
  Seattle city gun ban, state law override of, challenge to, 54
Second Amendment Foundation v. Renton (Washington), 51–52
Secondary school shootings, 115–118
"Seeking Gun or Selling One, Web Is a Land of Few Rules" (Luo, McIntire, & Palmer), 28
Self-defense
  gun control laws, arguments against stricter, 141–142
  gun homicides, number of justifiable, compared to criminal, 94(t6.10)
  guns and, 93–95
  handguns used for, 6–7
  justifiable homicides, firearms used in, by type, 95t
  justifiable homicides, race of shooter/person killed in, 94(t6.11)
  safe storage of guns in home and, 95
  state laws protecting right to use deadly force in, 48t
  women's use of guns for, 132
Semiautomatic Assault Weapons Ban, 29
Semiautomatic firearms
  assault weapons ban, 29
  Connecticut ban on, 49–50
  LaPierre, Wayne, on ban of, 139–141
  overview of, 12–14
September 11, 2001, terrorist attacks, 31
Shaffer, Catherine, 54
Shall-issue laws, 35–36
Sherrill, Patrick Henry, 65
Shootings. See Mass shootings; School shootings
Shotguns, 41–42
Sickmund, Melissa
  on youth gangs, 108
  on youth homicide, 99–102
  on youthful homicide victims, 104–107
Silberman, Laurence H., 47
Silencers, 44
Silveira v. Lockyer, 5, 44

"6 New Gun Control Laws Enacted in California, as Gov. Brown Signs Bills" (Chappell), 33
Sklar, Jerome, 53
Sklar v. Byrne (Illinois), 53
Small Arms Survey, 11–12
Small Arms Survey 2015 (Small Arms Survey), 11–12
Smith, Erica L.
  on homicide trends, 58
  on youth homicide trends, 99
  on youthful homicide victims, 107–108
Smith, John Angus, 44
Smith, Tom W., 14
Smith & Wesson, 10–11
Smith v. United States, 44
Snedeker, Thomas E., 54
Solomon, Dan, 37–38
Son, Jaesok, 14
Sorenson, Susan, 131
"Source of Firearms Used by Students in School-Associated Violent Deaths—United States, 1992–1999" (CDC), 110
South
  gun ownership in, 124
  police deaths/injuries in, 70
  robberies involving firearms in, 76
South Carolina
  open carry law in, 38
  state constitution article concerning weapons, 145
South Dakota, 145
"Spending on School Security Rises" (Porter), 119–120
Spitzer, Robert J., 5, 6
Spousal abuse, 131–132
Standing armies, 3–4
State constitution articles, 143–145
State Constitutional Right to Keep and Bear Arms Provisions, by Date (Volokh), 4
State firearms control laws
  California, San Francisco handgun ban preemption, 50–51
  on concealed weapons, 35–36
  Connecticut, child access prevention challenge, 50
  Connecticut, semiautomatic firearms ban in, 49–50
  firearms purchase laws, by state, 32*t*
  Indiana, right to possess handgun in, 48–49
  Maine, Portland Housing Authority provision against gun possession, 49
  on minors, 38
  New Jersey, assault weapons ban challenge, 50
  Oklahoma, guns on corporate property challenge, 50
  on open carry, 36–38
  Oregon, mental illness limits on right to bear arms, 49

  overview of, 32–34, 48
  protecting right to use deadly force in self-defense, 48*t*
  right-to-carry weapons laws, 36*t*
  states with loosest gun laws, 34*f*
  states with strictest gun laws, 35*f*
State v. Blocker (Oregon), 51
State v. Boyce (Oregon), 51
State v. Kessler (Oregon), 51
State v. Owenby, 49
State v. Wilchinski (Florida), 50
States
  child access prevention laws, by state, 96*f*
  constitutions, post–Revolutionary War, 4
  federal firearms licensees, trends in, 31
  federal firearms licenses, by state, 31*t*
  gun control laws, gun violence rates and, 135
  gun trafficking in, 79–80
  guns, numbers of in, 9
  guns recovered from youths in, 110
  local ordinances and, 38–39
  murder, by state/type of weapon, 60
  NICS background checks by, 24–25
  NICS participation map, 26*f*
  Permanent Brady permit chart, 37*t*
  ratification of Constitution, militias and, 3–4
  registered weapons by, 13*t*–14*t*
  Second Amendment rules, states adherence to, 48
  state firearms control laws, 32–38
  state laws, 48
  state protections *vs.* federal, 41
  student reports of threats/injuries, 112–113
  students who carried weapon/gun, by sex/state/urban area, 114*t*
  students who reported being threatened/injured with weapon on school property, by state, 116*t*
  with universal background checks, 29
  weapon-carrying by students in, 111
Statistical information
  active shooter incidents, number of, annually, 70*f*
  active shooters, number of casualties (killed or wounded) by, 71*f*
  aggravated assault, by type of weapon used, percentage distribution by region, 78(*t*5.18)
  arrest trends, by offense/age, 81*t*
  arrests, by race, 82*t*
  BB/pellet gun injuries among children aged 0–19 years, rates per 100,000, 86(*t*6.4)
  child access prevention laws, by state, 96*f*
  crime trends, by population group, 59*t*
  death, 10 leading causes of, for 10–24-year-olds, 122*f*

  death rates for firearm-related injuries, by sex, race, Hispanic origin, age, 87*t*–89*t*
  death rates for suicide, by sex/age, 90*t*–91*t*
  deaths/death rates for 10 leading causes of death among young people, 121*t*–122*t*
  fatal gunshot (homicide and suicide) deaths/rates per 100,000 people, 91(*t*6.7)
  federal firearms licensees, 30*t*
  federal firearms licenses, by state, 31*t*
  federal officers, assaults on, by extent of injury/type of weapon, 75(*t*5.13)
  firearm deaths of children aged 0–19 years, 101*t*
  firearm deaths per 100,000 people, 86*f*
  firearm violence, 76*t*
  firearms exported, numbers of, 12*t*
  firearms imported, by country, 15(*t*2.5)
  firearms imported, numbers of, 11*t*
  firearms manufactured, numbers of, 10*t*
  firearms purchase laws, by state, 32*t*
  gun homicides, number of justifiable, compared to criminal, 94(*t*6.10)
  gun owners by selected demographic characteristics, 16*t*
  gun ownership, prevalence of, among selected groups, 125(*f*8.2)
  gun ownership, reasons for, 17(*t*2.8)
  guns at home, non-gun households attitudes toward having, 17(*t*2.9)
  guns recovered after use in crimes by state/age of possessor, 109*t*
  homicide circumstances, by relationship of victim to offender, 64*t*
  homicide circumstances, by weapon, 62*t*
  homicide firearm deaths of children aged 12/under, 107*t*
  homicide victims by age, sex, race, ethnicity, 61*t*
  homicides, by number of victims/weapon type, 65*t*
  homicides, by state/weapon type, 63*t*
  homicides, by victim/offender relationship/weapon use, 101*f*
  homicides, number/rate of, by victim demographics, 60*t*
  homicides committed by juveniles, by firearm involvement, 103(*f*7.4)
  homicides involving guns by age of victim, 106(*f*7.7)
  homicide/suicide/gun-related deaths among youths aged 15–19, 100*f*
  injury deaths, 10 leading causes of, by age group, 92*t*
  justifiable gun homicides, race of shooter/person killed in, 94(*t*6.11)
  justifiable homicides, firearms used in, by type, 95*t*

justifiable homicides committed by law enforcement officers, by weapon used, 78(t5.19)
justifiable homicides committed by private citizens, by weapon used, 79t
juvenile arrests, 104t
juvenile homicide offenders, by race, 105f
juvenile homicide offenders, by relationship to victim, 103(f7.3)
juvenile homicide offenders, demographics of, 102t
juvenile homicide victims by race, 107f
juvenile homicide victims killed with/without firearms, 106(f7.6)
juveniles murdered 1980–2010, by age/weapon type, 100t
juveniles murdered by age/relationship to offender, 105t
law enforcement officers accidentally killed, by circumstance at scene of incident, 74t
law enforcement officers assaulted, by circumstance at scene of incident/type of weapon, 75(t5.14)
law enforcement officers killed, by circumstance at scene of incident, 73t
law enforcement officers killed, by type of weapon/region, 72t
mass shootings, notable, 1949–2008, 66t–67t
mass shootings, notable, 2009–2016, 68t–69t
mass shootings as percentage of all U.S. homicides, 124f
National Firearms Act registered weapons, by state, 13t–14t
NICS checks, total, 24t
NICS firearms denials, by reason, 27(t3.2)
NICS Index, active records in, 27(t3.3)
nonfatal firearm injuries among children aged 0–19, rates per 100,000, 86(t6.3)
nonfatal firearm violence, 76f
nonfatal firearm violence, by race/Hispanic origin, 77f
nonfatal gunshot injuries/rates per 100,000 people, 84t
nonfatal injuries treated in hospital emergency departments, leading causes of, by age group, 85t
public opinion, Americans' dissatisfaction with gun laws, by party, 128t
public opinion, Americans' satisfaction with gun laws, 127f
public opinion, Americans' views on carrying concealed weapons, by demographic group, 130t
public opinion, Americans' views on controlling gun ownership vs. protecting right of Americans to own guns, 125(f8.3)

public opinion, Americans' views on potential handgun ban, 129f
public opinion, Americans' views on strictness of gun laws, 128f
public opinion, views of gun ownership as matter of protection vs.. safety risk, 126t
public opinion on background checks, 127t
robberies/aggravated assaults, by type of weapon used, 77t
robbery, by type of weapon used, percentage distribution by region, 78(t5.17)
Second Amendment, poll respondents on meaning of, 2t
states with loosest gun laws, 34f
states with strictest gun laws, 35f
students grades 9–12 who carried weapon/gun, by sex/race/ethnicity/grade, 113t
students grades 9–12 who reported carrying weapon by location/sex, 111f
students grades 9–12 who reported carrying weapon by location/student characteristics, 112t
students grades 9–12 who reported carrying weapon by race/ethnicity/location, 113f
students who carried weapon/gun, by sex/state/urban area, 114t
students who reported being threatened/injured with weapon on school property, 115t
students who reported being threatened/injured with weapon on school property, by state, 116t
unintentional firearm deaths/rates per 100,000 people, 91(t6.8)
U.S. households with guns, percentage of, 15(t2.6)
violent crime offenses reported to law enforcement, 58f
Statute of Northampton, 2
Stengl, Robert, 52
Sterling, Alton, 71
Stevens, John Paul, 47
Stewart, Robert Wilson, Jr., 47
Stewart, United States v., 47
Stockton, California, 29
Storage
  AAP on gun storage, 134, 135
  safe storage of guns in home, 95–96
Stout, David, 56
Straw purchasing
  background checks and, 28
  California's laws regarding, 33
  criminals' acquisition of firearms, 79
Students
  grades 9–12 who carried weapon/gun, by sex/race/ethnicity/grade, 113t

grades 9–12 who reported carrying weapon by location/sex, 111f
grades 9–12 who reported carrying weapon by location/student characteristics, 112t
grades 9–12 who reported carrying weapon by race/ethnicity/location, 113f
guns and, 110–111
threats/injuries, student reports of, 110–113, 115
who carried weapon/gun, by sex/state/urban area, 114t
who reported being threatened/injured with weapon on school property, 115t
who reported being threatened/injured with weapon on school property, by state, 116t
"Study Finds Vast Online Marketplace for Guns without Background Checks" (Rucker), 28
A Study of Active Shooter Incidents in the United States between 2000 to 2013 (FBI), 65, 70
Sturm, Ruger & Co., Inc., 10–11
Suffocation, 96, 120
Sugarmann, Josh, 30
Suicide
  CAP laws for prevention of, 95, 96
  death rates for, by sex/age, 90t–91t
  fatal gunshot (homicide and suicide) deaths/rates per 100,000 people, 91(t6.7)
  firearm suicides vs. homicides, 90–91
  firearm-related deaths, demographics of, 86
  firearm-related deaths attributable to, 89–90
  guns used in, numbers of, 15
  homicide/suicide/gun-related deaths among youths aged 15–19, 100f
  Native American teens and, 118
  youth suicide, firearms and, 120
"Suicide Prevention: Youth Suicide" (CDC), 95, 120
Supplemental Homicide Reports (FBI), 131, 132
"Survey Research and Self-Defense Gun Use: An Explanation of Extreme Overestimates" (Hemenway), 6–7
Swaine, Jon, 22
Swanson, Jeffrey, 129
Swift, Art, 126–127
Synnes, United States v., 44

T

Taichman, Darren B., 83
Taxes, 19
Tennessee, 145
Terrorism, 31

Texas
  FFLs in, 31
  firearms used in homicides in, 60
  mass shooting in Killeen, 65
  open carry incidents in, 37–38
  school shooting at University of Texas, Austin, 65
  state constitution article concerning weapons, 145
Thieu, Matt, 119
Third Way, 28
Thomas, Pierre, 70
Thurman, Russ, 10–11
Thurston High School, Springfield, Oregon, 115–116
Tiahrt, Todd, 9, 28
Tiahrt Amendment
  criminals' acquisition of firearms and, 78
  gun trace statistics restricted by, 9
  NICS records, destruction of, 28
Tilley, Edward A., III, 54
Time-to-crime (TTC) data, 80
Title I (U.S. federal firearms laws), 4
Title II (U.S. federal firearms laws), 4
"Top Ten Issues Regarding Guns in the US Workplace" (Association of Corporate Counsel), 50
Tot, Frank, 43
Tot, United States v., 43
"Total NICS Firearm Background Checks: November 30, 1998–October 31, 2016" (FBI), 5
"Toy Gun Law," 22–23
Toy guns, 22–23
Toy Guns: Involvement in Crime & Encounters with Police (Department of Justice), 22–23
Toys "R" Us, 22
Trace the Guns: The Link between Gun Laws and Interstate Gun Trafficking (MAIG), 79–80
Trade, international, 11
Trafficking. *See* Gun trafficking
Transportation, 21
Treasury and General Government Appropriations Act, 21
Tribe, Lawrence, 133
Trotter, Gayle S., 141–142
Truman, Jennifer L., 73, 84
Trump, Donald, 29, 126
TTC (time-to-crime) data, 80
"2014 Child Gun Deaths" (Children's Defense Fund), 120
"2016 Mid-year Law Enforcement Officer Fatalities Report" (NLEOMF), 71

U

Umpqua Community College, Roseburg, Oregon, 118–119
Undetectable Firearms Act, 22

Uniform Crime Reports (UCR), 57–58
United States
  firearm deaths in, 16
  firearms exports of, 11–12
  firearms imports of, 11
  firearms manufacturing in, 10–11
  households with guns, percentage of, 15(t2.6)
  number of guns in, 9–10
  right to bear arms/militia and, 3
  See also Federal court cases; Federal government
United States, Bailey v., 45
United States, Cases v., 43
United States, Muscarello v., 45
United States, Printz v., 5, 46–47
United States, Smith v., 44
United States v. Cruikshank, 42
United States v. Edwards, 46
United States v. Emerson, 5, 43
United States v. Harris, 44–45
United States v. Hayes, 20
United States v. Lopez, 23, 45–46
United States v. Miller, 41–43
United States v. Phelps, 44
United States v. Robinson, 45
United States v. Stewart, 47
United States v. Synnes, 44
United States v. Tot, 43
United States v. Warin, 44
"Universal Background Checks" (Law Center to Prevent Gun Violence), 29
Universities, shootings at, 118–119
University of Texas, Austin, 65
Unruh, Howard, 63, 65
U.S. Census Bureau, 80
U.S. Conference of Mayors, 132–134
U.S. Congress, 4
U.S. Constitution
  Eighteenth Amendment, 19
  Fourth Amendment, 137
  gun control, constitutionality of, 5
  right to bear arms, history of, 3–4
  Second Amendment, 19, 123, 137, 142
U.S. Court of Appeals for the Fifth Circuit, 5
U.S. Department of Education
  on Gun-Free Schools Act, 115
  school safety efforts post-Newtown, 119–120
U.S. Department of Homeland Security, 24
U.S. Department of Justice
  on retention of records, 27
  study on assault weapons ban, 133
  Toy Guns: Involvement in Crime & Encounters with Police, 22–23
"U.S. Firearms Industry Today 2016" (Molde & Thurman), 10–11
U.S. Government Accountability Office (GAO), 27–28, 31

U.S. Health in International Perspective: Shorter Lives, Poorer Health (Woolf & Aron), 15–16
U.S. International Trade Commission, 11
U.S. Postal Service, 19
U.S. Supreme Court
  District of Columbia v. Heller, 128
  on five-day waiting period of interim Brady, 23
  on handgun background checks, 5
  United States v. Lopez, 23
  See also Federal court cases
Utah, 145

V

Velazquez, Jose Cases, 43
Vermont
  concealed weapons law of, 35–36
  state constitution article concerning weapons, 145
Vernick, Jon S., 25–26
Victimization
  homicide victimization rate, 58
  youth homicide victimization, 100–102
Victims
  criminal advantages of guns and, 61
  of homicide, demographics of, 58, 60
  of homicide, relationship to murderer, 60–61
  homicide circumstances, by relationship of victim to offender, 64t
  homicides, by number of victims/weapon type, 65t
  homicides, by victim/offender relationship/weapon use, 101f
  homicides, number/rate of, by victim demographics, 60t
  homicides involving guns by age of victim, 106(f7.7)
  juvenile homicide offenders, by relationship to victim, 103(f7.3)
  juvenile homicide victims by race, 107f
  juvenile homicide victims killed with/without firearms, 106(f7.6)
  juveniles murdered by age/relationship to offender, 105t
  of nonfatal firearm violence, 73–74
  youth homicide trends, 99
  youthful homicide victims, 104–108
Village of Morton Grove, Kalodimos v. (Illinois), 52
Village of Morton Grove, Quilici v. (Illinois), 52
Violence Policy Center (VPC)
  on assault weapons, 133
  on Eddie Eagle GunSafe Program, 96–97
  on guns used for self-defense, 93–94
  on homicide victims, relationship to offender, 61
  as interdiction advocate, 124

on murders of women, 131
on states' gun control laws, 135
Violent crime
consequences for using gun in, 31
firearm violence, 76*t*
nonfatal firearm violence, 76*f*
nonfatal firearm violence, by race/Hispanic origin, 77*f*
nonfatal violent crimes involving firearms, 72–74, 76–77
RTC laws and, 36
trends in, 57–58
violent crime offenses reported to law enforcement, 58*f*
See also Crime; Homicide/murder
Violent Crime Control and Law Enforcement Act
assault weapons ban of, 13, 29
ban on armor-piercing ammunition, 21
FFLs, new regulations to obtain, 30–31
juvenile possession of handguns/ammunition, ban on, 29–30
Virginia
gun control laws of, 33
state constitution article concerning weapons, 145
Virginia Polytechnic Institute and State University (Virginia Tech)
NICS Improvement Amendments Act in response to shooting at, 31–32
shooting at, 65, 118
Virginia Youth Violence Project, 102
Volokh, Eugene, 4
VPC. *See* Violence Policy Center

W

Waiting period
elimination of, 26–27
five-day waiting period of interim Brady, 23
states that require, 33
Wal-Mart, 55
Wal-Mart Stores, Inc. v. Coker (Florida), 55
Warin, Francis J., 44
Warin, United States v., 44
Warren, Earl, 41
Warren, James L., 51
Washington (state)
Renton, guns not permitted where alcohol is served, 51–52
Seattle city gun ban, state law override of, 54
state constitution article concerning weapons, 145
"We Have Three Different Definitions of 'Mass Shooting,' and We Probably Need More" (Ingraham), 63
Weapons
aggravated assault, by type of weapon used, percentage distribution by region, 78(*t*5.18)

federal officers, assaults on, by extent of injury/type of weapon, 75(*t*5.13)
firearms used in homicides, 60–61
homicide circumstances, by weapon, 62*t*
homicides, by state/weapon type, 63*t*
justifiable homicides committed by law enforcement officers, by weapon used, 78(*t*5.19)
justifiable homicides committed by private citizens, by weapon used, 79*t*
law enforcement officers assaulted, by circumstance at scene of incident/type of weapon, 75(*t*5.14)
offenses, 80
robberies/aggravated assaults, by type of weapon used, 77*t*
robbery, by type of weapon used, percentage distribution by region, 78(*t*5.17)
state constitution articles concerning weapons, 143–145
student reports of threats/injuries, 111–113, 115
students and guns, 110–111
used in robberies/aggravated assaults, 76
youth homicide weapons, 99–100
Weapons offenses
arrest trends, by offense/age, 81*t*
arrests, by race, 82*t*
trends in, 80
Web-Based Injury Statistics Query and Reporting System (WISQARS), 83–84
Webber, Reagan, 119
Webster, Daniel W.
on impact of Brady law on gun violence, 25–26
on youth acquisition of guns, 110
Weibe, Douglas, 131
Weise, Jeffrey, 117
West
gun ownership in, 124
police deaths/injuries in, 70
robberies involving firearms in, 76
West Hollywood, California, 53–54
West Nickel Mines Amish School, Nickel Mines, Pennsylvania, 119
West Virginia, 145
"What a Balancing Test Will Show for Right-to-Carry Laws" (Lott), 36
"What Does Gun Violence Really Cost?" (Follman), 93
"What Led Jaylen Fryberg to Commit the Deadliest High School Shooting in a Decade?" (Kutner), 118
When Men Murder Women: An Analysis of 2014 Homicide Data (Violence Policy Center), 61
White House Taskforce on Gun Violence Prevention, 134

Whites
arrests for weapons offenses by, 80
firearm fatalities among, 87
gun ownership by, 125
gun use for self-defense, 93–94
homicide rate for, 58, 60
weapon-carrying by students, 110, 111
youth homicide offenders, 102, 104
youth homicide victims, 107
Whitman, Coalition of New Jersey Sportsmen v., 50
Why Own a Gun? Protection Is Now Top Reason (Pew Research Center), 16
Wiener, Malcolm, 108
Wilchinski, Joseph, 50
Wilchinski, State v. (Florida), 50
William III, King of England, 2
Williams, Daniel, 56
Williams v. Beemiller (New York), 56
Winchester Henry, 141
Wintemute, Garen J.
on crime guns recovered, 110
"Guns, Fear, the Constitution, and the Public's Health," 15
Wisconsin, 145
WISQARS (Web-Based Injury Statistics Query and Reporting System), 83–84
Wolf, Anna D., 131–132
Women
gun control laws, arguments against stricter, 137–138, 141–142
gun ownership among, 16
See also Females; Gender
"Women Under the Gun" (Center for American Progress), 137
Woolf, Steven H., 15–16
Wyoming
concealed weapons law of, 35–36
gun control laws of, 33
state constitution article concerning weapons, 145

Y

Youth
criminally active nongang groups, 108
deadly assaults by, 99–102
death, 10 leading causes of, for 10–24-year-olds, 122*f*
deaths/death rates for 10 leading causes of death among young people, 121*t*–122*t*
firearm deaths of children aged 0–19 years, 101*t*
gangs, 108
guns and, 99
guns recovered after use in crimes by state/age of possessor, 109*t*
homicide firearm deaths of children aged 12/under, 107*t*

homicides, by victim/offender relationship/weapon use, 101*f*

homicides committed by juveniles, by firearm involvement, 103(*f*7.4)

homicides involving guns by age of victim, 106(*f*7.7)

homicide/suicide/gun-related deaths among youths aged 15–19, 100*f*

injured/killed by gunfire, 120

juvenile arrests, 104*t*

juvenile homicide offenders, by race, 105*f*

juvenile homicide offenders, by relationship to victim, 103(*f*7.3)

juvenile homicide offenders, demographics of, 102*t*

juvenile homicide victims by race, 107*f*

juvenile homicide victims killed with/ without firearms, 106(*f*7.6)

juveniles murdered 1980–2010, by age/ weapon type, 100*t*

juveniles murdered by age/relationship to offender, 105*t*

offenders, acquisition of guns by, 108–110

school shootings, 115–116

school shootings, postsecondary, 118–119

school shootings, secondary, 116–118

schoolchildren shot by adults, 119–120

schools, barring guns from, 115

students, guns and, 110–111

students grades 9–12 who carried weapon/gun, by sex/race/ethnicity/ grade, 113*t*

students grades 9–12 who reported carrying weapon by location/sex, 111*f*

students grades 9–12 who reported carrying weapon by location/student characteristics, 112*t*

students grades 9–12 who reported carrying weapon by race/ethnicity/ location, 113*f*

students who carried weapon/gun, by sex/state/urban area, 114*t*

students who reported being threatened/ injured with weapon on school property, 115*t*

students who reported being threatened/ injured with weapon on school property, by state, 116*t*

suicide, firearms and, 120

threats/injuries, student reports of, 110–113, 115

youth gangs, 108

youthful offenders, 102, 104

youthful victims, 104–108

See also Children

Youth Crime Gun Interdiction Initiative, 109

Youth Risk Behavior Surveillance System, 110–111

Z

Zhang, Alexandria, 36

Zhang, Anlan, 116

CPSIA information can be obtained
at www.ICGtesting.com
Printed in the USA
FFOW05n1830100517

NOV - - 2017

9 781410 325495